O.-R. Kaaden
C.-P. Czerny
W. Eichhorn (eds.)

Viral Zoonoses and Food of Animal Origin

A Re-Evaluation of Possible Hazards for Human Health

SpringerWienNewYork

Prof. Dr. Oskar-Rüger Kaaden
Dr. Claus-Peter Czerny
Dr. Werner Eichhorn
Institute of Medical Microbiology, Ludwig-Maximilians-University,
Munich, Federal Republic of Germany

Printed in Austria

Typesetting: Thomson Press (India) Ltd., New Delhi
Printing: Adolf Holzhausens Nfg., A-1070 Wien
Binding: Fa. Papyrus, A-1100 Wien

Graphic design: Ecke Bonk

Printed on acid-free and chlorine-free bleached paper

SPIN: 10558738

With 43 Figures

ISBN 3-211-82927-X Springer-Verlag Wien New York
ISBN 3-211-83014-6 (Softcover) Springer-Verlag Wien New York

"Viral Zoonoses and Food of Animal Origin: A Re-Evaluation of Possible Hazards for Human Health"

Opening Speech

On behalf of the Federal Minister for Health, Mr. Horst Seehofer, and of the Federal Minister of Food, Agriculture and Forestry, Mr. Jochen Borchert, I would like to give you a warm welcome today to the 9th Munich Symposium on Microbiology from the „WHO Collaborating Centre for Collection and Evaluation of Data on Comparative Virology", Institute for Medical Microbiology, Infectious and Epidemic Diseases, Veterinary Faculty, Ludwig-Maximilians-University of Munich on the subject "Viral Zoonoses and Food of Animal Origin: A Re-Evaluation of Possible Hazards for Human Health".

In Central Europe and especially in Germany, it is an old tradition to consume various kinds of food of animal origin raw. Minced meat made of raw beef (*"Tatar"*) and uncooked pork (*"Hackepeter"*) are hardly ever missing from a country-style buffet. Certified milk, i.e. cows' milk intended for raw consumption and soft cheese manufactured from raw milk are also popular specialities. Live oysters or sabajone and *"Bienenstich"* cake with cream made from raw eggs are now standard dishes in every Italian restaurant in Germany.

It is reasonable to presume that statutory bans will hardly alter these old traditions of food consumption. Furthermore, a prohibition of selling such raw food would be evaded and the sale would be continued without any control. Moreover, it has to be expected that recommendations for the consumers to change their dietary habits will not have any effect.

On the basis of these practical reflections the legislator permitted, by virtue of legal provisions and within a clearly restricted framework, the manufacture and marketing of raw food of animal origin and subjected them to stringent precautions and requirements. In doing so, however, he also assumes responsibility for possible negative effects of this permission, e.g. for diseases arising from animal-borne zoonoses communicated to man via food.

However, general dietary habits also carry risks with regard to possible secondary and cross-contamination. Additionally, attention should be given to the relevance of direct contact between animals and people to the infection chain.

Therefore, prior to every statutory food-hygiene-admission and also in future, the risk involved must be estimated as exactly as possible and the best possible measures of restricting this risk must be ascertained.

It is mainly scientists and practicians who must deal with these questions. Thus the Federal Government is very grateful to the WHO-Collaborating Centre for Collection and Evaluation of Data on Comparative Virology for taking up this important subject.

We particularly appreciate the fact that you, ladies and gentlemen, have come to Munich to discuss this issue and to formulate recommendations. You are assisting us – the law-giving bodies – in judging to what extent food of animal origin can be consumed and where the limits are with regard to the risks involved in doing so. But you are also helping the producers of these foods and the controlling bodies to realize how and by what measures these risks can be minimized.

Each year for over 20 years now, the Federal Government has, in collaboration with the WHO, been supporting individual programmes aimed at clarifying the behaviour of zoonoses and identifying means of controlling them. The reports of these conferences are available from the Headquarter of the World Health Organization in Geneva.

The Federal Ministry for Health has contributed 55.000 DM to help make this conference a success. I am convinced that this was a worthwile investment since the composition of this assembly gives us reason to assume that your results will have a meaningful influence on the legislative work in your countries and in ours.

This event was organised by the WHO-Collaborating Centre, represented today by Professor Kaaden. I would like to give our special thanks to you and your colleagues for this achievement.

The Federal Government hopes to contribute to improving world health not only through its annual contribution to the World Health Organization of about 55 million DM, but also through the concrete measures it takes to find solutions to problems, these measures being financed with the aid of supplementary funds. One example of the latter is, for instance, the sum of approximately 2.5 million DM allocated annually for the WHO Collaboration Centers and for this kind of meeting.

Our goal is, through the exchange of opinions and knowledge which is taking place here today, to be able to identify health hazards in advance and ward off possible threats to human and animal health.

The governments of various countries including those comprising the European Union are awaiting the results of this conference with great interest. They are looking forward to obtaining results which they can use as a basis on which to examine their standards and, where necessary, make adjustments.

Not least due to this, I wish you the greatest posssible success in your endeavours today and tomorrow here in Munich and hereby declare this conference open.

Dr. Hans Böhm
Federal Ministry for Health, Bonn

Acknowledgement

The Editors gratefully acknowledge the competent and skilled secretarial assistance of Mrs. Elke Bongers to organize the 9th Munich Symposium on Microbiology and to publish the scientific papers as supplement to *Archives of Virology*.

Contents

Cowpox: a re-evaluation of the risks of human cowpox based on new epidemiological information

D. Baxby[1] and **M. Bennett**[2]

[1]Alder Hey Childrens' Hospital and Department of Medical Microbiology, Liverpool University, Liverpool, U.K.
[2]Department of Veterinary Clinical Science, Liverpool University, Liverpool, U.K.

Summary. Human cowpox is a rare but relatively severe infection of interest because of its links with Edward Jenner and the introduction of smallpox vaccine and, more recently, because of re-evaluation of the epidemiology of the infection. This indicates that cowpox is not enzootic in cattle, relegates the cow to a minor role, and emphasizes the importance of feline cowpox as a source of human infection and of wildlife as virus reservoirs. The evidence available suggests that the virus is of low infectivity for humans and should not become an increasing problem despite the cessation of smallpox vaccination and increasing numbers of immunocompromised individuals.

Introduction

Attention was first drawn to cowpox in Edward Jenner's famous *Inquiry* of 1798, when he reported his cowpox inoculation (vaccination) experiments, the first of which was done 200 years ago on May 14, 1796 [4, 6, 30]. Jenner provided a coloured plate illustrating natural human cowpox and, particularly in a later monograph [31], distinguished between 'true' and 'spurious' cowpox. The latter, which would not protect against smallpox, included what we would now recognize as pseudocowpox (milker's nodes) caused by a parapoxvirus antigenically unrelated to cowpox virus [4, 24].

During the early 19th century, when epidemiological studies were largely unknown, the germ theory of disease not established and laboratory help not available, cowpox was regarded as a natural disease of cattle with a bovine reservoir; hence the name, cowpox. The view that cowpox was enzootic in cattle persisted until the 1970s, although it was recognized that bovine cowpox was uncommon [16, 42]. It was also natural that, reflecting these views, human cowpox was regarded as an occupational hazard of dairyworkers etc. However, studies since then and reviewed here indicate that cowpox virus, which has a wide host range, has wildlife reservoirs and is seen most often in domestic cats.

This paper re-evaluates the risk of human cowpox in the light of this information, the declining residual immunity provided by smallpox vaccination, and the increasing proportion of immunocompromised individuals in the community.

Human cowpox

Human cowpox is rare, and although individual cases were described, there was no attempt to collate clinical and epidemiological information on different cases until the authors' survey of ten human cases in 1977 [2]. This was extended to 54 cases by 1993, by which time more certain information was available on severity, age, sex, number and location of lesions, and particularly on the source of infection [5].

Infection is through a break in the skin and this determines the location of the primary lesion (Table 1). The hand is most commonly affected, but a relatively high proportion of patients (39%) had facial lesions and in 16% the eye was involved. There was a preponderance of adolescent males with facial lesions which may be connected with poor shaving technique or puberty.

The majority of patients have only one lesion. Additional lesions may be coprimaries [9, 48] or occasional secondary lesions caused by accidental transfer from a developing primary lesion [9, 48]. More rarely, multiple lesions may be due to lymphatic or viraemic spread [9, 43]. Generalized infections are rare and occur only in patients with predisposing factors such as immunodefiency, eczema or dermatitis [9, 22, 32].

The lesions which are invariably painful start as inflamed macules and pass through papules and haemorrhagic vesicles over a period of 7–12 days. Cases are rarely seen at this stage and by the time they are, usually after 2–3 weeks, the lesion will have progressed to a hard black eschar (for colour illustrations see [9]). Inflammation, oedema, and lymphadenopathy are common and most

Table 1. Human cowpox: location of lesions in relation to age and sex of patient (54 cases)[a]

Location	Age (years)[b]						
	< 12		12–18		>18		
	m	f	m	f	m	f	Total (%)
Hands only	2	5	0	2	8	8	25 (46)
Face only	1	3	4	2	5	3	18 (33)
Hands & face	0	1	0	0	0	0	1 (2)
Misc[c]	0	2	0	1	3	1	7 (13)
Generalized	0	0	2	0	1	0	3 (6)
Total	3	11	6	5	17	12	54

[a]From data in [9]
[b]Age range 1–64 years
[c]For example elbow, knee, anogenital

patients report some systemic reaction such as pyrexia, malaise, lethargy, sore throat, and/or 'flu-like' symptoms. Sixteen out of 54 patients (30%) of whom 9 were <18 years were admitted to hospital for periods of between 3 and 24 days, and most others were absent from school or work and required bed-rest.

The majority of patients recover in 6–8 weeks although in some, complete healing may take 10–12 weeks or longer. Scarring is common and permanent, and cosmetic surgery may be considered for facial scars. Eye infections may result in conjunctivitis sometimes with corneal involvement; recovery is complete though often slow [9]. Rare deaths due to encephalitis or generalized infection have been recorded [9, 22]. In general, treatment is supportive with antibiotics used to control any secondary bacterial infection. Corticosteroids exacerbate the infection and Acyclovir is ineffective against poxvirus infection.

The most important differential diagnoses are: milker's nodes which is usually less painful and in which a bovine source can often be traced; herpes simplex, usually less haemorrhagic and erythematous and which should be recognized if it is a reactivation; and cutaneous anthrax, relatively painless and rare in Britain [9]. Electron microscopy can be of great value. All but one of 24 cases from which suitable material was available were diagnosed by this method [9]. Epidemiological features which might suggest cowpox include; involvement of all age groups, but perhaps more commonly young girls (Table 1); presentation particularly during July to November; and, as discussed below, contact with domestic cats [9, 12].

Epidemiology of cowpox

Bovine cowpox

Although there is evidence that bovine cowpox was uncommon during the 19th century [4] the idea that cows were the reservoir host persisted. Consequently when viruses were isolated from human and bovine cases and shown to be different from other poxviruses [23, 24] the term cowpox virus was adopted. However there is now strong evidence that cows are not the reservoir host of cowpox virus (Table 2).

Table 2. Sources of 56 cases of human cowpox 1969–93[a]

Source	Total	%
Unknown[b]	26	46
Feline [b]	18	32
Occupational[c]	7	12.5
Bovine	3	5
Rodents[d]	2	3.6

[a]From data in [9, 50]
[b]For 1985–93 unknown = 30%, feline = 55%
[c]Veterinary, laboratory and zoo workers
[d]Circumstantial evidence only

Cowpox has been detected in Europe including Britain but not Ireland, and some of the states of the former USSR. There are no reliable reports from the Iberian Peninsula and it is not present in the Americas, Africa, and Australasia [8, 14]. If cowpox was enzootic in cattle it would have been exported along with infected animals, in the same way as pseudocowpox, in the days before quarantine was introduced. Surveys in Great Britain indicated that clinical bovine cowpox was very rare [29], and no case of bovine cowpox has been detected there since 1976 ([9], authors' unpublished data). Serological surveys also indicated a low prevalence of antibody to cowpox in the British herd; only 0.65% of 1 076 animals tested had antibodies and in most of those the titres were very low [2]. All this evidence suggested that cowpox is not enzootic in cattle. Further evidence comes from observations on the epidemiology of human cowpox. Although occasional cases of cowpox in farmworkers have a known bovine source [2], only 3 of 28 cases with a known source were acquired from cattle [9]. Further, the age- and sex-distribution of human cases, with 26% of 54 cases occurring in children <12 years and 20% in young girls (Table 1), suggests a non-bovine non-occupational source for most cases (see below).

It is difficult to see how a population of biologically-homogeneous strains of cowpox virus could be maintained in cattle, and the problem is compounded by the co-existence of stable, biologically-distinct strains of cowpox virus [1, 45]. The inescapable conclusion from analysis of these data was that cowpox virus is not enzootic in cattle and that most likely reservoirs were to be found among wildlife, probably rodents [2, 3, 8]. These conclusions are now generally accepted [25, 26, 41, 44] and evidence on wildlife reservoirs is discussed below.

Cowpox in non-native species

From 1960 onwards various isolations of orthopoxviruses have been made from infection in exotic species kept in Continental and British zoos or circuses. The animals concerned included okapi [57], lions, black panthers, cheetahs, pumas, jaguars, ocelots and anteaters in Moscow Zoo [40], black and white rhinoceros in Munster and Frankfurt zoos [50], cheetahs in English zoos [7], but particularly in African and Indian elephants in a number of European zoos and circuses [28, 33, 50]. Although the viruses isolated in England were clearly strains of cowpox virus [7, 45], the taxonomic position of the Continental isolates was less certain. They are very closely-related to typical strains of cowpox virus although minor differences can sometimes be detected, and the term 'cowpox-like viruses' was sometimes used to describe them [10, 11, 44, 49]. It is now agreed that these strains represent slightly different isolates of cowpox virus [24–26]. The animal species from which these virus strains were isolated are not native to Europe, and the obvious conclusion was that the virus was introduced into their environment from some indigenous wildlife reservoir [3].

Feline cowpox

Cowpox was first detected in the domestic cat in England in 1978 [58]. Since then many cases of feline cowpox have been investigated in Britain and Continental Europe [12, 13, 17, 37, 44, 45, 47], and the cat is now the most commonly-detected victim of cowpox virus. The infection in cats, though producing generalized skin lesions, is usually benign and most animals recover uneventfully; cat-to-cat transmission is very rare [12, 13]. All the evidence available from field and laboratory studies suggests that, as with cows and exotic species, the cat is not a reservoir host of cowpox virus [9, 12, 13, 37, 41, 44]. However, at least in Western Europe including Britain, the domestic cat plays an important role in the epidemiology of human cowpox [9]. Domestic cats were first recognized as a source of human cowpox in 1985 [48, 55]. Since then a number of cases of human cowpox have been shown to have a feline source [9, 15, 22, 32, 43, 44], and at present 50% or more human infections are acquired from cats (Table 2). Obviously the cat acts as the important link between the probable wildlife reservoirs and humans.

Wildlife reservoirs of cowpox virus

Investigations into possible wildlife reservoirs of cowpox virus are hindered by the extremely low rate of cowpox virus isolation from such animals, and the close antigenic relationship among orthopoxviruses [25, 26]. In this last respect it is of interest that ectromelia virus has recently been detected in wild mice in central Europe and which may complicate serological surveys there [38].

Although ectromelia (mousepox) virus has been isolated from laboratory mice cowpox is believed to be the only orthopoxvirus naturally indigenous to Britain. Antibody reacting with cowpox virus has been detected in bank voles. (*Clethrionomys glareolus*), field voles (*Microtus agrestis*) and woodmice

Table 3. Orthopoxvirus antibody in British wildlife[a]

Species	No. tested	No positive[b]	%
Field vole	24	8	33
Bank vole	63	25	40
Wood mouse	106	17	16
House mouse	44	1	2
Rat	65	0	<1.5[c]
Badger	31	0	<3.0
Squirrel	39	0	<2.9
Deer	27	0	<3.7

[a]From data [21] and authors' unpublished data

[b]Tested by virus neutralization and/or immunofluorescence

[c]Figure given is % represented by 1 positive specimen out of the number sampled

(*Apodemus sylvaticus*) ([21]; Table 3). Although virus has yet to be detected in these wild species, we believe this reflects cowpox infection in its British wildlife reservoirs, particularly as these species can be infected experimentally with low doses of cowpox virus but not with high doses of ectromelia virus [14].

Cowpox was first detected in Russia in 1973–74 when it was shown to be the cause of the Moscow Zoo outbreak [40, 41]. At about the same time antibody reactive with cowpox virus was detected in 18.6% of great gerbil (*Rhombomys opimus*) and 15% of yellow suslik (*Citellus fulvus*) populations in Turkmenia [39, 41]. Isolation of cowpox virus from small numbers of these species strengthens the conclusion that they represent wildlife reservoirs of the virus [41]. More recently virus and antibody to it was detected in gerbils (*Meriones spp.*) in Georgia former USSR [53, 54], and virus in root voles (*Microtus oeconomus*) in northern Russia [35].

Significant progress has been made in our understanding of the epidemiology of human cowpox, in particular by emphasizing the role of the domestic cat as a source of human infection. However preliminary evidence from experimental studies that virus strains may vary in their pathogenicity for cats needs further investigation [13, 56]. More work is also needed to confirm the identities and assess the relative importance of the different wildlife reservoirs. The development of tests for detecting cowpox-specific antibody and of sensitive tests for detecting virus, perhaps using the polymerase chain reaction, should be of value in such work.

The risks of human cowpox

The risks of human cowpox are a combination of the risks of acquiring infection together with the severity of infection if acquired. The former is determined by; the geographical distribution of the virus; its circulation among reservoir and other hosts; human contact with them which may depend on age, sex, and occupation; and the infectivity of the virus for humans which may be affected by the immune status. The severity may then be affected by such factors as age, number and location of lesions, and factors such as predispositions which may exacerbate, or immunity which may attenuate the infection.

The geographical range of cowpox virus is limited (Table 4), but has been extended by the discovery of the virus in some states of the former USSR. Future work may provide further evidence of the virus in this general area. The most important developments relevant to human cowpox have been studies which show that cowpox virus does not have a bovine reservoir (Table 4), that cows are an insignificant source of human infection, and that domestic cats are the most important source of human infection in western Europe including Britain (Table 2). Bovine cowpox was always rare [4] and there is no reason to suppose there has been a change in the epidemiology; more intensive investigations have simply revealed the true picture.

Table 4. Features of cowpox which suggest it has no bovine reservoir[a]

1.	Geographically restricted; not found in Australasia, Americas, Africa [8, 41]
2.	Clinical bovine cowpox uncommon; no cases detected in UK since 1976 [9, 29]
3.	Low antibody prevalence in cattle; 0.65% of 1 076 UK cattle [2]
4.	Cows are rare source of human infection; 10% of 28 cases with known source, 5.5% of total [9]
5.	Age distribution of human patients suggests non-occupational exposure [9]
6.	Above problems compounded by co-existence of stable biologically-distinct strains of virus [1, 46]

[a]From data in papers indicated by bracketed numbers

At present domestic cats are the direct source of about 55% of human cases, and a relatively high proportion (20%) in young girls is consistent with them handling infected pets (Table 1). However we know of no human case acquired from a cat after the feline case has been diagnosed and appropriate precautions taken. Human and feline cases are most common during July–November [9, 12]. This is when rodent populations are high and active, and bank and field voles and woodmice, the probable reservoirs in Britain, are among the most common prey of domestic cats [19].

Although human cowpox has occasionally been acquired from infected zoo animals it is otherwise extremely rare in the former USSR. It is possible that the circulation of cowpox virus among the different and larger rodents there does not readily permit transmission to feline and human hosts.

Although the historical risk to dairymaids is in fact very low, occupational infections may occur in veterinary, laboratory, and zoo workers (Table 2). However care when handling infected animals and materials should reduce this. No direct source of infection has been found for a high proportion of cases (Table 2). This may reflect poorly-taken histories but also possibly indirect infection via contaminated thorn bushes, barbed wire etc. No certain evidence of human infection acquired directly from a rodent is available but there is circumstantial evidence in two cases [34, 51].

Human cowpox is a relatively severe infection with about 30% of patients admitted to hospital, many requiring bed-rest, and virtually all unable to work or attend school for a few days. Multiple lesions, particularly in young children would be expected to add to the severity. However, generalized life-threatening or fatal infections have only occurred in those with underlying factors such as immunosuppression or atopic eczema [9, 15, 22, 32].

Human cowpox is rare and it is of interest briefly to discuss possible factors which may alter, and in particular increase, the incidence. The rarity is perhaps most simply explained by evidence which indicates that cowpox virus is of low infectivity for humans (Table 5). However, there have been suggestions that the incidence of human cowpox in the past has been controlled by smallpox vaccination, and that the cessation of vaccination and waning immunity in those vaccinated may lead to an increased incidence and/or severity of cowpox

Table 5. Evidence to suggest that cowpox virus is of low infectivity for humans

No case-to-case spread
Low cat-to-human spread
(None after diagnosis of feline case)
Low incidence in eczematous
Low incidence in immunocompromised
No cases reported in HIV positive individuals

Table 6. Incidence of 'primary' responses to smallpox vaccine in individuals revaccinated after various time intervals

Study	Reference	Interval[a]	No.	No. positive[b]	% positive[b]
UK, 1947	[18]	< 5	355	24	6.8
–		5–10	89	8	9
–		> 10	783	254	32
UK, 1961	[20]	NS[c]	520	26	5
–		>10	NS	NS	44.7
USA, 1970–71	[27]	0.5	213	22	10.3
Japan, 1975	[25]	1	714	NS	19

[a]Interval between primary and re-vaccination in years
[b]Number and % of individuals responding with a 'primary' take on re-vaccination
[c]Not stated

[15, 22, 32, 41, 46]. Vaccinia worked well against smallpox which was a systematic infection [25]. However, although it may reduce the severity, we believe it will provide only limited and uncertain protection against cowpox. We base this conclusion on the general, long-established practice of successful smallpox revaccination [25], and the results of trials (Table 6) which showed the relatively poor protection which primary smallpox vaccination offered to re-vaccination [5, 20, 25, 27]. This, and the often under-estimated morbidity of vaccination is what led British health authorities not to recommend smallpox vaccination for those working with cowpox and vaccinia viruses [5].

The small number of cowpox cases makes proper analysis difficult (Table 6). Routine smallpox vaccination was discontinued in Britain in 1971. Before then and up to abut 1981 only 1–2 cases of human cowpox were detected per year, despite particular interest being shown in cowpox from about 1969 onwards [2, 9]. Although there has been an increase to about 2.5 cases per year during 1985–95 (Fig. 1), this may be as much due to increased attention paid to developments in our understanding of the epidemiology as to waning immunity. For example, slight increases during 1978–81 (8 cases), and 1987–90 (11 cases), followed increased publicity, with only 5 cases in the intervening period. Case numbers have declined since 1992 and only one case was detected in 1996 (Fig. 1).

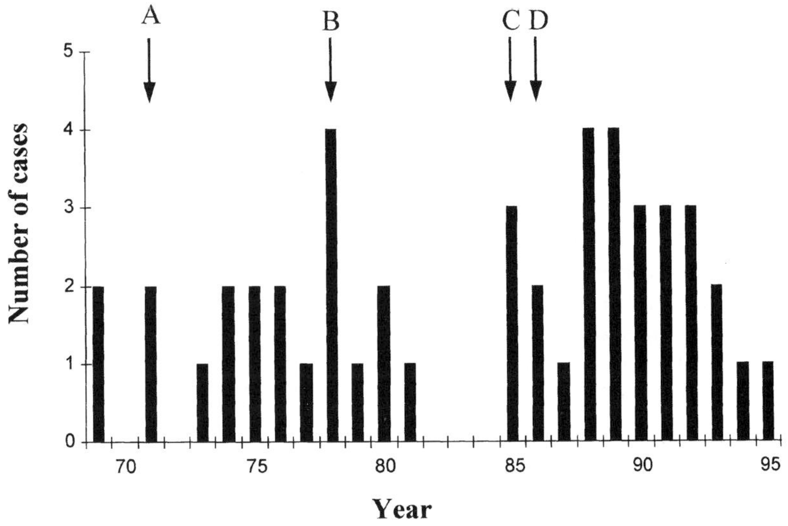

Fig. 1. Human cowpox in Britain 1969–1995. **A** Smallpox vaccination discontinued 1971. **B** Cowpox first detected in domestic cats, 1978. **C, D** Attention first focused on cats as a source of human infection 1985–1986

Human cowpox is severe in those with eczematous conditions and in the immunocompromised, but in the past has been rare in such patients; part of the evidence that cowpox virus is of low human infectivity. Progress in cancer therapy and transplant surgery has increased the number of individuals receiving immunosuppressive therapy, and to this population must be added the increasing number of HIV-positive individuals. However, although the number of immunocompromised individuals will increase the risk of cowpox for each person should remain the same. Similarly the risk to the eczematous individuals, for whom vaccination would have been contra-indicated, will remain the same.

Finally, despite increased interest and continued work on cowpox and its reservoir hosts, increasing financial pressure and restraints in the British health services may mean that fewer human cases will be investigated. If so, future trends in incidence, severity, and epidemiology could go unexplored.

References

1. Baxby D (1975) Laboratory characteristics of British and Dutch strains of cowpox virus. Zentralbl Vet Med B22: 480–487
2. Baxby D (1977) Is cowpox misnamed? a review of ten human cases. Br Med J 1: 1 379–1 381
3. Baxby D (1977) Poxvirus hosts and reservoirs. Arch Virol 55: 169–179
4. Baxby D (1981) Jenner's smallpox vaccine; the riddle of vaccinia virus and its origin. Heinemann, London
5. Baxby D (1993) Indications for smallpox vaccine; policies still differ. Vaccine 11: 395–396

6. Baxby D (1996) Edward Jenner's role in the introduction of smallpox vaccination. In: Plotkin S, Fantini B (eds) Vaccinia, vaccination, vaccinology: Jenner, Pasteur and their successors. Elsevier, Paris, pp 59–64
7. Baxby D, Ashton DG Jones DM, Thomsett LR (1982) An outbreak of cowpox in captive cheetahs; virological and epidemiological studies. J Hyg 89: 365–372
8. Baxby D, Bennett M (1994) Cowpox virus. In: Webster RG, Granoff A (eds) Encyclopedia of virology. Academic Press, Orlando, pp 261–267
9. Baxby D, Bennett M, Getty B (1994) Human cowpox; a review based on 54 cases, 1 969–1 993. Br J Dermatol 131: 598–607
10. Baxby D, Ghaboosi B (1977) Laboratory characteristics of poxviruses isolated from captive elements in Germany. J. Gen Virol 37: 407–414
11. Baxby D, Shackleton WB, Wheeler J, Turner A (1979) Comparisons of cowpox-like viruses isolated from European zoos. Arch Virol 61: 337–340
12. Bennett M, Gaskell CJ, Baxby D, Gaskell RM, Kelly DF, Naidoo J (1990) Feline cowpox virus infection. J Small Anim Pract 31: 167–173
13. Bennett M, Gaske RM, Gaskell CJ, Baxby D, Kelly DF (1989) Studies on poxvirus infection in cats. Arch Virol 104: 19–33
14. Bennet M, Crouch A, Begon M, Duffy B, Feore S, Gaskell RM, Kelly DF, McCracken CM, Vicary L, Baxby D (1997) Cowpox in British voles and mice. J Comp Pathol 116: 35–44
15. Blackford S, Roberts DL, Thomas PD (1993) Cowpox infection causing a generalized eruption in a patient with atopic dermatitis, Br J Dermatol 129: 625–627
16. Blood DC, Henderson JA (1974) Veterinary medicine, 4th ed. Balliere and Tyndall, London
17. Bomhard D, Plfeghaar S, Mahnel H (1992) Zur Epidemiologie, Klinik, Pathologie und Virologie der Katzenpocken-Infektion. Kleintierpraxis 37: 219–230
18. Broom JC (1947) Revaccination in adults. Lancet i: 364–366
19. Churcher PB, Lawton JH (1987) Predation by domestic cats in an English village. J Zool Lond 212: 439–455
20. Cross RM (1961) Observations on the classification and interpretation of reactions to smallpox vaccination. Bull World Health Organ 25: 7–17
21. Crouch AC, Baxby D, McCracken CM, Gaskell RM, Bennett M (1995) Serological evidence for the reservoir hosts of cowpox virus in British wildlife. Epidemiol Infect 115: 185–191
22. Czerny C-P, Eis-Hübinger AM, Mayr A, Schneweiss KE, Pfeiff B (1991) Animal poxviruses transmitted from cat to man; current event with lethal end. J Vet Med B38: 421–431
23. Downie AW (1939) A study of the lesions produced experimentally by cowpox virus. J Pathol Bacteriol 48: 361–379
24. Esposito JJ, Baxby B, Black D, Dales S, Darai G, Dumbell KR, Granados RR, Joklik WK, McFadden G, Moss B, Moyer RW, Pickup DJ, Robinson AJ, Tripathy DN (1995) Poxviruses. In: Murphy FA, Fauquet CM, Bishop DHL, Ghabrial SA, Jarvis AW, Martelli GP, Mayo MA, Summers MD (eds) Virus Taxonomy. Classification and Nomenclature of Viruses. Sixth Report of the International Committee on Taxonomy of Viruses. Springer, Wien New York, pp 79–91 (Arch Virol [Suppl] 10)
25. Fenner F, Henderson DA, Arita I, Jezek Z, Ladnyi ID (1988) Smallpox and its eradication. WHO, Geneva
26. Fenner F, Wittek R, Dumbell KR (1989) The orthopoxviruses. Academic Press, San Diego

27. Galasso GJ, Karzon DT, Katz SL, Krugman S, Neff J, Robbins FC (1977) Clinical and serologic study of four smallpox vaccines comparing variations of dose and route of administration. J Infect Dis 135: 131–186
28. Gehring H, Mahnel H, Mayer H (1972) Elephantenpocken. Zentralbl Vet Med B19: 258–261
29. Gibbs EPJ, Johnson RH, Collings DH (1973) Cowpox in a diary herd in the United Kingdom. Vet Rec 92: 56–64
30. Jenner E (1798) Inquiry into the causes and effects of the variolae vaccinae. Sampson Low, London
31. Jenner E (1799) Further observations on the variolae vaccinae. Sampson Low, London
32. Klingebiel T, Vallbracht A, Döller G, Steirhof YD, Gerth HJ, Glasshauser E, Herzau V (1988) A severe human cowpox infection in Germany. Pediatr Infect Dis 7: 883–885
33. Kubin GO, Kölbl O, Gerstl F (1975) Charakerisierung eines vom Elefanten isolierten Pockenvirusstammes. Wien Tierarztl Monatschr 62: 271–276
34. Lewis-Jones MS, Baxby D, Cefai C, Hart CA (1993) Cowpox can mimic anthrax. Br J Dermatol 129: 625–627
35. Lvov SD, Gromashevskyi VL, Marennikova SS, Bogoyavlenskyi GV, Bajluk FN, Butenko AM, Guschina E, Shelukhina EM, Thukova OA, Morozova T (1988) Poxvirus isolation from *Microtus oeconomus* Pal 1776 in Colsky peninsula. Vop Virusol 1: 92–94 (Russian).
36. Maenhout T, Declercq J, de Keuster T, Hoorens (1991) Drie gevallen van koepokken-infektie bij de kat in Belgie. Vlaams Diergeneeskd Tijdschr 60: 64–67
37. Mahnel H (1991) Katzenpocken in Deutschland Tierarztl Prax 19: 419–422
38. Mahnel H, Holejsovsky J, Bartek P, Czerny C-P (1993) Kongenitale 'Ektromelie' bei Pelztieren durch *Orthopoxvirus muris*. Tierärztl Praxis 21: 469–472
39. Marennikova SS (1979) Field and experimental studies of poxvirus infections in rodents. Bull World Health Organ 57: 461–464
40. Marennikova SS, Maltseva NN, Korneeva VI, Garanina NM (1977) Outbreak of pox disease among Carnivora (*Felidae*) and Edentata. J Infect Dis 135: 358–366
41. Marennikova SS, Shelukhina EM, Efremova EV (1984) New outlook on the biology of cowpox virus. Acta Virol 28: 437–444
42. Miller WL, West GP (1970) Black's veterinary dictionary, 9th ed. Black, London
43. Motley RJ, Holt PJA (1990) Cowpox presenting with sporotrichoid spread: a case report. Br J Dermatol 122; 705–708
44. Munz E, Lickh S, Renner-Müller ICE (1992) Infektionen mit originärem Kuhpockenvirus und kuhpockenähnlichen Erregern bei Mensch und Tier: Eine Literaturübersicht. J Vet Med B39: 209–225
45. Naidoo J, Baxby D, Bennett M, Gaskell RM, Gaskell CJ (1992) Characteristics of orthopoxviruses isolated from feline infections in Britain. Arch Virol 125: 261–272
46. Nasemann T, Mayr A, Schaeg G, Kimmig W, Mahnel H (1987) Infektion eines Mädchens mit Kuhpockenvirus. Hautarzt 38: 414–418
47. Novotny N, Fischer OW, Schilcher F, Schwendenwein I, Loupal G, Schorzmann T, Meyer J, Hermanns W (1994) Pockenvirusinfektionen bei Hauskatzen; klinische, pathohistologische, virologische und epizootiologische Untersuchungen. Wien Tierärztl Monatschr 8: 362–369
48. Pether JVS, Trevains PH, Harrison SRB, Baxby D, Bennett M (1986) Cowpox from cat to man. Lancet i: 38–39
49. Pilaski J, Rösen A, Darai G (1986) Comparative analysis of the genomes of orthopoxviruses isolated from elephant, rhinoceros, and okapi by restriction enzymes. Arch Virol 88: 135–142

50. Pilaski J, Rösen-Wolff A (1988) Poxvirus infection in zoo-kept mammals. In: Darai G (ed) Virus diseases in laboratory and captive animals. Martinus Nijhoff, Boston, pp 84–100
51. Postuma BH, Diepersloot RJA, Niessen GJC, Droog RP (1991) Cowpoxvirus-like infection associated with rat bite. Lancet i: 733–734
52. Thomsett LR, Baxby D, Denham EM (1978) Cowpox in the domestic cat. Vet Rec 108: 567
53. Tsanava SA, Marennikova SS, Sakvarelidze LA, Shelukunia EM (1989) Isolation of cowpox virus from the red-tailed Libyan jird. Vop Virusol 1: 95–97 (in Russian)
54. Tsanava SA, Sakvarelidze LA, Shelukhina EM (1989) Serologic survey of wild rodents in Georgia for antibodies to orthopoxviruses. Acta Virol 33: 91
55. Willemse A, Egberink HF (1985) Transmission of cowpox infection from domestic cat to man. Lancet i: 1 515
56. Zhukova OA, Tsanava SA, Marennikova SS (1992) Experimental infection of domestic cats by cowpox virus. Acta Virol 36: 329–331
57. Zwart P, Gispen R, Peters JC (1970) Cowpox in okapis *Okapia johnstoni* at Rotterdam zoo. Br Vet J 127: 20–24

Authors' address: Dr. D. Baxby, Department of Medical Microbiology, Liverpool University, Liverpool L693GA, U.K.

Characterization of a cowpox-like orthopox virus which had caused a lethal infection in man

C.-P. Czerny[1,2], C. Zeller-Lue[1], A. M. Eis-Hübinger[3], O.-R. Kaaden[1,2], and H. Meyer[4]

[1]Institute for Medical Microbiology, Infectious and Epidemic Diseases,
[2]WHO Collaborating Centre for Collection and Evaluation of Data on Comparative Virology, Veterinary Faculty, Ludwig-Maximilians-University, Munich, Germany
[3]Institute for Medical Microbiology and Immunology, University of Bonn, Bonn, Germany, [4]Institute of Microbiology, Federal Armed Forces Medical Academy, Munich, Germany

Summary. In August 1990 an orthopox virus (OPV) had been isolated from a severe case of a generalized infection with lethal outcome in an immunosuppressed 18-year-old man. In this communication we present a detailed characterization of the causative virus strain. Based on distinct epitope configurations detected by various monoclonal antibodies the isolate could be differentiated from other OPV species and was classified as a cowpox virus (CP). This classification was confirmed by a species-specific PCR assay and by establishing physical maps for the restriction enzymes HindIII and XhoI. Based on serological data of neutralization assays, blocking-ELISAs and Western blotting analysis evidence is provided that the infection had beed acquired from a stray cat.

Introduction

Human cowpox is a relatively rare zoonosis and cows are currently not known to be involved [6]. In recent years virus strains drew medical attention causing localized and generalized skin affections in humans [12, 18, 19, 29, 39, 40, 45, 49–51] as well as in cats [7–9, 21, 30, 31, 38, 46, 48] or zoo-animals like large felides or elephants [4, 5, 32, 52]. Usually, a low infectivity for healthy persons is observed and benign skin lesions are found mainly localized on fingers, hands, arms, face or neck. The clinical features are often misinterpreted. In predisposed and immunocompromised individuals the infection can lead to severe diseases which may end lethally [19]. The virus strains isolated belong to the genus *Orthopoxvirus* and owing to their biological qualities they are classified as "cowpox-like virus" [7, 29, 38]. Based on serological surveys it is assumed that these virus strains are maintained in wildlife rodents with humans, cats, elephants, and cows as final hosts [6, 14, 16, 32]. Most of the human cowpox infections described so far were acquired from cats or elephants [6, 12, 14, 18, 39, 40, 45, 49, 51].

In this paper we are reporting about the detailed characterization of the human isolate OPV-90/2 based on antigenic sites and DNA analysis. It was isolated in 1990 from lesions of a generalized fatal infection of an 18-year-old, immunosuppressed and non-vaccinated man of Lüdenscheid, Germany. The clinical picture resembled smallpox [14, 19, 41]. The virus has been typed as a cowpox virus based on the pock morphology after infection of the chorioallantoic membrane (CAM) and based on skin lesions after intradermal inoculation of rabbits [14]. According to investigations described herein, including the determination of antigenic sites, PCR, and DNA analysis, this classification was confirmed.

Materials and methods

Cells and viruses

The permanent African green monkey kidney cell line MA-104 cultured in minimum essential medium (MEM) and supplemented with 5% fetal calf serum was used to propagate the cowpox virus reference strain Brighton (BR), the Lister strain (LS) of vaccinia virus [28], elephantpox virus EP-1 as well as the isolate OPV-90/2. Infectivity titres were determined on 24-well plates (Nunc, Wiesbaden) and calculated as plaque forming units (pfu). The virus preparations were purified and concentrated by sucrose gradient centrifugation [13, 24]. Protein concentrations were determined by the method of Lowry and co-workers [27].

Preparation of polyclonal hyperimmune sera

Polyclonal hyperimmune sera against purified vaccinia virus MVA, cowpox virus BR, and OPV-90/2 were raised in rabbits (Chinchilla Bastard) by four subcutaneous injections of 175 µg antigen/animal at intervals of three weeks.

Detection of OPV-specific antibodies in sera

Antibodies against OPV were determined in human and feline sera using a previously established blocking-ELISA [16]. In short, the 96-well microplates (Immunoplate II, Nunc, Wiesbaden) were directly coated for 4h at 37 °C with 1 µg/ml of vaccinia virus MVA diluted in carbonate/bicarbonate-buffer (pH 9.6; 100 µl/well). Log_2-dilutions (in PBS containing 10% FCS) of feline or reference sera were titrated in duplicate wells (0.1 ml/well) and incubated for 30 min at 37 °C. Subsequently, tissue culture fluids of the MAbs were adjusted to an optimal working dilution in PBS containing 10% FCS. Plates were incubated for 30 min at 37 °C (0.1 ml/well), washed four times before horseradish peroxidase- (HRP) conjugated anti-mouse-IgG-globulins (1:1 000 in PBS+10% FCS; Dako, Hamburg) were added to each well (0.1 ml/well). After an incubation of 30 min at 37 °C and four washings, the reaction was made visible by using 3,3′,5,5′-tetramethylbenzidine (TMB; Serva, Heidelberg) as indicator (0.1 ml/well). Ten minutes later the reaction was stopped by 2M H_2SO_4 (0.05 ml/well) and measured in a Titertek photometer (Flow, Bonn) at a wavelength of 450 nm (optical density $OD_{450\,nm}$). Reduction of the photometer extinction of the MAbs by the competing serum antibodies was calculated by the formula: %-inhibition = [1−$OD_{450\,nm}$ (MAb+MAb-conjugate)/$OD_{450\,nm}$ MAb-conjugate]× 100. A reduction of the photometer extinction ⩾ 50% was regarded as a significant inhibition.

Plaque-reduction test

The plaque-reduction test (PRT) was performed by standard methods [33] without addition of complement. Sera were diluted 1:2 with EMEM, titrated in log_2-steps, before the equal volume of vaccinia virus (VV) Munich 1 or cowpox virus BR was added, each containing 100 pfu. After an incubation for 2 h at 37 °C, the virus/serum mixtures were transferred to 24-well Linbro plates (100 µl/well) containing a monolayer of 5×10^5 MA-104 cells/well. The plates were left for 1 h at room temperature, washed twice, and EMEM supplemented with 2% FCS and 1% dimethylsulfoxide (DMSO) was added. The plates were incubated for 48 h at 37 °C and stained with crystal violet.

Species-specific differentiation of orthopox viruses by an antigen-capture-ELISA

Orthopox viruses can be differentiated by a previously established monoclonal species-specific antigen-capture-ELISA (Czerny et al., paper in prep.). In short, monoclonal antibodies against VV MVA, as well as against cowpox virus BR, monkeypox virus (MP) Kopenhagen, and ectromelia virus (EM) Munich 1 were established as described previously [13, 15]. Eight MAbs offered species-specific characteristics in detection of their antigenic sites localized on the 14kD fusion protein, the 32kD adsorption protein, and the A-type inclusion body (ATI) protein encoded by the open reading frames ORF A27L, D8L, and A25L [22]. These Mabs were chosen to capture orthopox virus reference strains and isolates derived from clinical samples. Antigen detection was performed with an equivalent mixture of rabbit hyperimmune sera raised against vaccinia, monkeypox, and ectromelia virus. Staining was achieved with horseradish conjugated anti-rabbit immunoglobulins, 3,3′,5,5′-tetramethylbenzidine (TMB) and 2M H_2SO_4.

SDS-PAGE and Western blotting

Proteins of purified OPV strains (10 µg/slot) were fractionated on vertical 12% sodium dodecyl sulfate (SDS)-polyacrylamide gels [26]. The proteins were silver-stained [34] or subsequently transferred to nitrocellulose membranes [47]. Immunodetection was performed by standard techniques using either human, feline and rabbit sera (1:150) or monoclonal antibodies (1:2 diluted tissue culture fluids, HRP-conjugated anti-IgG antisera (1:250), and HRP colour developing reagent (Bio-Rad, Munich)). The M_r of stained viral proteins was determined in a Beckman gel mate 1 000 sonic digitizing system (Beckman, Munich) on the basis of M_r standards from 250 kD to 14.3 kD (Sigma, Munich).

Orthopox virus-specific polymerase chain reaction

Identification and differentiation of OPV were achieved by PCR using a single primer pair based on the gene encoding the acidophilic inclusion (ATI) protein [36]. Amplification of OPV-DNA resulted in amplicons of different size depending on the species. Briefly, 5 µl of 10x reaction buffer (1x=50 mM KCl, 10 mM Tris-HCl, pH 8.3, and 2.5 mM $MgCl_2$), 1 µl of sample DNA, 200 µM of each dNTP, 100 ng of each primer ATI-up-1 (5′-AATACAAG-GAGGATCT-3′) and ATI-low-1 (5′-CTTAACTTTTTCTTTCTC-3′), 2.5 U Taq polymerase

(Boehringer, Mannheim) and water were combined to give a total volume of 50 µl. Reactions were thermocycled 25 times. Each cycle included a denaturation step at 94 °C for 1 min, an annealing step at 40 °C for 1 min and an extension step at 72 °C for 2.5 min. To differentiate the amplicons more clearly and to prove specificity, 5 U of BglII (Boehringer, Mannheim) were added to 30 µl of the cycled PCR mixtures. Digests were incubated at 37 °C for 2 h followed by electrophoresis in 3% NuSieve-GTG agarose containing 1% Seakem-GTG agarose (FMC Corp., Hameln).

Southern blot hybridization

DNA of cowpox virus BR, vaccinia virus LS and isolate OPV-90/2 was digested with HindIII or XhoI, respectively. DNA-fragments were separated on 0.5% agarose gels, transferred to nylon membranes (Hybond N+, Amersham-Buchler, Braunschweig) and fixed by drying at 80 °C. DNA probes were prepared from either VV LS or CP BR cloned DNA-fragments [43] using a kit (DNA labeling and detection kit, non radioactive; Boehringer, Mannheim). After prehybridization, membranes were hybridized in the presence of labeled probes. The detection of filter-bound labeled DNA was done according to the manufacturer's protocol. The same Hybond-N+ membranes were probed in succession after removing DNA probes.

Identification of cross-linked restriction fragments

The termini of OPV are cross-linked covalently and renature rapidly on release from denaturating conditions [17]. This property can be exploited to identify the terminal restriction fragments. In the present study denaturation was achieved by treatment with 50% (v/v) formamide at 60 °C for 6 min followed by chilling prior to gel analysis.

Results

Characterization of OPV-90/2 based on species-specific antigenic sites

The isolate OPV-90/2 with a typical orthopox virus-like morphology in the electron microscope reacted positive in a genus-specific antigen-capture-ELISA [14, 23]. To classify the species-specificity within the genus *Orthopoxvirus* the isolate was tested in a newly established antigen-capture-ELISA based on eight distinct monoclonal antibodies reactive with epitopes on the 14 kD fusion-, the 32 kD adsorption-, and the ATI-protein. The isolate OPV-90/2 was compared to several OPV reference strains and virus isolates. OPV-90/2 had an identical reaction pattern with the eight MAbs (key 12–5 678) as cowpox virus strains of the antigenic group I (Fig. 1).

Characterization of OPV-90/2 by PCR

PCR experiments with primer pair ATI-up-1/ATI-low-1 and template DNA derived from cowpox virus BR and vaccinia virus LS and OPV-90/2 produced amplicons of the expected size of 1673 bp and 1603 bp, respectively [36]. The amplicon obtained with OPV-90/2 was of similiar size as compared to that of cowpox virus BR. In order to prove specificity and to more clearly differentiate the OPV species cowpox virus and vaccinia virus, the amplicons were digested

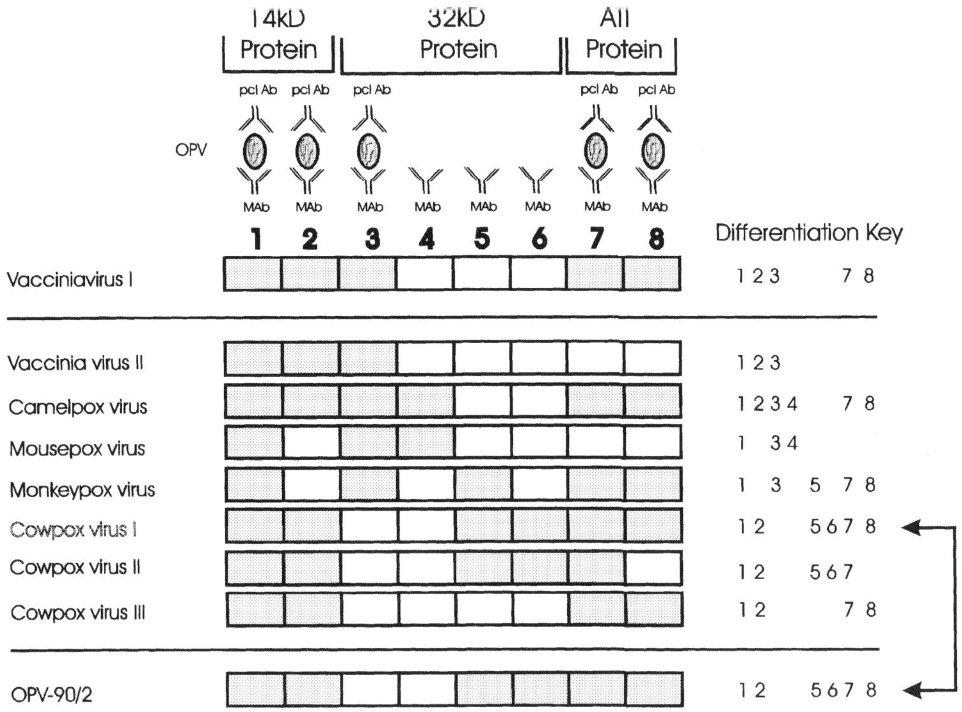

Fig. 1. Antigen capture-ELISA for the species-specific differentiation of orthopox viruses. Eight monoclonal antibodies were used as capture antibodies (*1* MAb anti-EM-5B1; *2* MAb anti-VV-5B4; *3* MAb anti-EM-1F7; *4* MAb anti-EM-4F1; *5* MAb anti-MP-1F8; *6* MAb anti-KR2-3E5; *7* MAb anti-MP-5F7; *8* MAb anti-MP-2C3 [15]. MAbs 1 and 2 are directed against the 14 kD fusion protein, MAbs 3, 4, 5, and 6 against the 32 kD adsorption protein, and MAbs 7 and 8 against the ATI-protein. Virus detection was achieved with an equivalent mixture of hyperimmune sera against VV, MP, and EM. The isolate OPV-90/2 reacted in the same way as cowpoxvirus strains of antigenic group I. ▒ Positive OPV-detection by the MAbs results in the species-specific differentiation key. ☐ Negative reaction

with BglII. This resulted in two different patterns. The pattern obtained for OPV-90/2 was identical to that of cowpox virus BR (data not shown).

Restriction endonuclease pattern of OPV-90/2

DNA-fragments resulting from cleavage of genomic DNA from vaccinia virus LS, cowpox virus BR, and OPV-90/2 with HindIII and XhoI were separated by electrophoresis (data not shown). Their sizes were determined by comparison with molecular weight standards. The total size of the genomes was calculated by summation. Independently, the summation of HindIII and XhoI-fragments resulted in similar values for OPV-90/2 and the cowpox virus reference strain BR (219–222 kb), whereas vaccinia virus reference strain LS displayed a 20 kb

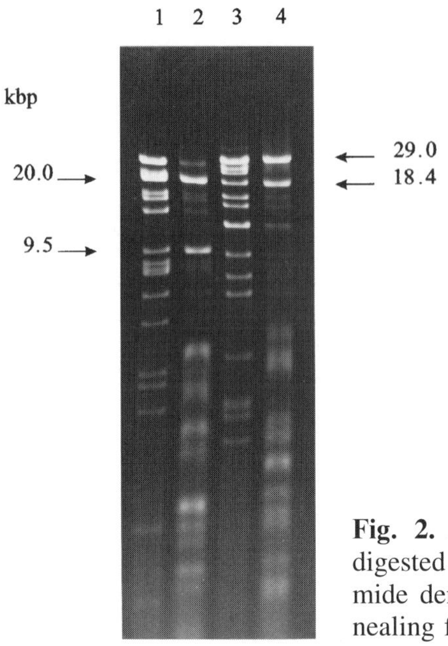

Fig. 2. Agarose gel electrophoresis of OPV-90/2 DNA digested with HindIII (*1*, *2*) or XhoI (*3*, *4*), either formamide denatured (*2*, *4*) or undenatured (*1*, *3*). Rapidly annealing fragments in *2*, *4* represent the terminal fragments

smaller genome. The sizes obtained for the reference strains correspond well with data of the literature [20].

In order to determine the terminal fragments OPV-90/2-DNA was digested with either HindIII or XhoI, denatured with formamide, and immediately electrophoresed in an agarose gel (Fig. 2). Two HindIII fragments (20.0 and 9.5 kb, resp.) comigrating with two fragments of an undenaturated HindIII digest of the same DNA were assigned as terminal fragments. A similar result was obtained with the XhoI digest (29.0 and 19.4 kb).

Mapping of restriction fragments of OPV-90/2

The HindIII and XhoI sites in the genome of OPV-90/2 could be mapped by cross-hybridization with Digoxigenin-labeled probes derived from vaccinia virus LS and cowpox virus BR. Published maps of the reference strains and the general assumption that cross-hybridizing fragments represent equivalent regions in the genome enabled construction of physical maps for the OPV-90/2 genome for HindIII and XhoI (Fig. 3). Based on the overall size of the genome and on the organization of the terminal regions OPV-90/2 can be regarded as a cowpox virus. Results here are consistent with earlier studies on OPV digests and maps [20], demonstrating a high degree of midregion DNA sequence conservation. Interspecies variance is reflected by differences in the sequence and the length of the termini.

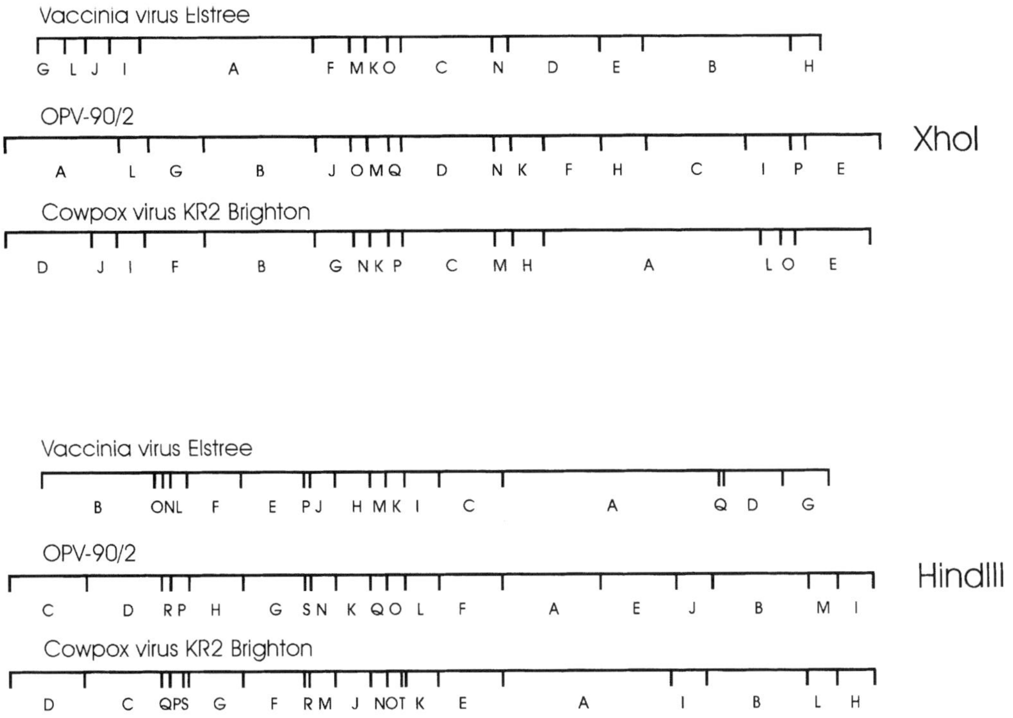

Fig. 3. Physical maps (HindIII, XhoI) established for cowpox virus BR and vaccinia virus LS [20] are compared with those obtained for OPV-90/2

Detection of OPV-specific antibodies

The serum from the 18-year-old man and from a stray cat were tested in a previously established monoclonal blocking-ELISA [16]. Antibody titres were 1:80 in the human and 1:640 in the feline serum. Neutralizing activities of the ELISA-positive sera were analyzed in a plaque-reduction test (PRT) using 100 pfu of vaccinia virus Munich 1 and cowpox virus BR. The titre of the human serum mounted to 1:40 against VV and cowpox virus BR, the titre of the feline serum was 1:320 against both virus strains. An anti-OPV-90/2 hyperimmune serum showed an ELISA titre of 1:3 200 and a PRT titre of 1:1 600 against VV Munich 1 or cowpox virus BR.

The feline serum had eight-fold higher titre levels in both tests than the human serum. This could also be confirmed by immuno-blotting analysis using purified VV MVA, cowpox virus BR, elephantpox virus EP1, and OPV-90/2 as the test antigens (Fig. 4). The rabbit anti-VV MVA (Fig. 4A) and rabbit anti-cowpox virus BR hyperimmune sera (Fig. 4E) served as positive controls and reacted well with virus proteins of 61kD, 42kD, 38kD, 37kD, 35kD, 32kD, 16kD, and 14kD. A similar pattern was detected by the anti-OPV-90/2 rabbit hyperimmune serum (Fig. 4D). The serum of the transmitter cat (Fig. 4B) reacted specifically but with lower intensity than the hyperimmune sera. The

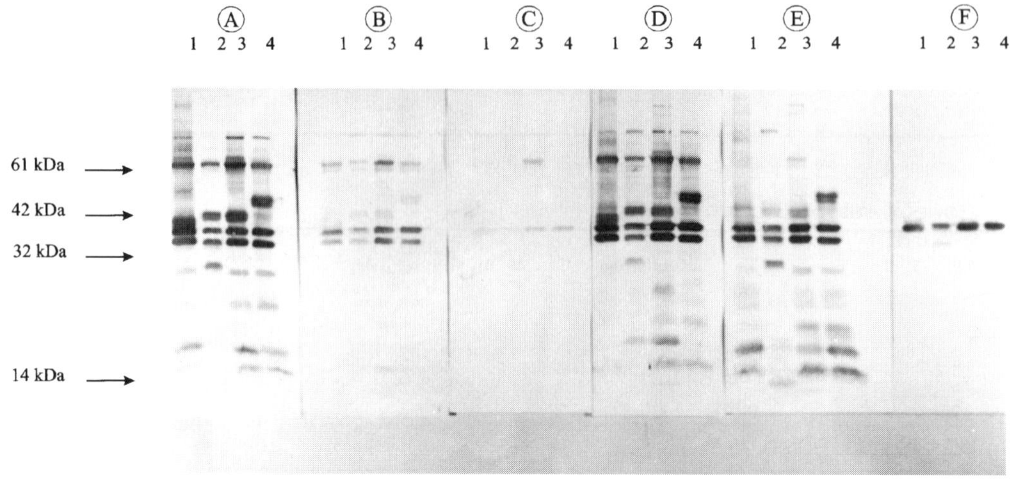

Fig. 4. Electrophoresis and separation of gradient-purified vaccinia virus MVA (*1*), elephantpox virus EP1 (*2*), OPV-90/2 (*3*) and cowpox virus KR2 Brighton (*4*) on 12% SDS-polyacrylamide gel under reducing conditions. Detection of immunogenic proteins in Western blotting was performed by anti-VV MVA rabbit-hyperimmune serum (**A**), serum of the transmitter cat (**B**), serum of the 18-year-old man (**C**), anti-OPV-90/2 rabbit-hyperimmune serum (**D**), rabbit anti-CP KR2 Brighton hyperimmune serum (**E**), and anti-CP KR2 Brighton monoclonal antibody 4C4 (**F**). The apparent M_r of immunoreactive proteins are given in kilodaltons (kDa). Negative human, feline or rabbit sera offered no reaction (data not shown)

61kD, 35kD, and 32kD proteins were the most prominent bands detected. The best reaction was seen with the causative agent OPV-90/2. The human serum (Fig. 4C) was weekly positive. It mainly detected the 35 kD envelope protein. The best reaction was seen with OPV-90/2 and cowpox virus BR. A monoclonal antibody against the 35kD envelope protein (Fig. 4) was positive with all antigen preparations.

Discussion

The fatal infection caused by the isolate OPV-90/2 is the only known case worldwide where a cowpox virus transmitted by a cat is responsible for a lethal disease in a human being. This event confirms the admonitions of some investigators not to ignore OPV after abolition of the compulsory vaccination [30, 31]. During the smallpox eradication campaign immunosuppressed children have not been vaccinated due to the well-known post-vaccinal complications like encephalitis or allergic eczema. A possible risk even of vaccinia virus for immunosuppressed persons was later demonstrated by a lethal VV-infection following vaccination of an HIV-infected recruit [42]. The 18-year-old male from Luedenscheid had not been vaccinated in his infancy because he suffered from atopic dermatitis and allergic asthma bronchiale requiring continuously steroid therapy since he was six years old [19, 41].

The involvement of cats in the epidemiology of OPV is already well-known for several years. Numerous isolates were obtained from ulcerative skin lesions mainly without tendency to generalize [7, 29, 31, 38]. Serological investigations in Germany confirmed that between 2 and 10% of the cats had serum antibodies against OPV without clinical signs [16, 51]. In the same way as other mammals and humans, cats obviously are final targets in the infective course. A transmission of OPV strains from cat to man by a close and intensive contact was suspected by several investigators [6, 16, 25, 30] because the clinical lesions were often seen in humans and animals living in the same household [10, 51]. In our case the isolate OPV-90/2 was clearly transmitted by a stray cat with a skin lesion at the anterior paw. Serum antibodies were detected by a monoclonal blocking-ELISA, as well as by a plaque-reduction test and Western blotting analysis. Titres of the feline serum were eight-fold higher than in the 18-year-old male. However, it has to be taken into account that the serum sample was drawn 14 days after the admission to hospital and approximately 18 days after first initial clinical signs. The feline serum titer was similar to that of cats with clinically apparent OPV infections [16].

The severity of this human OPV-90/2 infection rises the question about a potential hazard of OPVs which are present in European wildlife. All means of laboratory diagnosis valuable to differentiate OPV indicate that no new virus strain caused the disease. In the antigen-capture-ELISA established quite recently OPV-90/2 could be identified as cowpox virus. Due to close cross-reactivities within the species of the genus *Orthopoxvirus* a serological differentiation has not been successful in the past [23]. The combination of eight MAbs intensively characterized in their antigenic profile [13, 15, 35] now enables the differentiation of OPV-isolates into the species vaccinia virus (two subgroups), cowpox virus (three subgroups), monkeypox virus, camelpox virus, and ectromelia virus (Czerny et al., paper in prep.) based on species-specific epitope configurations. The differentiation can be performed with crude material and needs only 2.5 hours. Up to now more than 100 OPV strains and isolates have been investigated and in every case the ELISA classification agreed with the classification by phenotype markers or PCR.

Analysis of the viral genome using restriction enzymes also allows the differentiation of OPV-species [20, 28]. Whereas the internal part of the OPV-genome displays a very similar distribution of restriction endonucleases sites, differences among OPV-species are located near the ends of the genome. Comparison of HindIII and XhoI restriction enzyme patterns of OPV-90/2 with those of the cowpox virus reference strain BR provided no clear cut identification as a cowpox virus. This is in contrast to findings of Naidoo and co-workers [38] who investigated 12 cowpox viruses isolated in Britain and found a very close relationship between all isolates and the BR strain. However, by determining the genome size and by establishing physical maps OPV-90/2 could be clearly identified as a cowpox virus. Further studies will be needed to demonstrate whether the differences observed among cowpox viruses may prove to be of epidemiological value. The classification of OPV-90/2 as a cowpox virus by establishing physical maps was confirmed by the ATI-PCR amplicon restriction fragment length polymorphism assay [36, 37] and successfully applied for the identification and differentiation of 95 strains and isolates belonging to eight OPV species.

The lethal outcome of the OPV-90/2-infection obviously was also dependent on the immune status of the 18-year-old man. He was nearly life-long medicinally immunosuppressed. OPV-90/2 caused a pyrexia and multiple, partially confluent vesicles in his face. His body showed severe and deep necroses. The carrier cat was only affected with skin lesions mainly localized at the paws. On the other side, it is known that a parallel infection with FIV can cause a severe poxvirus disease in cats [11]. It is not fully understood what happens in an immunocompromised organism. However, there is evidence that poxviruses can also use several mechanisms of immune evasion to outwit the immune system [1, 2, 44]. In cowpox virology we are at the beginning to identify the corresponding mechanisms. Human OPV-infections have to be supervised even if they still remain singular cases and represent only a low risk for healthy persons [3, 6]. Meanwhile good diagnostic tools are available.

Acknowledgements

We are grateful to Gudrun Zöller, Andrea Jahnke, and Barbara Köttgen for their excellent technical assistance and Ursula Wolf who assisted us with the photographic documentation. We also thank Elke Bongers for the review of the manuscript. This work was partly supported by funds provided by the Fraunhofer Gesellschaft Munich and by the German Federal Ministry of Health.

References

1. Alcami A, Smith GL (1995) Vaccinia, cowpox, and camelpox viruses encode soluble gamma interferon receptors with novel broad species specificity. J Virol 69: 4633–4639
2. Alcami A, Smith GL (1995) Cytokine receptors encoded by poxviruses: a lesson in cytokine biology. Immunol Today 16: 474–478
3. Baxby D (1988) Human poxvirus infection after the eradication of smallpox. Epidemol Infect 100: 321–334
4. Baxby D, Ashton DG, Jones DM, Thomsett LR (1982) An outbreak of cowpox in captive cheetahs: virological and epidemiological studies. J Hyg Camb 89: 365–372
5. Baxby D, Shackleton WB, Wheeler J, Turner A (1979) Comparison of cowpox-like viruses isolated from European zoos. Arch Virol 61: 337–340
6. Baxby D, Bennett M, Getty B (1994) Human cowpox: a review based on 54 cases, 1969–93. Br J Dermatol 131: 598–607
7. Bennett M, Gaskell RM, Gaskell CJ, Baxby D, Kelly DF (1989) Studies on poxvirus infection in cats. Arch Virol 104: 19–33
8. Bennett M, Gaskell CJ, Baxby D, Gaskell RM, Kelly DF, Naidoo J (1990) Feline cowpox virus infection. J Small Anim Pract 31: 167–173
9. Bennett M, Gaskell CJ, Gaskell RM, Baxby D, Gruffydd-Jones TJ (1986) Poxvirus infection in the domestic cat. Some clinical and epidemiological observations. Vet Rec 118: 387–390

10. Bomhard D, Pfleghaar S, Mahnel H, Schneekloth-Dücker I (1991) Fallbericht: Katzenpockeninfektion als Zoonose für Hund und Mensch. Kleintierpraxis 36: 511–514
11. Brown A, Bennett M, Gaskell CJ (1989) Fatal poxvirus infection in association with FIV infection. Vet Rec 124: 19–20
12. Casemore DP, Emslie ES, Whyler DK, Baxby D, Bennett M, Davies AB (1987) Cowpox in a child, acquired from a cat. Clin Exp Dermatol 12: 286–287
13. Czerny C-P, Mahnel H (1990) Structural and functional analysis of orthopox virus epitopes with neutralizing monoclonal antibodies. J Gen Virol 71: 2341–2352
14. Czerny C-P, Eis-Hübinger AM, Mayr A, Schneweis KE, Pfeiff B (1991) Animal poxviruses transmitted from cat to man: current event with lethal end. J Vet Med B 38: 421–431
15. Czerny C-P, Johann S, Hölzle L, Meyer H (1994) Epitope detection in the envelope of intracellular mature orthopox viruses and identification of encoding genes. Virology 200: 764–777
16. Czerny C-P, Wagner K, Gessler K, Mayr A, Kaaden O-R (1996) A monoclonal blocking-ELISA for detection of orthopox virus antibodies in feline sera. Vet Microbiol 52: 185–200
17. De Fillipes FM (1976) Restriction enzyme mapping of vaccinia virus DNA. J Virol 43: 136–149
18. Egberink HF, Willemse A, Horzinek MC (1986) Isolation and identification of a poxvirus from a domestic cat and a human contact case. J Vet Med B 33: 237–240
19. Eis-Hübinger AM, Gerritzen A, Schneweis KE, Pfeiff B, Pullmann H, Mayr A, Czerny C-P (1990) Fatal cowpox-like virus infection transmitted by cat. Lancet 336: 880
20. Esposito JJ, Knight JC (1985) Orthopoxvirus DNA: comparison of restriction profiles and maps. Virology 143: 230–251
21. Gaskell RM, Gaskell CJ, Evans RJ, Dennis PE, Bennett AM, Udall ND, Voyle C, Hill TJ (1983) Natural and experimental pox virus infection in domestic cat. Vet Rec 112: 164–170
22. Goebel SJ, Johnson GP, Perkus ME, Davis SW, Winslow JP, Paoletti E (1990) The complete DNA sequence of vaccinia virus. Virology 179: 247–266
23. Johann S, Czerny C-P (1993) A rapid antigen capture ELISA for the detection of orthopox viruses. J Vet Med B 40: 569–581
24. Joklik WK (1962) The purification of four strains of poxvirus. Virology 18: 9–18
25. Klingebiel T, Vallbracht A, Döller G, Stierhof YD, Gerth HJ, Glashauser E, Herzau V (1988) A severe human cowpox infection in South germany. Pediatr Infect Dis J 7: 883–885
26. Laemmli UK (1970) Cleavage of structural proteins during the assembly of the head of bacteriophage T4. Nature 227: 680–685
27. Lowry ON, Rosebrough NJ, Farr L, Randall RJ (1951) Protein measurement with the folin phenol reagent. J Biol Chem 193: 265–278
28. Mackett M, Archard LC (1979) Conservation and variation in orthopox virus genome structure. J Gen Virol 45: 683–701
29. Mahnel H (1986) Identifizierung eines Kuhpockenvirus, isoliert von einem Kind. J Vet Med B 33: 362–370
30. Mahnel H (1991) Katzenpocken in Deutschland. Tierärztl Praxis 19: 419–422
31. Mahnel H, Czerny C-P, Mayr A (1989) Nachweis und Identifizierung von Pockenvirus bei Hauskatzen. J Vet Med B 36: 231–236
32. Marennikova SS, Maltseva NN, Korneeva VI, Garania NM (1977) Outbreak of pox disease among carnivora (felidae) and edentata. J Infect Dis 135: 358–366
33. Mayr A, Bachmann PA, Bibrack B, Wittmann G (1977) Virologische Arbeitsmethoden Band II: Serologie. G. Fischer, Stuttgart

34. Merril CR, Goldman D, Sedman SA, Ebert MH (1981) Ultrasensitive stain for proteins in polyacrylamide gels shows regional variation in cerebrospinal fluid proteins. Science 211: 1437–1438
35. Meyer H, Osterrieder N, Czerny C-P (1994) Identification of binding sites for neutralizing monoclonal antibodies on the 14-kDa fusion protein of orthopox viruses. Virology 200: 778–783
36. Meyer H, Pfeffer M, Rziha HJ (1994) Sequence alterations within and downstream of the A-type inclusion protein genes allow differentiation of orthopox virus species by polymerase chain reaction. J Gen Virol 75: 1975–1981
37. Meyer H, Ropp SL, Esposito JJ (1997) Gene for A-type inclusion protein is useful for a polymerase chain reaction assay to differentiate orthopoxviruses. J Virol Methods 64: 217–221
38. Naidoo J, Baxby D, Bennett M, Gaskell RM, Gaskell CJ (1992) Characterization of orthopoxviruses isolated from feline infections in Britain. Arch Virol 125: 261–272
39. O'Connor GM, Thiru AAA, Caul OE (1990) Ocular cowpox: transmission from the domestic cat to man. Br J Ophthalmol 74: 245–246
40. Pether JVS, Trevains PH, Harrison SRB, Baxby D, Bennett M, Gibb APR (1986) Cowpox from cat to man. Lancet 1: 38–39
41. Pfeiff B, Pullmann H, Eis-Hübinger AM, Gerritzen A Schneweis KE, Mayr A (1991) Lethale Tierpockeninfektion bei einem Atopiker unter dem Bild einer Variola vera. Hautarzt 42: 292–297
42. Redfield RR, Wright DC, James WD, Jones TS, Brown C, Burke DS (1987) Disseminated vaccinia in a military recruit with human immunodeficiency virus (HIV) disease. N Engl J Med 316: 673–676
43. Renner-Müller ICE, Meyer H, Munz E (1995) Characterization of camelpox virus isolates from Africa and Asia. Vet Microbiol 45: 371–381
44. Smith GL (1993) Vaccinia virus glycoproteins and immune evasion. The sixteenth Fleming Lecture. J Gen Virol 74: 1725–1740
45. Stolz W, Götz A, Thomas P, Ruzicka T, Süss R, Landthaler M, Mahnel M, Czerny C-P (1996) Characteristic but unfamiliar – the cowpox infection transmitted by a domestic cat. Dermatology. 193: 140–143
46. Thomsett LR, Baxby D, Denham EH (1978) Cowpox in the domestic cat. Vet Rec 103: 567
47. Towbin H, Staehelin T, Gordon J (1979) Electrophoretic transfer of proteins from polyacrylamide gels to nitrocellulose sheets: Procedure and some applications. Proc Natl Acad Sci USA 76: 4350–4354
48. Vestey JP, Yirrell DL, Aldridge RD (1991) Cowpox/catpox infection. Br J Dermatol 124: 74–78
49. Willemse A, Egberink HF (1985) Transmission of cowpox virus infection from domestic cat to man. Lancet: 1515
50. Yirrel DL, Vestey JP, Norval M, Aldridge RD (1990) Human catpox/cowpox infection. Vet Rec 127: 24
51. Zimmer K, Bogantes JC, Herbst W, Räther W (1990) Poxvirus infection in a cat and its owner. Schweiz Arch Tierheilk 132: 409–484
52. Zwart P, Gispen R, Peters C (1971) Cowpox in Okapi "Okapi johnstoni" at Rotterdam Zoo. Br Vet J 127: 20–24

Authors' address: Dr. C.-P. Czerny, Institute für Medizinische Mikrobiologie, Veterinärstr. 13, D-80539 München, Federal Republic of Germany.

Molecular genetic analyses of parapoxviruses pathogenic for humans

A. Mercer[1], **S. Fleming**[1], **A. Robinson**[1,*], **P. Nettleton**[2], and **H. Reid**[2]

[1]Virus Research Unit and Centre for Gene Research, University of Otago, Dunedin, New Zealand
[2]Moredun Research Institute, Edinburgh, U.K.

Summary. The current members of the genus parapoxvirus are orf virus (ORFV), bovine papular stomatitis virus (BPSV), pseudocowpoxvirus (PCPV) and parapoxvirus of red deer in New Zealand (PVNZ). BPSV and PCPV are maintained in cattle while ORFV is maintained in sheep and goats, but all three are zoonoses. Only the recently reported PVNZ has yet to be recorded as infecting humans. Tentative members of the genus are camel contagious ecthyma virus, chamois contagious ecthyma virus and sealpoxvirus.

The separation of the parapoxviruses into 4 distinct groups has been based on natural host range, pathology and, more recently, on restriction endonuclease and DNA/DNA hybridisation analyses. The latter studies have shown that the parapoxviruses share extensive homology between central regions of their genomes, but much lower levels of relatedness within the genome termini.

The high G+C content of parapoxvirus DNA is in contrast to most other poxviruses and suggests that a significant genetic divergence from other genera of this family has occurred. DNA sequencing of portions of the genome of ORFV, the type species of the genus, has allowed a detailed comparison with the fully sequenced genome of the orthopoxvirus, vaccinia virus (VACV). These studies have provided a genetic map of ORFV and revealed a central core of 88 kbp within which the genomic content was strikingly similar to that of VACV. This conservation is not maintained in the genome termini where insertions, deletions and translocations have occurred.

The characterisation of specific ORFV genes may lead to the construction of attenuated vaccine strains in which genes such as those with the potential to interfere with the immune response of the host have been deleted. The current ORFV vaccines are living unattenuated virus and vaccination lesions produce virus which contaminates the environment in a manner similar to natural infection. The virus in scab material is relatively resistant to inactivation and

*Present address: CSIRO Division of Wildlife and Ecology, Lyneham, Australia.

this virus both perpetuates the disease in sheep and provides the most likely source of human infections. A vaccine which immunises animals without perpetuating the disease could be the best way of reducing the incidence of ORFV infection of humans. It is likely that protection against infection by ORFV is cell mediated and will require the endogenous production of relevant antigens. We have recently constructed a series of VACV recombinants each of which contains a large multigene fragment of ORFV DNA. Together the recombinants represent essentially all of the ORFV genome in an overlapping manner. Vaccination of sheep with the recombinant library provided protection against challenge with virulent ORFV. Further studies with this library may enable dominant protective antigens of ORFV to be identified and lead to their incorporation into a subunit vaccine.

Introduction

Reports of infections by what we now recognise as parapoxviruses appeared in the scientific literature in the latter part of last century (reviewed in [30]). These reports of a pustular dermatitis of sheep, cattle and humans described infection by three members of the genus: orf virus (ORFV), bovine papular stomatitis virus (BPSV) and pseudocowpoxvirus (PCPV). BPSV and PCPV are maintained in cattle while ORFV is maintained in sheep and goats, but all three are zoonoses. More recently a fourth member has been added to the genus: parapoxvirus of red deer in New Zealand (PVNZ) [32]. PVNZ has yet to be recorded as infecting humans. Tentative members of the genus are camel contagious ecthyma virus, chamois contagious ecthyma virus and sealpox virus. Sealpox virus can be transmitted to humans.

Parapoxviruses cause a contagious pustular dermatitis that follows a characteristic course of vesicle-pustule-scab formation over a period of around 4 to 6 weeks. There is little evidence of systemic spread by parapoxviruses; rather lesions tend to be confined to the epithelium and oral mucosa. Mortality rates associated with infection of animals by parapoxviruses are low (1% or less) and probably relate to secondary infections. Infection around the mouth or teats can interfere with feeding and may retard growth especially of young animals.

Definitive reports of the susceptibility of humans to infection by parapoxviruses appeared in the earlier parts of this century (reviewed in [30]) although it seems very probable that the human disease was well known from much earlier times. One of the more intriguing historical references can be found in a letter of Pope Leo X written at the time of the excommunication of Martin Luther. The letter deplores the actions of Luther and states that "we cannot suffer the scabby sheep longer to infect the flock" (cited in [2]). This was written in 1 520 and perhaps is the earliest documented reference to ORFV.

In immune competent individuals infection is generally benign and confined to proliferative lesions on the skin at the points of infection. The lesions persist for one month or more after which time they heal usually leaving no scar. Erythema multiformae reactions in the form of rashes on the backs of the hands and on the legs and ankles are common [1, 3, 7] and cases of Stevens-Johnson syndrome (a severe form of erythema multiformae involving skin and mucous membranes) have been reported [6]. Severe progressive disease has been reported in immunocompromised individuals infected with parapoxviruses [15, 34, 39]. Severe reactions have also been recorded in otherwise normal individuals [41], in cases of burns [14], and in cases of atopic dermatitis [5].

Molecular characterisation

Restriction endonuclease cleavage profiles and, in some cases, cleavage site maps of genomic DNA strains of ORFV, BPSV, PCPV and PVNZ have been published [11, 27, 29, 32]. When these profiles are compared the four species of parapoxvirus can easily be distinguished [30]. The distinction between these four sub-types of the *Parapoxvirus* genus has also been borne out by DNA:DNA hybridisation studies [11, 32]. In one of these studies DNA probes from internal genomic locations hybridised to representative isolates of ORFV, BPSV and PCPV whereas DNA probes from terminal regions of the genome hybridised only to the DNA of other members of the same species [11]. Similarly, the total genome of PVNZ hybridised strongly to DNA fragments of internal regions of the genomes of each of the other parapoxviruses, but only weakly or not at all to fragments from terminal regions of the other species [32]. This pattern of a conserved genomic center in conjunction with greater variation in near-terminal regions has also been observed in comparisons of members of the *Orthopoxvirus* genus.

Parapoxviruses share numerous features with other members of the poxvirus family. These features include their large double stranded DNA genomes with cross-linked ends and inverted terminal repeats, the cytoplasmic site of replication, conserved transcriptional regulatory signals and aspects of their pathology. But the distinctive "ball of yarn" morphology of parapoxvirus virions and the high G+C content of their genomes [42] suggests that a significant genetic divergence from other genera of this family has occurred. The sequence of blocks of DNA of the type species of the genus, ORFV, have been reported but these total less than 20% of the 139 kb genome [8, 10, 19–21, 23, 36–38]. In contrast the entire genomic sequences of the orthopoxviruses, vaccinia virus (VACV) and variola virus have been reported. This has enabled a limited comparison of the genomic content and organisation of ORFV and VACV. The comparison revealed numerous points of homology between the two genomes. This homology was evident within predicted amino acid sequences but because of the large difference in G+C content of the two genomes, it was not generally apparent in the DNA sequences. The data allowed the mapping of 32 ORFV genes with homologues in VACV and revealed that the order, orientation and spacing of these genes was strikingly conserved between the two genera. However there was much less evidence of homology within the near-terminal regions of the genomes and it was suggested that as much as 30 kb at one end of the ORFV genome might be composed of genes not found in

VACV and which might contribute to the distinguishing features of ORFV. Within this region we have identified two such genes which have not been reported in other poxviruses. One of these is a homologue of mammalian vascular endothelial growth factor (VEGF) [18].

A homologue of vascular endothelial growth factor (VEGF)

VEGF is a mitogen specific for vascular endothelial cells and induces angiogenesis in several in vivo models. It is believed that VEGF plays an important role in the formation of new blood vessels during normal physiological processes such as embryonic development, formation of the corpus luteum and wound healing. It has also been implicated in pathological processes such as the vascularisation and rapid growth of solid tumours. VEGF also promotes leakage from blood vessels and is alternatively called vascular permeability factor. It has yet to be shown that the ORFV homologue has VEGF-like activity but the literature contains references to extensive vascularisation associated with ORFV and PCPV lesions [12] and it seems probable that VEGF-like activity by the viral homologue is responsible for this feature. In addition the tumour-like appearance of some lesions, particularly those occurring in immunoincompetent or immunosuppressed individuals may in part be the result of the activity of this gene. Definition of the precise role of the viral VEGF in the pathogenesis of ORFV will be provided by assays of the purified protein and analysis of a mutant virus in which the VEGF gene has been inactivated.

Two independent New Zealand isolates of ORFV have been shown to have a single copy of the VEGF-like gene. The genes share the same relative genomic location and a similar level of relatedness to mammalian VEGF, but the two genes are unexpectedly different from one another (only 41% amino acid identity) [18]. Analyses of further ORFV isolates from New Zealand, Australia and United Kingdom have indicated that each has a copy of the VEGF-like gene. These genes are similar, but generally not identical, to one or other of the two versions first identified. In the isolates examined so far this divergence from the general pattern ranges up to 20% mismatch in the amino acid sequence (unpublished). The significance and origin of this unexpected sequence diversity remains unclear.

A homologue of interleukin-10

A second gene found in the near-terminal region of the ORFV genome and which has not been reported in other poxviruses encodes a predicted protein with extensive amino acid identity to interleukin-10 (IL-10) (Fleming and Mercer, unpubl.). IL-10 is a multifunctional cytokine that has many stimulatory and inhibitory activities. IL-10 acts as a suppressor of macrophage function, as an inhibitor of T helper 1 type responses and as a stimulator of T helper 2 type responses. It is likely that expression of this viral IL-10 might significantly influence the immune response to ORFV infection and might relate to the rather short-lived protection against re-infection. Epstein-Barr virus and equine herpes virus type 2 have also been shown to encode a homologue of IL-10.

It is intriguing that the equivalent near-terminal region of the PVNZ genome showed little cross-hybridisation with other parapoxviruses. DNA sequencing of parapoxviruses other than ORFV will be required to determine the full extent of the sequence variation in this region of parapoxvirus genomes.

Infection of humans by parapoxviruses is most commonly seen in individuals working with or living in close proximity to the relevant animals. For example, a survey of workers in the New Zealand sheepmeat processing industry detected 500 cases of ORFV infection in a one year period [33]. It is clear that the vast majority of human infections are acquired either directly or indirectly from animals and human-to-human transmission is rare. Consequently, the incidence of human infection could be reduced by reducing the incidence of animal infection. This could be achieved by an effective vaccine.

Vaccination

Vaccines for ORFV are available but these consist of live unattenuated virus. It is common farming practice to inoculate young lambs with these vaccines at a site on the animal (usually the hind leg) where the ensuing infection does not interfere with feeding. This procedure induces an immunity to reinfection that lasts for some months and thereby avoids the production losses associated with natural ORFV infection around the mouth. However, the vaccination lesions produce virus which contaminates the environment just as surely as natural infection. The virus in scab material that falls from the animal after resolution of the lesion, either natural or vaccine-derived, is relatively resistant to inactivation and it is this virus which is a source of infection for subsequent lamb crops and contributes to the perpetuation of the disease.

A means of immunising against ORFV infection without the production of infectious virus would be highly desirable. Such a vaccine should reduce the level of ORFV in sheep flocks and in the environment, so reducing infections in humans. The range of possible vaccine formulations includes a subunit vaccine, a specifically attenuated ORFV or a vector expressing the appropriate ORFV genes. The choice between these options will be influenced by the type of immune response that is desired, however that choice is currently difficult because the nature of the protective immune response to parapoxvirus infection is unclear.

Protective immunity

Experimental studies of various viral infections of mice indicate that the major components of the T-cell response to viral infection are $CD8^+$ cytotoxic T lymphocytes (CTLs), $CD4^+$ helper 1 cells, γ interferon and interleukin-2 (IL-2). The primary T-cell response is likely to be initiated in lymphoid tissue where circulating naive T-cells are brought into contact with antigen presenting cells.

In this environment, with the participation of lymphokines and generally, $CD4^+$ T helper cells, they undergo clonal expansion and differentiation. The proliferating cells eventually acquire effector function and various homing molecules which can direct their migration to specific tissues, such as the skin. Effector $CD8^+$ T cells kill viral infected cells following interaction between the T cell receptor of the CTL and a viral peptide presented on the surface of an infected cell in association with a MHC-I glycoprotein. Most cells of the body express MHC-I molecules and are therefore subject to surveillance by $CD8^+$ CTLs.

It is clear that CTL responses against MHC-I presented antigens are involved in the protective response against poxviruses in inbred mice [25, 35] and humans with impaired cellular immunity have been shown to develop severe disease after inoculation with VACV [17]. In the case of ORFV there is evidence that antibody has only a limited or perhaps no role in protection. For example, circulating antibody as measured in neutralisation tests or ELISA can be found in ewes that are immune to OV. Although this antibody can be found in colostrum and in lambs after suckling, the lambs are fully susceptible to infection when they are conventionally reared [4] or hysterectomy-procured, colostrum-fed and barrier-maintained [22]. This would tend to rule out a major role for antibody in protection unless the relevant class of antibody was selectively excluded from entering the colostrum. There are two earlier reports that claim that protection can be passed on in colostrum [16, 26] but neither Buddle and Pubfort [4] nor we have been able to reproduce those results.

Our analysis in hysterectomy-procured, barrier-maintained lambs showed that ORFV infection induces a protective response to secondary challenge: primary lesions lasted an average of 17.75 days while reinfection 4 weeks later produced lesions that resolved in 7.5 days and showed neither the reddening nor pustule formation seen in primary lesions [22]. This pattern was essentially the same as that seen when primary and secondary infections with VACV were compared. Furthermore, the same experiments showed that this memory response is specific, in that when ORFV-immunised lambs were challenged with VACV the resulting lesions were indistinguishable from a primary exposure to that virus. Clearly ORFV infection induces a specific immunological memory that results in an infection that is less severe and is more quickly resolved than that seen in naive individuals. These results are similar to those reported by others in both sheep and humans (see [28] for references).

With the above observations in mind, we believe it is likely that a cellular immune response, probably involving CTL recognition of ORFV antigens associated with MHC-I, is responsible for recovery from ORFV infection and for providing protection against subsequent challenge. A $CD8^+$ CTL response to ORFV infection of sheep was reported recently [13]. Stimulation of cell mediated responses will require the endogenous production of the relevant antigen(s). Consequently, procedures such as screening expression libraries using anti-ORFV sera or fractionating virions and using these as immunogens are unlikely to lead to the identification of those antigens that are necessary for inducing protection.

Identification of protective antigens

Our approach to identifying the relevant antigens is to use VACV as an expression vector to screen ORFV genes for their ability to induce protection in sheep. The use of VACV to express foreign genes has become commonplace. In nearly all cases the sequence of the foreign gene has been determined and the gene placed so as to be transcribed from a strong VACV promoter. In contrast to this we have attempted to circumvent the necessity to first identify and then screen individual genes by inserting large multigene fragments of the ORFV genome into VACV. We have constructed 16 recombinant VACVs (Lister strain) that together contain an overlapping set of ORFV fragments covering the entire 135 kb coding region except for a 5 kb region which has defied our attempts to clone as a single fragment. The average size of the ORFV DNA fragments recombined into VACV is 12 kb and the average overlap between the fragments is 4.3 kb (manuscript submitted). The strategy is to inoculate sheep with these recombinants and determine which are able to immunise against ORFV challenge. In this way we hope to dissect out the ORFV genes encoding the peptides that stimulate specific and protective immune responses.

Clearly for our approach to be successful there are two main prerequisites. Firstly, as the fragments inserted are greater than the size of an average gene and no attempt has been made to use VACV promoters, transcription of ORFV genes would require that the VACV RNA polymerase complex faithfully initiate transcription from the ORFV promoters. We have extensive data indicating that many, perhaps all ORFV promoters are recognised by the VACV RNA polymerase complex: (a) we have used S1 and primer extension analyses to map the 5′ ends of 3 ORFV early genes and have shown that these genes are faithfully transcribed in a VACV recombinant and that the same transcriptional start sites are used in the recombinant as in ORFV [9, 40]; (b) we have linked an ORFV late promoter to the β-galactosidase reporter gene and detected strong expression of this gene in VACV-infected cells (unpublished); (c) cells infected with each of the 16 members of the VACV-ORFV recombinant library give clear ORFV-specific immunofluorescence (submitted); (d) radio-immune precipitation using hyperimmune ORFV serum detects 10 antigens in extracts of cells infected with ORFV and at least 7 of these are also precipitated from various recombinant extracts (submitted). The second prerequisite was that there should be no cross-protection between VACV and ORFV. We have performed two experiments which clearly show that despite a degree of antigenic cross-recognition between ORFV and VACV antisera, these two viruses do not cross-protect [22, 31].

Of major interest is whether or not members of the recombinant library will protect sheep against challenge with ORFV. Two lambs were immunised with all 16 recombinants and 2 other lambs immunised with a VACV recombinant containing only the β-galactosidase reporter gene. Three weeks later when the lesions had resolved all animals were challenged with ORFV. The resulting lesions were examined daily and scored on a scale of 0 to 3 for each of erythema, vesicule/pustule formation and scab formation [24]. These scores were summed each day and at the end of the experiment all scores for each pair of animals were combined to give a total clinical score. In lambs vaccinated with the β-gal recombinant, OV lesions were typical of a primary ORFV leison and the total clinical score was 236. In contrast, lambs immunised with VACV-ORFV recombinants had a much milder and shorter course of disease after challenge with ORFV and their total clinical score was only 52 (submitted). These results show that the full set of recombinants is able to present the ORFV antigens required to induce a protective immune response and demonstrate the validity of our approach. The degree of protection provided by the recombinant library was comparable to that recorded in earlier experiments in which lambs were immunised with ORFV [24]. It is hoped that further experiments using subsets of the library will identify single recombinants and thereafter single genes whose products are able to immunise against ORFV infection. The identification of single proteins able to induce protective immunity might lead to a vaccine able to deliver these relevant immunogens in the absence of infectious virus.

It is 200 years since Jenner demonstrated that inoculation with material from a cowpox lesion could protect against smallpox. That observation established the principle of vaccination. It also led to the widespread use of a VACV vaccine against smallpox and eventually to the global eradication of that devastating disease. Despite this distinguished history and extensive knowledge of VACV, the viral proteins that are the targets of the protective immune response have not been identified for VACV or any poxvirus. The strategy described here is likely to lead to the identification of the protective antigens of one poxvirus and could be adapted to other poxviruses.

References

1. Agger WA, Webster SB (1983) Human orf infection complicated by erythema multiformae. Cutis 31: 334–338
2. Bainton RH (1950) Here I stand. The life of Martin Luther. Abingdon-Cokesbury Press, New York
3. Blakemore F, Abdussalam M, Goldsmith WN (1948) A case of orf (contagious pustular dermatitis): identification of the virus. Br J Dermatol 60: 404–409
4. Buddle BM, Pulford HD (1984) Effects of passively acquired antibodies and vaccination on the immune response to contagious ecthyma virus. Vet Microbiol 9: 515–522
5. Dupre A, Christol B, Bonafe JL, Lassere J (1981) Orf and atopic dermatitis. Br J Dermatol 105: 103–104
6. Erickson GA, Carbrey EA, Gustafson GA (1975) Generalised contagious ecthyma in a sheep rancher: diagnostic considerations. J Am Vet Med Assoc 166: 262–263
7. Fastier LB (1957) Human infections with the virus of ovine contagious pustular dermatitis. N Z Med J 56: 121–123

8. Fleming SB, Blok J, Fraser KM, Mercer AA, Robinson AJ (1993) Conservation of gene structure and arrangement between vaccinia virus and orf virus. Virology 195: 175–184
9. Fleming SB, Mercer AA, Fraser KM, Lyttle DJ, Robinson AJ (1992) In vivo recognition of orf virus early transcriptional promoters in a vaccinia virus recombinant. Virology 187: 464–471
10. Fraser KM, Hill DF, Mercer AA, Robinson AJ (1990) Sequence analysis of the inverted terminal repetition in the genome of the parapoxvirus, orf virus. Virology 176: 379–389
11. Gassmann U, Wyler R, Wittek R (1985) Analysis of parapoxvirus genomes. Arch Virol 83: 17–31
12. Groves RW (1991) Human orf and milkers' nodule: a clinicopathological study. J Am Acad Dermatol 25: 706–711
13. Haig DM, Hutchison G, Thomson J, Yirrell D, Reid HW (1996) Cytolytic activity and associated serine protease expression by skin and afferent lymph $CD8^+$ T cells during orf virus reinfection. J Gen Virol 77: 953–961
14. Hooser SB, Scherba G, Morin DE, Whiteley HE (1989) A typical contagious ecthyma in a sheep after extensive cutaneous thermal injury. J Am Vet Med Assoc 195: 1255–1 256
15. Hunskaar S (1986) Giant orf in a patient with lymphocytic leukaemia. Br J Dermatol 114: 631–634
16. Jan CL, L'Haridon R, Madelaine MF, Cornu C, Asso J (1978) Transfer of antibodies against CPD virus through colostrum and milk. Ann Rech Vet 9: 343–346
17. Kempe CH (1960) Studies on smallpox and complications of smallpox vaccination. Pediatrics 26: 176–189
18. Lyttle DJ, Fraser KM, Fleming SB, Mercer AA, Robinson AJ (1994) Homologs of vascular endothelial growth factor are encoded by the poxvirus orf virus. J Virol 68: 84–92
19. Mercer AA, Fraser KM, Stockwell PA, Robinson AJ (1989) A homologue of retroviral pseudoproteases in the parapoxvirus, orf virus. Virology 172: 665–668
20. Mercer AA, Green G, Sullivan JT, Robinson AJ, Drillien R (1996) Location, DNA sequence and transcriptional analysis of the DNA polymerase gene of orf virus. J Gen Virol 77: 1 563–1 568
21. Mercer AA, Lyttle DJ, Whelan EM, Fleming SB, Sullivan JT (1995) The establishment of a gene map of orf virus reveals a pattern of genomic organisation that is highly conserved among divergent poxviruses. Virology 212: 698–704
22. Mercer AA, Yirrell DL, Reid HW, Robinson AJ (1994) Lack of crossprotection between vaccinia virus and orf virus in hysterectomy-procured, barrier-maintained lambs. Vet Microbiol 41: 373–382
23. Naase M, Nicholson BH, Fraser KM, Mercer AA, Robinson AJ (1991) An orf virus sequence showing homology to the 14 K 'fusion' protein of vaccinia virus. J Gen Virol 72: 1 177–1 181
24. Nettleton PF, Brebner J, Pow I, Gilray JA, Bell GD, Reid HW (1996) Tissue culture-propagated orf virus vaccine protects lambs from orf virus challenge. Vet Rec 138: 184–186
25. O'Neill HC, Brenan M (1987) A role for early cytotoxic T cells in resistance to ectromelia virus infection in mice. J Gen Virol 68: 2 669–2 673
26. Poulain J, Gourreau J-M, Dautigny A (1972) Ecthyme contagieux du mouton; anticrops seriques neutralisants. Ann Rech Vet 3: 561–579
27. Rafii F, Burger D (1985) Comparison of contagious ecthyma virus genomes by restriction endonucleases. Arch Virol 84: 283–289

28. Robinson AJ, Balassu TC (1981) Contagious pustular dermatitis (orf). Vet Bull 51: 771–781
29. Robinson AJ, Barns G, Fraser K, Carpenter E, Mercer AA (1987) Conservation and variation in orf virus genomes. Virology 157: 13–23
30. Robinson AJ, Lyttle DJ (1992) Parapoxviruses: their biology and potential as recombinant vaccines. In: Binns M, Smith GL (ed) Recombinant poxviruses. CRC Press, Boca Raton, pp 285–327
31. Robinson AJ, Mercer AA (1988) Orf virus and vaccinia virus do not cross protect in sheep. Arch Virol 101: 255–259
32. Robinson AJ, Mercer AA (1995) Parapoxvirus of red deer: evidence for its inclusion as a new member in the genus *Parapoxvirus*. Virology 208: 812–815
33. Robinson AJ, Peteresen GV (1983) Orf virus infection of workers in the meat industry. N Z Med J 96: 81–85
34. Savage J, Black MM (1972) "Giant orf" of finger in a patient with a lymphoma. Proc R Soc Med 64: 766–768
35. Sharma DP, Ramsay AJ, Maguire DJ, Rolph MS, Ramshaw IA (1996) Interleukin-4 mediates down regulation of antiviral cytokine expression and cytotoxic T-lymphocyte responses and exacerbates vaccinia virus infection in vivo. J Virol 70: 7 103–7 107
36. Sullivan JT, Fleming SB, Robinson AJ, Mercer AA (1995) Sequence and transcriptional analysis of a near-terminal region of the orf virus genome. Virus Genes 11: 21–29
37. Sullivan JT, Fraser KM, Fleming SB, Robinson AJ, Mercer AA (1995) Sequence and transcriptional analysis of an orf virus gene encoding ankyrinlike repeat sequences. Virus Genes 9: 277–282
38. Sullivan JT, Mercer AA, Fleming SB, Robinson AJ (1994) Identification and characterisation of an orf virus homologue of the vaccinia virus gene encoding the major envelope antigen p37k. J Virology 202: 968–973
39. Tan ST, Blake GB, Chambers S (1991) Recurrent orf in an immunocompromised host. Br J Plastic Surg 44: 465–467
40. Vos JC, Mercer AA, Fleming SB, Robinson AJ (1992) In vitro recognition of an orf virus early promotor in a vaccinia virus extract. Arch Virol 123: 223–228
41. Wilkinson JD (1977) Orf: a family with unusual complications. Br J Dermatol 97: 447–450
42. Wittek R, Kuenzle CC, Wyler R (1979) High G+C content in parapoxvirus DNA. J Gen Virol 43: 231–234

Authors' address: Dr. A.A. Mercer, Virus Research Unit, Department of Microbiology, University of Otago, P.O. Box 56, Dunedin, New Zealand.

Recent advances in molluscum contagiosum virus research

J. J. Bugert and **G. Darai**

Institut für Medizinische Virologie der Universität Heidelberg, Heidelberg, Federal Republic of Germany

Summary. Molluscum contagiosum virus (MCV) and variola virus (VAR) are the only two poxviruses that are specific for man. MCV causes skin tumors in humans and primarily in children and immunocompromised individuals. MCV is unable to replicate in tissue culture cells or animals. Recently, the DNA sequence of the 190 kbp MCV genome was reported by Senkevich et al. MCV was predicted to encode 163 proteins of which 103 were clearly related to those of smallpox virus. In contrast, it was found that MCV lacks 83 genes of VAR, including those involved in the suppression of the host response to infection, nucleotide biosynthesis, and cell proliferation. However, MCV possesses 59 genes predicted to code for novel proteins including MHC-class I, chemokine and glutathione peroxidase homologs not found in other poxviruses. The MCV genomic data allow the investigation of novel host defense mechanisms and provide new possibilities for the development of therapeutics for treatment and prevention of the MCV infection.

Introduction

Poxviruses are a highly versatile group of viruses that can adapt easily to specific host environments. They cause generalized as well as locally limited infections in a broad range of host organisms, including mammals, birds, and insects. The most notable and historically well known example for a generalized poxvirus infection is the human smallpox disease caused by smallpox virus (variola). Variola infections are acute and systemic with the cell mediated host response clearing the infection rapidly or not at all, in the latter case usually with fatal consequences. However, some poxviruses are limited to specific host tissues and the diseases caused by these poxviruses are usually benign but difficult to eradicate. An example for a chronic and localized poxviral infection in humans are the benign epidermal tumors caused by Molluscum contagiosum virus (MCV). Like other poxviruses MCV has a brick shape morphology and a large DNA genome that carries all the genetic information needed to complete the viral replicative cycle in the host cell cytoplasm. Recently, the complete DNA genome sequence of MCV type 1 was determined. It potentially encodes 163 proteins, 103 of which have homologs in smallpoxvirus (variola). Other poxviruses are known to encode proteins that perturb host growth, differentiation and defense processes at both the local and the systemic level. Among the 59 predicted MCV genes that encode previously unknown proteins are a major histocompatibility complex class 1 homolog, a beta chemokine homolog, and a highly conserved homolog of glutathione peroxidase, which suggests there are specific MCV immunevasive mechanisms [72].

Molluscum contagiosum

MCV is a member of the family *Poxviridae* (subgenus *Molluscipoxvirus*) and after the eradication of smallpox, the only remaining poxvirus that causes disease in humans. MCV is the causative agent of Molluscum Contagiosum (MC). MC patients, mainly preadolescent children, sexually active adults and individuals with impaired cellular immunity, present with benign tumors of the skin. The appearance of the typical molluscum lesion was first described by Thomas Bateman: The tumors are hard, smooth and nearly the color of skin, but with a shiny surface and a slight appearance of transparency. They are generally of a globular form, sometimes ovate and sessile upon a contracted base. A milky fluid can be expressed from an imperceptible aperture in the apex of the larger tumors. The infection is spread through contact with this fluid [2]. The tumors are classified as acanthomas because of their strict limitation to the epidermis and have a striking resemblance to hair follicles of the human skin. In immunocompetent hosts MC is a benign and self limiting viral infection of the skin [23, 67]. In HIV infected patients and otherwise immunocompromised individuals, however, MC is a marker of late stage disease and, although not per se life threatening, can lead to severly disfiguring cutaneous lesions, that are prone to bacterial superinfection [6]. The incidence of molluscum contagiosum in the HIV infected population was found to be between 5.2 and 24% [20, 47]. In HIV infected patients MC is a cutaneous correlate of cellular immunodeficiency, is not selflimiting, difficult to eradicate, possibly latent, and usually widespread [44, 71]. Eczema molluscum with up to 700 lesions has been described [19]. Immunocompromised patients may develop hundreds of lesions, that persist over prolonged periods of time and MCV infections of the eyelid with secondary conjunctivitis is common [17]. Severe MC cases were observed after splenectomy [45]. Association of MCV with hair follicles [35], epidermoid cysts [1, 22, 58], ossification [52] and hemorrhagic lesions [56, 62] are described. The MCV infection fulfills the criteria of a local infection: mass effect of the original inoculum and self limitation.

A MCV reservoir other than man is not known, and it is commonly believed that MCV only infects the epidermis of the human skin, therefore and in contrast to most other poxviruses presenting a very limited hostrange and a highly specific tissue tropism. Reports of molluscum contagiosum in various animals are not substantiated by molecular analyses, but have never been positively ruled out (e.g. [43]). The lack of positive knowledge about the viral reservoir is an impediment for the eradication of the disease in the future.

MCV has not been grown in cell or tissue culture and there is no established animal model. At the beginning of the century, Chamberland filtrates of homogenized MCV lesion material were inoculated into human volunteers and the first Koch postulate for a transmissible pathogen fulfilled. Incubation times of 14 to 50 days were observed [36, 86]. However, subsequent attempts to cultivate the virus in cell or tissue culture including human primary keratinocyte cultures were frustrated. Cytopathic effects (CPE), particularly cell aggregation and degeneration, could be observed within 6 to 24 h p.i. but disappeared after 2 to 3 days without signs of viral replication [14, 16, 40, 41, 42]. It was concluded that MCV behaves like a host dependent conditionally lethal mutant, that transcribes and expresses early gene products responsible for the CPE in vitro but completes the viral replicative cycle only under the specific conditions of human skin in vivo [48, 73].

Recently, MCV replication has been described in human epidermal tissue explants [13]. In MCV infected human foreskin grafts to the dorsolateral thorax of athymic mice, MCV induced morphological changes that were similar to patient biopsies and included the development and migration of molluscum bodies containing mature virions to the epidermal surface. This effect was observed between 74 and 142 days p.i. However, the xenotransplantation of human tissue caused a 50 to 75% rate of rejected grafts even in athymic mice. Not more than 24 mice were transplanted at one given time and only three of these showed graft survival and MCV maturation at the same time. Findings were confirmed by in situ hybridization. A passage of the MCV viral progeny to secondary animals was not attempted.

MCV lifecycle

The typical MC lesion is histologically organised in closely clustered lobuli of MCV infected keratinocytes, that are separated by hyperplastic basal epidermal cells. Separating basal cells do not contain molluscum inclusion bodies and are similar to cells found in talkum glands accompanying and differentiated from embryonic hair follicles [28, 68].

Electron microscopically it is possible to identify MCV cores in all layers of the epidermis, also in areas that appear uninfected in light microscopy [76]. The basal layer of the epidermis shows an increased rate of mitoses. MCV early transcription is assumed to occur at the same time in the same cells. MCV infected cells in the neighboring stratum spinosum show premature keratinization, an effect called dyskeratinization. MCV DNA replication is presumed to occur simultaneously [24, 25, 84, 85]. New virions mature inside poxviral factories in keratinocytes that are on their way up to the epidermal horn layer and therefore still seem to follow their natural migration pattern. However, the migration time of infected keratinocytes to the horn layer is decreased, indicating accelerated differentiation. MCV factories are surrounded and sealed off intracellularly by a collagen/lipid rich layer forming a distinctive sac-like structure in each infected keratinocyte. It is postulated that this sac favours replication of the virions by providing a site that is both anatomically and immunologically privileged. The protected MCV viral colony becomes first visible even by light microscopy about four layers away from the basal membrane in the stratum spinosum as MCV inclusion bodies or Henderson-Paterson bodies [68, 74, 82]. MCV infected cells grow in size, while internal organelles are dislocated and eventually completely obliterated by the growing molluscum bodies. Rupture and discharge of the virus packed cells occurs in a process similar to holocrine secretion. Membrane debris and infectious MCV virons accumulate in a crater-like ostium formed by the histologically lobulated lesion towards the hornlayer and the epidermal surface. The MCV infection is then spread venerally by contact with this infectious debris.

The above described MCV lifecycle within the human epidermis is unique in the poxvirus family. MCV would therefore be expected to have a gene complement very different from that of other poxviruses.

MCV is a cytoplasmatically replicating virus with typical poxvirus brick shape morphology [50]. Because of its lack of crosshybridization or immunological cross-reactivity with other poxviruses it has been classified as sole member of the subgenus molluscipox. MCV can neither nongenetically reactivate other poxviruses, a typical feature of the poxvirus family [7, 41], nor can it be reactivated by them, indicating a specific genetic distinctiveness contrasting the apparent morphological similarity [48]. On the other hand the large doublestranded MCV DNA genome shares certain structural characteristics with the DNA genomes of other poxviruses, e.g covalently closed termini and inverted terminal repeated sequences [12, 59, 66]. Two main genetic types (MCV type 1 and 2; MCV type 1: estimated 188 kbp; MCV type 2: estimated 195 kbp), and several variant subtypes were described by fingerprinting techniques based on BamHI restriction patterns of their viral DNAs [21, 59, 69, 70]. MCV of all genetic types causes the same benign epidermal tumors. For a virus that cannot be characterized by classical virological methods, the molecular analysis of the viral gene complement by DNA sequence analysis is the only approach that promises clues about viral pathogenic strategies. Genome sequencing provides the opportunity for further characterization of specific viral gene products and by these means to learn some biological lessons from this interesting human pathogen.

MCV genome analysis

DNA fragments of the MCV type 1 genome were cloned into bacterial plasmid vectors [9, 12, 21] and physical maps of the viral genome were constructed [12, 64, 66]. Parts of the viral genome were characterized by DNA nucleotide sequence analysis. The boundaries of the terminal inverted repeats were determined in the MCV-1 HindIII fragments J-1 and K [10]. A large cluster of three repetitive DNA elements R1, R2 and R3, with a complex structural arrangement was detected in MCV-1 HindIII DNA fragment C [32]. This cluster of repetitive DNA sequences, encoding a matched set of Q/A repeats, was found in a large open reading frame (ORF number 17, corresponding to MC006L [72, 32], that comprises 1, 175 amino acids with a predicted molecular weight of 126 kD. This ORF has a promoter signal with similarity to described poxviral early promoters 21 nucleotides upstream from the start codon. A MCV homolog to the vaccinia virus 37 kDa major envelope protein was found in the MCV-1 BamHI DNA fragment J and an identical copy was found on its MCV-2 counterpart. Orf virus and fowlpoxvirus encode another homologue of this vaccinia virus protein [15, 78].

MCV like other poxviruses, the African swine fever virus (ASFV) and iridoviruses, replicates in the cytoplasm of infected cells. This implies that it must have a specific transcription and RNA modification system within the core of its virions that enables it to transcribe early mRNA independently from the host nuclear transcription machinery. Genes encoding the poly(A) polymerase, a small (22 kDa) and the largest subunit (147 kDa) of the DNA-dependent RNA polymerase (DdRP) of molluscum contagiosum virus were identified using PCR technology and oligonucleotide primers, corresponding to two conserved domains (RQP[T/S]LH and NADFDGDE) of the known largest subunits of DdRPs from other DNA viruses. The oligonucleotide primers were designed according to the coding usage statistics of known open reading frames of the viral genome [77]. The MCV-1 gene for the largest subunit DdRP and the adjacent genes were localized within the DNA sequences of a part of the BamHI DNA fragment A (BamHI/HindIII DNA fragment A8a; 13.5 kbp, 0.454 to 0.525 viral map units; corresponding to MC076R to MC079R of ref [72] of the MCV-1 genome.

The definitive description of the MCV type 1 genome complement was achieved through DNA nucleotide sequencing of the complete MCV type 1 genome based on the cloned DNA fragments of the type 1 prototype genome and terminal DNA sequences from several individual MCV-1 isolates [12, 72]. The DNA sequence was determined to be 190, 289bp, comprising the entire MCV-1 genome with the exception of the covalently closed terminal hairpin loops (Genbank accession number U60315). The MCV genome has as expected from earlier estimates a G+C content of 63.8% [21, 59]. A similar G+C content is found within the poxvirus family only in the ORF virus (subgenus Parapoxvirus), outside the poxvirus family in herpesviruses, particularly herpes simplex virus.

The arrangement of MCV type 1 terminal repetitive sequences reported by Senkevich et al. support earlier observations of MCV-1 inverted terminal repeats and terminal restriction fragment polymorphisms [9, 12] that are also described for other poxviruses and that may be related to reiterative processes taking place during replication of the MCV viral genome. Variability up to 1.5 kbp was previously observed and occurred symmetrically on both ends of the genome [12]. A DNA sequence that resembles a late poxviral promoter was found in close proximity to the termini of the MCV type 1 genome. A conserved sequence of this type was previously described as the DNA component of a poxvirus telomere resolution site (TRT; [49]). The DNA nucleotide sequence of the terminal hairpins was not determined.

The genomes of poxviruses are known to encode many proteins that perturb host processes at both the cellular and systemic levels. For example, a viral homolog of epidermal growth factor is active in vaccinia virus infections of cultured cells, rabbits, and mice. Several poxviral proteins with homology to the serine protease inhibitor family have been identified. Vaccinia and cowpoxvirus encode a secreted IL 1 receptor homolog that binds and inhibits IL 1 beta (reviewed in [51]). MCV lacks almost all the genes that encode proteins known to be involved in virus host interactions of other poxviruses. Instead it uses its own set of genes that seems specifically designed to adapt the virus to its epidermal environment. In contrast the core gene complement needed for transcription and replication of the viral DNA and structural components of the virion is located mainly in the center part of the MCV genome and highly conserved between the poxviruses. The genomic localization of conserved genes is colinear between MCV and variola and vaccinia virus with only two exceptions [72].

A gene for an epidermal growth factor as described for other poxviruses was proposed for MCV [63] but could not be found by sequence analysis of the MCV type 1 genome [72]. However, some of the predicted MCV proteins appear to be secretory and may represent previously unknown growth factors.

One of these predicted secreted MCV-1 proteins (ORF MC148R [72]) has structural homology to the beta or CC family of chemokines (reviewed in [4]). Chemokines are involved in leukocyte trafficking. They specifically mediate attachment of leukocytes to the vascular endothelium, leukocyte activation and the leukocyte diapedesis process required for tissue invasion. Beta or CC chemokines attract mononuclear leucocytes into sites of tissue injury. Alpha or CXC chemokines (e.g. IL-8) attract neutrophil leukocytes. Intracellular pathogens usually attract mononuclear cells, therefore CC chemokine signalling would be expected to take place in early MCV infection. The typical location and arrangement of 4 cysteine residues within the amino acid sequence of the MCV-1 ORF 148R sterically favors the disulfide bonding pattern and the general conformation of a chemotactic factor. However, the absence of 5 from 10 amino acid residues at the NH_2 terminal end of the protein that are required for target cell activation in known CC chemokines may define the role of the predicted MCV-1 protein as a receptor blocking agent, that ultimately inhibits leukocyte immigration into the site of MCV infection [57].

The predicted MCV-1 chemokine has a typical early promoter as described for other poxviruses. This suggests its expression at the earliest possible timepoint in the MCV infective cycle, still in the basal cell layer of the epidermis, close to the vascularized dermis, where the MCV infection would be positionally and developmentally most vulnerable to interference by the host immune system. An antiinflammatory activity gradient in the transition zone between epidermal and dermal tissue would guarantee the survival of the MCV infection until other protective mechanisms take over. MCV is the first virus described to encode a viral chemokine homolog.

The function of a MCV-1 MHC class I heavy chain homolog encoded by gene MC080R [72] is less clear. The MCV protein lacks conserved amino acids important for peptide binding. Possibly the MCV protein prevents cellular MHC-1 from binding antigenic MCV peptides, that are then not quantitatively presented on the infected cell membrane. Display of mock viral MHC-1 molecules on the cell surface may deceive components of the first line epidermal cellular defense systems, e.g. Langerhans cells and natural killer cells. The MCV MHC-1 homolog has an early promoter and three potential signal peptides instead of one found in cellular MHC-1 heavy chain genes. MCV is the second virus after human cytomegalovirus [3, 8, 26] to encode a MHC-1 gene homolog.

The discovery of these predicted MCV gene products correlates well with earlier findings, describing the nature of the MCV specific local immune response in MCV infected tissue.

Recently, T cell subpopulations were studied in MCV infected skin biopsies [33, 83]. T lymphocytes and NK cell subpopulations were completely absent in the dermis underlying the majority of MCV lesions, in contrast to the findings in papilloma virus infected skin. In some lesions, however, activated T cells could be observed in the underlying dermis and Langerhans cells were detectable perifocally. The patients were not systemically immunodeficient. It was postulated that intact MCV infected cells do not present MCV specific glycoproteins on their cell surface because the intracellular MCV colony is sealed off by a collagen/lipid screen as described before [74]. Damaged lesions may present virus specific antigen and therefore attract immune cells. It was furthermore observed that MCV infected keratinocytes react with a monoclonal antibody against MHC class 1 antigen. This may lead in some cases to autoimmunity or immune tolerance, both previously described for MC lesions [75]. MCV infected keratinocytes did not bind beta 2 microglobulin, activation of EGF and TGF receptors in the vicinity of MC lesions was also observed elsewhere [55].

It is postulated that early MCV lesions secrete the predicted MCV chemokine blocking factor, preventing dermal vascular T lymphocytes from becoming activated and entering the nascent MCV lesion. Later in the infectious cycle, the intracellular MCV colony screens itself off by collagen/lipid membranes, thereby preventing leakage of MCV proteins from the infected cell. Additionally, the predicted MCV MHC heavy chain homolog may by competition with the cellular homolog keep MCV peptides from being presented on keratinocyte cell surfaces. Viral MHC might be presented that lacks peptide binding activity and deceives lingering immunocompetent cells as to the presence of virus. As long as the MCV infected site is not mechanically damaged or otherwise exposed, the lesions will be conspicuous for the absence of immune cells. Trauma inflicted by squeezing, needle picking or curettage/

cauterization will result in immediate exposure and clearance of the lesion. Minor trauma to the MC lesion would also explain spontanous resolution that is common in immunocompetent hosts. In immunodeficient hosts even traumatized lesions are not instantely cleared and lead to the uncontrolled spread and the clinical picture of eczema molluscum. The MCV infection promises to be an extremely usefull model system for the study of immune mechanisms employed in human skin as highlighted by MCV viral countermeasures.

The availability of MCV genome sequences may provide the necessary information for the development of future MCV specific diagnostic and therapeutic reagents.

Treatment

MC therapy is symptomatic. However, prophylaxis of bacterial secondary infections is critical in immunocompromised patients. The recommended method is curettage of the entire primary lesion; needle pricks, electrocauterization and laser therapy are less reliable [30, 31]. Curettage works best if the number of MCV lesions is limited. In immunocompromised patients with hundreds of lesions it is practically impossible to remove all of them surgically. In these cases systemic therapy with a variety of immunomodulating substances and systemically administred antiviral drugs has been attempted with little success. MCV is strictly localized to epidermal tissue and the architecture of its lesions is designed to shield it against systemic and local host defenses. To its possible detriment, however, it seems that this localization makes it an excellent candidate for topical treatment with substances that cannot be used systemically because of their severe toxic side effects. Adenosine N1-oxide (ANO), a potent and highly selective inhibitor of vaccinia virus replication, blocks viral DNA replication and viral late protein synthesis completely. The findings suggest a novel antiviral mechanism whereby incorporation of a modified nucleotide into viral mRNAs might selectively block viral gene expression at the level of translation. It was suggested that ANO merits consideration as an antipoxvirus drug for topical treatment of molluscum contagiosum in humans [37].

Diagnostics

The clinical diagnosis of MC is in most cases easily made by light microscopic examination of MC biopsies [46, 74]. Under certain circumstances, especially in HIV patients with cutaneous cryptococcosis [18, 61] or histoplasmosis [27] a unambigous diagnosis is more difficult and state of the art diagnostics is required. Established is a MCV specific immunosorbent assay based on rabbit antibody to MCV viral antigen for serological testing [39]. Furthermore protocols for in situ hybridization of MCV infected tissue with MCV specific probes were established in two independent laboratories [29, 79]. Protocols for dot blot hybridization of MCV lesion material [34] and immunohistochemical characterization of MCV biopsies exist [60]. Detection of MCV specific DNA sequences for diagnostic purposes were described recently as side results of partial genome sequencing projects [11].

MCV molecular epidemiology

In Europe and in Caucasian populations MCV genotype 1 and variants are with an incidence of 98% predominant over MCV type 2 and variants. A rare MCV type 3 is described that is identical to a previously observed MCV type 2 variant [64, 65, 66, 69, 70, 79, 80]. The MCV genotype 2 prevailed with 32% in a group of 75 Australian patients, 31 of which were HIV infected. Double infections with different MCV strains and genetic types were observed in 4% of the cases. In a separate investigation 69% of 31 HIV patients were infected with MCV type 2 [81]. MCV isolates in Japan show a different epidemiological pattern: Among 477 Japanese patients MCV type 1 variant and subvariants thereof prevail over the MCV type 1 prototype described for Europe. MCV types 2 and 3 are more prevalent than in Europe and a new, extremely rare MCV type 4 was observed [53, 54].

Acknowledgements

The authors thank Dr. Bernard Moss and Dr. Tania Senkevich, Laboratory of Viral Diseases, NIH, NIAID, for helpful discussions and for critically reading the manuscript.

References

1. Aloi FG, Pippione M (1985) Molluscum contagiosum occurring in an epidermoid cyst. J Cutan Pathol 12: 163–165
2. Bateman T (1814) A practical synopsis of cutaneous diseases, 3rd ed. Longman Hurst Rees Orme & Brown, London
3. Beck S, Barrel BG (1988) Human cytomegalovirus encodes a glycoprotein homologous to MHC class-I antigens. Nature 331: 269–272
4. Ben-Baruch A, Michiels DF, Oppenheim J (1995) Signals and rezeptors involved in recruitment of inflammatory cells. J Biol Chem 270: 11 703–11 706
5. Blake NW, Porter CD, Archard LC (1991) Characterization of a molluscum contagiosum virus homolog of the vaccinia virus p37K major envelope antigen. J Virol 65: 3 583–3 589
6. Brandrup F, Asschenfeldt P (1989) Molluscum contagiosum-induced comedo and secondary abscess formation. Pediatr Dermatol 6: 118–121
7. Brown T, Butler P, Postlethwaite R (1973) Nongenetic reactivation studies with the virus of molluscum contagiosum. J Gen Virol 19: 417–421
8. Browne H, Chrucher M, Minson T (1992) Construction and characterization of a human cytomegalovirus mutant with the UL18 (class I homolog) gene deleted. J Virol 66: 6784–6787
9. Bugert J, Rösen-Wolff A, Darai G (1989) Genomic characterization of molluscum-contagiosum virus type 1; Identification of the repetitive DNA sequences in the viral genome. Virus Genes 3: 159–173
10. Bugert JJ, Raab K, Rösen-Wolff A, Janssen W, Darai G (1993) Determination of the position of the boundaries of the terminal repetitive sequences within the genome of molluscum contagiosum virus type 1 by DNA nucleotide sequence analysis. Virology 192: 391–396

11. Bugert JJ, Darai G (1995) Detection of mollucum contagiosum virus by polymerase chain reaction In: Becker Y, Darai G (eds) Frontiers of virology: PCR: protocols for diagnosis. Springer, Berlin Heidelberg New York, Tokyo, pp 303–306
12. Bugert JJ, Darai G (1991) Stability of molluscum contagiosum virus DNA among 184 patient isolates: evidence for variability of sequences in the terminal inverted repeats. J Med Virol 33: 211–217
13. Buller RM, Burnett J, Chen W, Kreider J (1995) Replication of molluscum contagiosum virus. Virology 213: 655–659
14. Burnett J, Neva FA (1966) Studies on the mechanism of the molluscum contagiosum cytotoxicity. J Invest Dermatol 46: 76–93
15. Calvert JG, Ogawa R, Yanagida N, Nazerian K (1992) Identification and functional analysis of the fowlpox virus homolog of the vaccinia virus p37K major envelope antigen gene. Virology 191: 783–792
16. Chang TW and Weinstein L (1961) Cytopathic agents isolated from lesions of molluscum contagiosum. J Invest Dermatol 37: 433–439
17. Charteris DG, Bonshek RE, Tullo AB (1995) Ophthalmic molluscum contagiosum: clinical and immunopathological features. Br J Ophthalmol 79: 476–481
18. Concus AP, Helfand RF, Imber MJ, Lerner EA, Sharpe RJ (1988) Cutaneous cryptococcosis mimicking molluscum contagiosum in a patient with AIDS. J Infect Dis 158: 897–898
19. Cotton DW, Cooper C, Barrett DF, Leppard BJ (1987) Severe atypical molluscum contagiosum infection in an immunocompromised host. Br J Dermatol 116: 871–876
20. Dann FJ, Tabibian P (1995) Cutaneous diseases in human immunodeficiency virus-infected patients referred to the UCLA immunosuppression skin clinic: reasons for referral and management of select diseases. Cutis 55: 85–88, 93–98
21. Darai G, Reisner H, Scholz J, Schnitzler P, Lorbacher de Ruiz H (1986) Analysis of the genome of molluscum contagiosum virus by restriction endonuclease analysis and molecular cloning. J Med Virol 18: 29–39
22. Egawa K, Honda Y, Ono T (1995) Multiple giant molluscum contagiosa with cyst formation. Am J Dermatopathol 17: 414–416
23. Epstein WL (1992) Molluscum contagiosum. Semin Dermatol 11: 184–189
24. Epstein WL, Fukuyama K (1973) Maturation of Molluscum Contagiosum Virus (MCV) in vivo: Quantitative electron microscopic microscopic autoradiography. J Invest Dermatol 60: 73–79
25. Epstein WL, Conant MA, Krasnobrod H (1966) Molluscum Contagiosum: normal and virus infected epidermal cell kinetics. J Invest Dermatol 46: 91–103
26. Fahnestock ML, Johnson JL, Feldman RM, Neveu JM, Lane WS, Bjorkman PJ (1995) The MHC class I homolog encoded by human cytomegalovirus binds endogenous peptides. Immunity 3: 583–590
27. Feuilhade de Chauvin M, Revuz J, Deniau M (1983) Histoplasma duboisii histoplasmosis. Skin lesions simulating molluscum contagiosum. Ann Dermatol Venereol 110: 715–716
28. Fivenson DP, Weltman RE, Gibson SH (1988) Giant molluscum contagiosum presenting as basal cell carcinoma in an acquired immunodeficiency syndrome patient. J AM Acad Dermatol 19: 912–914
29. Forghani B, Oshiro LS, Cham CS, Hurst JW, Dennis J, Darai G, Warford AL, Cohen RM (1992) Direct detection of Molluscum contagiosum virus in clinical specimens by in situ hybridization using biotinylated probe. Mol Cell Probes 6: 67–77
30. Ginsburg CM (1986) Management of selected skin and soft tissue infections. Pediatr Infect Dis 5: 735–774

31. Gonnering RS and Kronish JW (1988) Treatment of periorbital Molluscum contagiosum by incision and curettage. Ophthalmic Surg 19: 325–327
32. Hadasch RP, Bugert JJ, Janssen W, Darai G (1993) Characterization of the genome of molluscum contagiosum virus type 1 between the genome coordinates 0.045 and 0.075 by DNA nucleotide sequence analysis of a 5.6-kb HindII/MluI DNA fragment. Intervirology 36: 32–43
33. Heng MC, Steuer ME, Levy A, McMahon S, Richman M, Allen SG, Blackhart B (1989) Lack of host cellular immune response in eruptive molluscum contagiosum. Am J Dermatopathol 11: 248–254
34. Hurst JW, Forghani B, Chan CS, Oshiro L, Darai G (1991) Direct detection of molluscum contagiosum virus in clinical specimen by dot blot hybridization. J Clin Microbiol 29: 1 959–1 962
35. Ive FA (1985) Follicular molluscum contagiosum. Br J Dermatol 113: 493–495
36. Juliusberg M (1905) Molluscum contagiosum. Dtsch Med Wochenschr 31: 1 598
37. Kane EM, Shuman S (1995) Adenosine N1-oxide inhibits vaccinia virus replication by blocking translation of viral early mRNAs. J Virol 69: 6 352–6 358
38. Katzman M, Carey JT, Elmets CA, Jacobs GH, Lederman MM (1987) Molluscum contagiosum and the acquired immunodeficiency syndrome: clinical and immunological details of two cases. Br J Dermatol 116: 131–138
39. Konya J, Thompson CH, De Zwart-Steffe RT (1992) Enzyme-linked immunosorbent assay for measurement of IgG antibody to Molluscum contagiosum virus and investigation of the serological relationship of the molecular types. J Virol Methods 40: 183–194
40. La Placa M (1966) On the mechanism of the cytopathic changes produced in human amnion cell cultures by the molluscum contagiosum virus. Arch Ges Virusforsch 18: 374–378
41. La Placa M, Portolani M, and Rosa A (1967a) Lack of "reactivating" activity in the molluscum contagiosum virus. Arch Ges Virusforsch 18: 251–255
42. La Placa M, Portolani M, Mannini-Palenzola A, Brabanti-Brodano G, Bernadini A (1967b) Further studies on the mechanism of the cytopathic changes produced by the molluscum contagiosum virus in human amnion cell cultures. J Microbiol 15: 205–216
43. Lange L, Marett S, Maree C, Gerdes T (1991) Molluscum contagiosum in three horses. J S Afr Vet Assoc 62: 68–71
44. Lim W, Sadick N, Gupta A, Kaplan M, Pahwa S (1990) Skin diseases in children with HIV infection and their association with degree of immunosuppression. Int J Dermatol 29: 24–30
45. Linberg JV and Blaylock WK (1990) Giant molluscum contagiosum following splenectomy. Arch Opthalmol 108: 1 076
46. Margolis S (1984) Genital warts and molluscum contagiosum. Urol Clin North Am 11: 163–170
47. Matis WL, Triana A, Shapiro R, Eldred L, Polk BF, Hood AF (1987) Dermatologic findings associated with human immunodeficiency virus infection. J Am Acad Dermatol 17: 746–751
48. McFadden G, Pace WE, Purres J, Dales S (1979) Biogenesis of poxviruses: expression of molluscum contagiosum early functions. Virology 94: 297–313
49. McFadden G, Stuart D, Upton C, Dickie P, and Morgan AR (1988) Replication and resolution of poxvirus telomers. In: Kelly T, Stillman B (eds) Cancer cells 6: eukaryotic DNA replication. Cold Spring Harbor Laboratory Press, Cold Spring Harbor, pp 77–85

50. Melnick JL, Bunting H, Banfield WG (1952) Electron microscopy of viruses of human papillomas: Molluscum contagiosum and vaccinia; including observations on the formation of virus within the cell. Ann NY Acad Sci 54: 1 214–1 225
51. Moss B (1996) Poxviridae: The viruses and their replication. In: Fields B (ed) Fields Virology, 3rd ed. Lippincott-Raven, Philadelphia, New York, pp 2 637–2 671
52. Naert F and Lachapelle JM (1989) Multiple lesions of molluscum contagiosum with metaplastic ossification. Am J Dermatopathol 11: 238–241
53. Nakamura J, Muraki Y, Yamada M, Hatano Y, Nii S (1995) Analysis of molluscum contagiosum virus genomes isolated in Japan. J Med Virol 46: 339–348
54. Nakamura J, Arao Y, Yoshida M, Yamada M, and Nii S (1992) Molecular epidemiological study of molluscum contagiosum virus in two urban areas of western Japan by the in-gel endonuclease digestion method. Arch Virol 125: 339–45
55. Nanney LB, Ellis DL, Levine J, King LE (1992) Epidermal growth factor receptors in idiopathic and virally induced skin diseases. Am J Pathol: 140 915–140 925
56. Ogino A, Ishida H (1984) Spontaneous regression of generalized molluscum contagiosum turning black. Acta Dermatol Venereol Stockh 64: 83–85
57. Omz H, Ben-Baruch A, and Oppenheim J (1996) Chemokines-progress toward identifying molecular targets for therapeutic agents. Trends Biotechnol 14: 46–51
58. Park SK, Lee JY, Kim YH, Kim SY, Cho BK, Houh W (1992) Molluscum contagiosum occurring in an epidermal cyst-report of 3 cases. J Dermatol 19: 119–121
59. Parr AP, Burnett JW, Garon CF (1977) Structural characterisation of the molluscum contagiosum virus genome. Virology 81: 247–256
60. Penneys NS, Matsuo S, Mogollon R (1986) The identification of molluscum infection by immunohistochemical means. J Cutan Pathol 13: 97–101
61. Picon L, Vaillant L, Duong T, Lorette G, Bacq Y, Besnier JM, Choutet P (1989) Cutaneous cryptococcosis resembling molluscum contagiosum: a first manifestation of AIDS. Acta Dermatol Venereol Stockh 69: 365–367
62. Pierard-Franchimont C, Legrain A, Pierard GE (1983) Growth and regression of molluscum contagiosum. J Am Acad Dermatol 9: 669–672
63. Porter CD, Archard LC (1987) Characterization and physical mapping of Molluscum Contagiosum Virus DNA and location of a sequence capable of encoding a conserved domain of epidermal growth factor. J Gen Virol 68: 673–682
64. Porter CD, Archard LC (1992) Characterization by restriction mapping of three subtypes of molluscum contagiosum virus. J Med Virol 38: 1–6
65. Porter CD, Blake NW, Archard LC, Muhlemann MF, Rosedale N, Cream JJ (1988) Molluscum contagiosum virus types in genital and non-genital lesions. Br J Dermatol 120: 37–41
66. Porter CD, Muhlemann MF, Cream J, Archard LC (1987) Molluscum contagiosum: characterization of viral DNA and clinical features. Epidemiol Infect 99: 563–567
67. Postlethwaite R (1970) Molluscum contagiosum: a review. Arch Environ Health 21: 432–452
68. Reed RJ, Parkinson RP (1977) The histogenesis of molluscum contagiosum virus. Am J Surg Pathol 1: 161–166
69. Scholz J, Rösen-Wolff A, Bugert J, Reisner H, White MI, Darai G, Postlethwaite R (1988) Molecular epidemiology of molluscum contagiosum virus. J Infect Dis 158: 898–900
70. Scholz J, Rösen-Wolff A, Bugert J, Reisner H, White MI, Darai G, Postlethwaite R (1989) Epidemiology of molluscum contagiosum using genetic analysis of the viral DNA. J Med Virol 27: 87–90
71. Schwartz JJ, Myskowski PL (1992) Molluscum contagiosum in patients with human immunodeficiency virus infection. A review of twenty-seven patients. J Am Acad Dermatol 27: 583–588

72. Senkevich TG, Bugert JJ, Sissler J, Koonin EV, Darai G, Moss B (1996) Genome sequence of a human tumorigenic poxvirus: Prediction of specific host response-evasion genes. Science 273: 813–816
73. Shand JH, Gibson P, Cooper GRJ, Keir HM, Postlethwaite R (1976) Molluscum contagiosum – a defective poxvirus? J Gen Virol 33: 281–295
74. Shelley WB and Burmeister V (1986) Demonstration of a unique viral structure: the molluscum viral colony sac. Br J Dermatol 115: 557–562
75. Shirodaria PV, Matthews RS (1977) Observations on the antibody responses in molluscum contagiosum. Br J Dermatol 96: 29–34
76. Smith KJ, Skelton HG, Yeager J, James W, Wagner KF (1992) Molluscum contagiosum. Ultrastructural evidence for its presence in skin adjacent to clinical lesions in patients infected with human immunodeficiency virus type 1. Arch Dermatol 128: 223–227
77. Sonntag KC, Clauer U, Bugert JJ, Schnitzler P, Darai G (1995) Identification and properties of the genes encoding the poly (A) polymerase and a small (22 kDa) and the largest subunit (147 kDa) of the DNA-dependent RNA polymerase of molluscum contagiosum virus. Virology 210: 471–478
78. Sullivan JT, Mercer AA, Fleming SB, Robinson AJ (1994) Identification and characterization of an orf virus homologue of the vaccinia virus gene encoding the major envelope antigen p37K. Virology 202: 968–973
79. Thompson CH, Biggs IM, De-Zwart-Steffe RT (1990a) Detection of molluscum contagiosum virus DNA by in situ hybridization. Pathology 22: 181–1863
80. Thompson CH, De-Zwart-Steffe RT, Biggs IM (1990b) Molecular epidemiology of Australian isolates of molluscum contagiosum. J Mod Virol 32: 1–9
81. Thompson CH, De-Zwart-Steffe RT, Donovan B (1992) Clinical and molecular aspects of molluscumm contagiosum infection in HIV-1 positive patients. Int J AIDS 3: 101–106
82. van Rooyen CE (1938) The micromanipulation and microdissection of the molluscum contagiosum inclusion body. J Pathol 46: 425–436
83. Viac J, Chardonnet Y (1990) Immunocompetent cells and epithelial cell modifications in molluscum contagiosum. J Cutan Pathol 17: 202–205
84. Vreeswijk J, Kalsbeck GL, Nanninga N (1977a) Envelope and nucleotide ultrastructure of molluscum contagiosum virus. Virology 83: 120–127
85. Vreeswijk L, Leene W, Kalsbeck GL (1977b) Early host cell molluscum contagiosum virus interactions. II. Viral interactions with the basal epidermal cells. J Invest Dermatol 69: 249–256
86. Wile UJ, Kingery LB (1919) The etiology of molluscum contagiosum: preliminary report of experimental study. J Cutan Dis 37: 431–446

Authors' address: Dr. G. Darai, Institut für Medizinische Virologie der Universität Heidelberg, INF 324, D-69120 Heidelberg, Federal Republic of Germany.

Molecular anatomy of lymphocystis disease virus

C. A. Tidona and **G. Darai**

Institut für Medizinische Virologie, Universität Heidelberg,
Heidelberg, Federal Republic of Germany

Summary. Lymphocystis disease (LD) has been reported to occur in over one hundred different species of fish worldwide. The disease is caused by lymphocystis disease virus (LCDV), a member of the iridovirus family. Numerous fish species that play an important role in fishery and fish farming are highly susceptible to LCDV infection. The infected animals develop disseminated clusters of aberrant hypertrophied cells within their connective tissue, the so-called lymphocystis cells. In the cytoplasm of these cells a massive accumulation of virions can be observed. As a first step towards understanding the mechanisms of viral infection and pathogenesis the complete genomic nucleotide sequence of lymphocystis disease virus type 1 (LCDV-1; flounder isolate) was determined. LCDV-1 is the type species of the genus *Lymphocystivirus* within the family *Iridoviridae*. The virions contain a single linear double-stranded DNA molecule that is circularly permuted, terminally redundant and heavily methylated. Since there is no convenient cell system for virus replication we determined the complete nucleotide sequence of the viral genome (102, 653 base pairs). Computer assisted analyses of 195 potential open reading frames resulted in the identification of a number of putative gene products with significant homology to functionally characterized proteins of other species.

Lymphocystis disease of fish

Lymphocystis disease (LD) was the first viral disease of fish to be described [18] and has been studied for more than one hundred years. The causative agent of LD was discovered by electron microscopy [32] and was called lymphocystis disease virus (LCDV). LCDV is a member of the family *Iridoviridae* representing large non-enveloped viruses with icosahedral symmetry. The iridoviruses can be subdivided into at least four genera [21] which either infect insects (genera *Iridovirus* and *Chloriridovirus*) or poikilothermic vertebrates (genera *Ranavirus* and *Lymphocystivirus*). It is remarkable that all host species of iridoviruses are associated with an aquatic environment which may be essential for virus transmission.

LD is a ubiquitous disease which affects a wide variety of fish species in fresh and salt water [35]. Although LD rarely causes mortalities the major impact to the infected animals is the production of unsightly external lesions which make the fish unsellable. This is a serious problem in aquarium fish as well as in wild or cultured food fish. The major target cells for LCDV infection have been shown to be fibroblasts or osteoblasts [23]. Upon infection LCDV causes massive hypertrophy and encapsidation of the host cells by a hyaline extracellular matrix. Individual cells may undergo a 50 000 to 100 000-fold increase in size and reach a diameter of up to two millimeters. These lymphocystis cells usually appear as disseminated white to gray nodules on fins and skin. There have been reports of ocular lesions [8] and involvement of internal organs [25] which implicate a much higher mortality rate.

LD can be diagnosed histologically. Infected cells have an enlarged nucleus, basophilic cytoplasmic inclusions and a hyaline capsule surrounding the individual cells [33]. As the infection progresses and the virus replicates the nucleus becomes vacuolated and undergoes necrosis while the capsule thickens. The disease does not usually elicit an inflammatory response. Often the infections are self-limiting and spontaneously disappear. In natural infections the disease is apparently spread when the lymphocystis cells are sloughed and burst to release the virus progeny. There seem to be many different host-specific LCDV strains since in most cases of LD the infectivity can only be transmitted between fish species of the same genus. The process of virus spread is still not understood in detail and remains to be elucidated. However, it has been shown that the susceptibility of flounders to LCDV-1 infection can be increased by stress factors, e.g. malnutrition and water pollution [14].

Molecular biology of iridoviruses

Iridoviruses have been isolated from a large variety of different poikilothermic animals from humid or aquatic habitats. The virions are non-enveloped icosahedral particles with a diameter of 120–300 nm. They contain a single linear double-stranded DNA molecule usually between 150 kbp and 200 kbp in size which is circularly permuted and terminally redundant [4, 6, 12]. The circular permutation and terminal redundancy of the iridoviral genome appears to be due to a unique replication strategy which has been studied in detail for frog virus 3 (FV3; type species of the genus *Ranavirus*). After virus entry a rapid shut-off of the host cell macromolecular synthesis can be observed. The viral replication and assembly takes place in the cytoplasm of the infected cell where the large concatemeric replication intermediates are processed and packaged into virus capsids [11, 22]. The mature virus progeny then accumulates in paracrystalline arrays in the cytoplasm of the infected cells. The virions of LCDV contain at least 33 different polypeptides with apparent molecular masses between 14 kD and 22 kD [9] including ten different glycoproteins [24]. Enzymatic activities such as protein kinase, nucleotide phosphohydrolase, RNAse, DNAse, and protein phosphatase have been reported to be present in the virions of some iridoviruses [9, 10]. There has been evidence for the presence of viral toxic polypeptides since solubilized viral proteins of FV3 and *Chilo* iridescent virus (CIV; types species of the genus *Iridovirus*) caused toxic degenerative hepatitis in mice shortly after intravenous or intraperitoneal administration [1, 17]. No viral progeny could be detected in moribund animals indicating that the toxicity is not correlated to virus replication. It was shown that the toxic effect could be inhibited by pretreatment of the viral isolates with heat, protease, or specific antisera [1, 17]. This indicates that at least one of the polypeptides present in virions is capable of producing a hepatotoxic effect in mice.

Lymphocystis disease virus type 1 (LCDV-1) which can be isolated from the skin lesions of infected flouders (*Platichthys flesus*) is the type species of the genus *Lymphocystivirus* within the iridovirus family. The genome of LCDV-1 has been characterized by molecular cloning and physical mapping [4, 5]. The virus particles contain a single linear double-stranded DNA molecule that is circularly permuted, terminally redundant and heavily methylated at cytosine residues. Comparison of cleavage patterns of LCDV-1 DNA after treatment with methyl-sensitive restriction endonucleases indicated that CCGG sequences are completely methylated at the inner C [4]. The content and distribution of the 5-methylcytosine in the DNA of LCDV-1 which was determined by HPLC and nearest neighbour analysis [31] revealed that 22% of all C residues are methylated, including 74% in CpG, about 1% in CpC, and 2–5% in CpA sequences. Recently the gene encoding the viral DNA (cytosine-5) methyltransferase was identified in the genome of LCDV-1 [29]. Although the function of extensive DNA methylation during viral replication is still unclear one can assume that the DNA methyltransferase might be part of a viral restriction modification system similar to that found in bacteria [19]. Genes encoding viral DNA methyltransferases and corresponding methyl-sensitive restriction endonucleases have been identified in the genome of *Paramaecium bursaria Chlorella* virus (PBCV) [36]. PBCV is a member of the family *Phycodnaviridae* and has many features in common with LCDV-1 including the large and highly methylated double-stranded DNA, the non-enveloped polyhedral virion morphology and the cytoplasmic replication strategy.

Genomic structure of LCDV-1

The complete nucleotide sequence of the LCDV-1 genome was determined (102, 653 base pairs) by automated sequencing [30] using a complete genomic library of LCDV-1 [5]. Computer assisted analyses of 195 potential open reading frames (ORFs) resulted in the identification of a variety of putative gene products with significant sequence homology to functionally characterized proteins of other species in the SWISS-PROT 33 databank (Fig. 1, Table 1).

A number of genes coding for proteins involved in viral replication and DNA modification were identified by sequence homology [30]. These include the DNA polymerase (ORF 135R) and the enzymes representing a putative viral

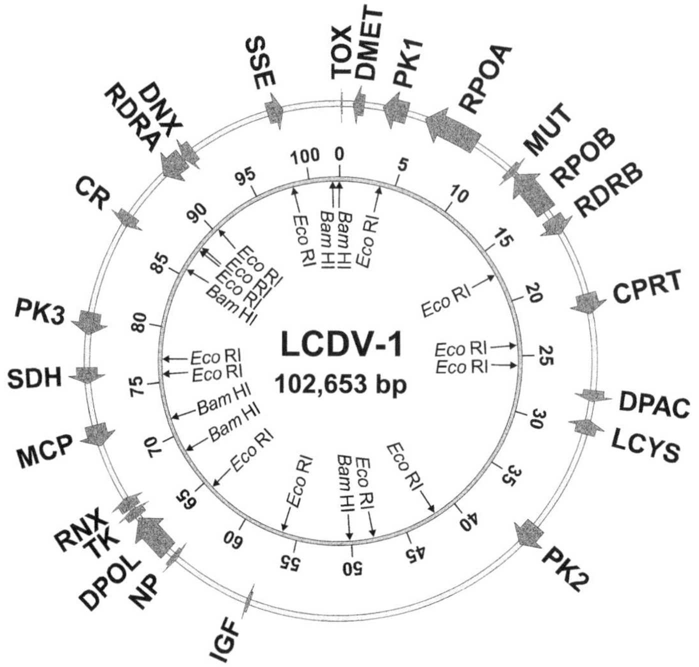

Fig. 1. Physical map of the LCDV-1 genome. Inner circle: the arrows indicate target sites for the restriction endonucleases *Eco*RI and *Bam*HI, and the figures indicate nucleotide positions given in kilo base pairs (kbp). Outer circle: the arrows represent putative genes identified by significant amino acid sequence homology in respect of their size, position and orientation. Abbreviations refer to Table 1

restriction modification system, i.e. the C-5 cytosine-specific DNA methylase (ORF 005L) [29], which is proposed to be responsible for the heavy methylation of the viral DNA at cytosines in CpG sequence, and a restriction endonuclease-like protein specific for a CCGG DNA target sequence (ORF 178L) [30]. There are two putative gene products, a structure-specific endonuclease (ORF 191R) and an ATP-dependent DNA packaging factor (ORF 054R), that are possibly involved in the processing of the large concatemeric replication intermediates and packaging of the viral DNA into virus particles [30]. The presence of a gene product that contains the bacterial mutator protein signature (ORF 022R) [16, 30] is of particular interest, since a variety of poxviruses and African swine fever virus encode homologous proteins.

Cytoplasmic DNA viruses code for a DNA-dependent RNA polymerase that is essential for the transcription of viral genes. Two ORFs encoding the largest subunits of the viral RNA polymerase (ORF 016L and ORF 025L) have been identified in the genome of LCDV-1 [20, 30]. With the exception of a putative ribonuclease (ORF 137R) there is no significant evidence for the presence of enzymes involved in the processing of viral RNA transcripts [30].

Table 1. Putative gene products of LCDV-1 identified by significant sequence homology to functionally characterized proteins of other species

ORF	Gene	A.a.[a]	Significant homologue
001L	TOX	40	neurotoxin TX2-9 of Brazilian armed spider
005L	DMET	228	C-5 cytosine-specific DNA methylase of frog virus 3
010L	PK1	517	serin/threonine protein kinase RCK2 of *Saccharomyces cerevisiae*
016L	RPOA	1 199	DNA-dependent RNA polymerase largest subunit of *Chilo* iridescent virus
022R	MUT	147	mutator mutT protein of *Streptococcus pneumoniae*
025L	RPOB	1 024	human DNA-dependent RNA polymerase II subunit 2
027R	RDRB	378	human ribonucleoside-diphosphate reductase M2 chain
043R	CPRT	430	human cathepsin B precursor
054R	DPAC	244	ATPase of frog virus 3
063L	LCYS	311	nonfibrillar collagen type IX subunit $\alpha 1$ of rat
080R	PK2	519	serine/threonine protein kinase of equine herpesvirus type 1
125R	IGF	80	insulin-like growth factor II of chicken
134L	NP	120	sperm-specific protein of blue mussel
135R	DPOL	932	DNA polymerase δ catalytic chain of *plasmodium falciparum*
136R	TK	195	thymidine kinase of channel catfish virus
137R	RNX	251	ribonuclease III of *Escherichia coli*
147L	MCP	459	major capsid protein of *Chilo* iridescent virus
153L	SDH	339	3-β-hydroxy-5-ene steroid dehydrogenase of vaccinia virus
158L	PK3	488	protein kinase UME5 of *Saccharomyces cerevisiae*
167L	CR	267	OX40 antigen precursor of rat
176L	RDRA	547	ribonucleoside-diphosphate reductase large chain of vaccinia virus
178L	DNX	234	type II restriction enzyme *Msp*I of *Moraxella sp.*
191R	SSE	333	DNA repair protein RAD2 of *Schizosaccharomyces pombe*

[a]A.a. Amino acids

Large DNA viruses like poxviruses and herpesviruses encode a number of proteins that interfere with the nucleic acid metabolism of the host cell. Accordingly it was found that LCDV-1 codes for the two subunits of the ribonucleoside-diphosphate reductase (ORF 176L and ORF 027R) and a thymidine kinase-like protein (ORF 136R) [3, 30]. Several putative gene products of LCDV-1 contain conserved active site signatures of enzymes involved in protein modification such as a cysteine proteinase (ORF 043R) [7] and three protein kinases (ORF 010L, ORF 080R, and ORF 158L) [15, 30]. Cysteine proteinases are known to cleave lamins which may lead to the disintegration of the nucleus observed in LCDV-1 infected cells.

The major structural component of the icosahedral virion has been shown to be the major capsid protein (ORF 147L) [26], which is highly conserved among iridoviruses, African swine fever virus and some algal viruses [13]. Therefore the major capsid protein gene can be considered a useful tool for comparative phylogenic studies and classification of iridoviruses [34]. LCDV-1 also encodes a small basic histone-like protein that might be involved in condensation of the large viral DNA thus functioning as a nucleocapsid protein (ORF 134L) [30].

It has been shown that LCDV-1 codes for a number of potential virulence factors that are also present in other large DNA viruses. For example the putative polypeptide derived from ORF 125R shows significant homology to a variety of insulins and insulin-like growth factors [30] and could therefore be responsible for the hypertrophic properties of the infected lymphocystis cells. The hyaline extracellular matrix of the virus-producing cells is assumed to consist at least in part of a putative secreted nonfibrillar collagen type IX homologue encoded by ORF 063L of LCDV-1 [30]. The putative protein derived from ORF 153L shows significant homology to members of the 3-β-hydroxysteroid dehydrogenase superfamily [2, 30] and could be involved in the alteration of the transcription activity of the host cell. Analogous to poxviruses [27] LCDV-1 encodes a tumor necrosis factor receptor homologue that contains the typical cysteine-rich repeat motif [28] but lacks a membrane anchor domain (ORF 167L) [30]. A soluble cytokine receptor homologue could have the function of a competitive inhibitor of cytokine-mediated signal transduction. This could be an explanation for the fact that the aberrant lymphocystis cells are not recognized and eliminated by the immune system of the infected host over a long period of time.

A unique finding is the detection of a gene coding for a polypeptide of 40 amino acids (ORF 001L) that shows significant homology to a small toxic polypeptide of the Brazilian armed spider [30]. The presence of a gene encoding a viral toxin is consistent with the observation that solubilized iridoviral proteins from isolated virions cause fatal hepatitis in mice shortly after intravenous or intraperitoneal administration [1, 17]. The evaluation of possible risks for humans eating LCDV-infected fish or fish products requires further analyses of potentially toxic virus components present in infected fish.

Acknowledgement

This work is dedicated to Dr. Yechiel Becker at the Hebrew University of Jerusalem, Israel on the occasion of his 65th birthday.

References

1. Aubertin A-M, Anton M, Bingen A, Elharrar M, Kirn A (1977) Solubilized viral proteins produce fatal hepatitis in mice. Nature 265: 456–457
2. Baker ME, Blasco R (1992) Expansion of the mammalian 3-beta-hydroxysteroid dehydrogenase/plant dihydroflavonol reductase superfamily to include a bacterial cholesterol dehydrogenase, a bacterial UDP-galactose-4-epimerase, and open reading frames in vaccinia virus and fish lymphocystis disease virus. FEBS Lett 301: 89–93
3. Blasco R, Lopez-Otin C, Munoz M, Bockamp E-O, Simon-Mateo C, Vinuela E (1990) Sequence and evolutionary relationship of African swine fever virus thymidine kinase. Virology 178: 301–304

4. Darai G, Anders K, Koch HG, Delius H, Gelderblom H, Samalecos C, Flügel RM (1983) Analysis of the genome of fish lymphocystis disease virus isolated directly from epidermal tumours of pleuronectes. Virology 126: 466–479
5. Darai G, Delius H, Clarke J, Apfel H, Schnitzler P, Flügel RM (1985) Molecular cloning and physical mapping of the genome of fish lymphocystis disease virus. Virology 146: 292–301
6. Delius H, Darai G, Flügel RM (1984) DNA analysis of insect iridescent virus 6: evidence for circular permutation and terminal redundancy. J Virol 49: 609–614
7. Dufour E (1988) Sequence homologies, hydrophobic profiles and secondary structures of cathepsins B, H and L: comparison with papain and actinidin. Biochimie 70: 1 335–1 342
8. Dukes TW, Lawler AR (1975) The ocular lesions of naturally occurring lymphocystis in fish. Can J Comp Med 39: 406–410
9. Flügel RM, Darai G, Gelderblom H (1982) Viral protein and adenosine triphosphate phosphohydrolase activity of fish lymphocystis disease virus. Virology 122: 48–55
10. Flügel RM (1985) Lymphocystis disease virus. Curr Top Microbiol Immunol 116: 133–150
11. Goorha R (1982) Frog virus 3 replication occurs in two stages. Virology 43: 519–528
12. Goorha R, Murti KG (1982) The genome of frog virus 3, an animal DNA virus, is circularly permuted and terminally redundant. Proc Natl Acad Sci USA 79: 248–252
13. Graves MV, Meints RH (1992) Characterization of the major capsid protein and cloning of its gene from algal virus PBCV-1. Virology 188: 198–207
14. Grinwis GCM, Wester PW, Kamstra A, Van de Brandhof EFJ, Van Dijk JE, Leewis RJ, Vaal MA, Vethaak AD, Vos JG (1995) Laboratory maintenance, topographical anatomy and histology of flounder, *Platichthys flesus*. National Institute of Public Health and Environmental Protection, Bilthoven, Netherlands, Rapp No. 732402 001
15. Hunter T (1991) Protein kinase classification. Methods Enzymol 200: 3–37
16. Koonin EV (1993) A highly conserved sequence motif defining the family of MutT-related proteins from eubacteria, eukaryotes and viruses. Nucleic Acids Res 21: 4 847
17. Lorbacher de Ruiz H, Gelderblom H, Hofmann W, Darai G (1986) Insect iridescent virus type 6 induced toxic degenerative hepatitis in mice. Med Microbiol Immunol 175: 43–53
18. Lowe K (1874) Fish disease. IV. Trans Norfolk Norwich Nat Soc Fish: 21–56
19. Meselson M, Yuan R, Heywood J (1972) Restriction modification enzyme systems in bacteria. Annu Rev Biochem 41: 447–466
20. Müller M, Schnitzler P, Koonin EV, Darai G (1995) Identification and properties of the largest subunit of the DNA-dependent RNA polymerase of fish lymphocystis disease virus: dramatic difference in the domain organization in the family *Iridoviridae*. J Gen Virol 76: 1 099–1 107
21. Goorha R (1995) Iridoviridae. In: Murphy FA, Fauquet CM, Bishop DHL, Ghabrial SA, Jarvis AW, Martelli GP, Mayo MA, Summers MD (eds) (1995) Virus Taxonomy. Classification and Nomenclature of Viruses. Sixth Report of the International Committee on Taxonomy of Viruses. Springer, Wien New York, pp 95–99 (Arch Virol [Suppl] 10)
22. Murti KG, Goorha R, Granoff A (1985) An unusual replication strategy of an animal iridovirus. Adv Virus Res 30: 1–19
23. Nigrelli RF, Ruggieri GD (1965) Studies in virus diseases of fishes. Spontaneous and experimentally induced cellular hypertrophy (lymphocystis disease) in fishes of the New York Aquarium, with a report of new cases and an annotated bibliography (1 874–1 965). Zoologica 50: 83–95

24. Robin J, Laperriere A, Berthiaume L (1986) Identification of the glycoproteins of lymphocystis disease virus (LDV) of fish. Arch Virol 87: 297–305
25. Russell PH (1974) Lymphocystis in wild plaice (*Pleuronectes platessa* L.) and flounder (*Platichthys flesus* L.) in British coastal waters. A histopathological and serological study. J Fish Biol 6: 771–778
26. Schnitzler P, Darai G (1993) Identification of the gene encoding the major capsid protein of fish lymphocystis disease virus. J Gen Virol 74: 2 143–2 150
27. Smith CA, Davis T, Wignall JM, Din WS, Farrah T, Upton C, McFadden G, Goodwin RG (1991) T2 open reading frame from the Shope fibroma virus encodes a soluble form of the TNF receptor. Biochem Biophys Res Commun 176: 335–342
28. Sprang SR (1990) The divergent receptors for TNF. Trends Biochem Sci 15: 366–368
29. Tidona CA, Schnitzler P, Kehm R, Darai G (1996) Identification of the gene encoding the DNA (cytosine-5) methyltransferase of lymphocystis disease virus. Virus Genes 12: 219–229
30. Tidona CA, Darai G (1996) The complete DNA sequence of lymphocystis disease virus. Virology 230: 207–216
31. Wagner H, Simon D, Werner E, Gelderblom H, Darai G, Flügel RM (1985) Methylation pattern of DNA of fish lymphocystis disease virus. J Virol 53: 1 005–1 007
32. Walker R (1962) Fine structure of lymphocystis disease virus in fish. Virology 18: 503–505
33. Weissenberg R (1951) Studies on lymphocystis tumor cells of fish. Granular structures of the inclusion substance as stages of the developmental cycle of the lymphocystis virus. Cancer Res 11: 608–613
34. Williams T (1996) The Iridoviruses. Adv Virus Res 46: 345–412
35. Wolf K (1988) Fish viruses and fish viral diseases. Cornell University Press, Ithaca
36. Zhang Y, Nelson M, Nletfeldt JW, Burbank DE, Van Etten JL (1992) Characterization of Chlorella virus PBCV-1 *Cvi* AII restriction and modification system. Nucleic Acids Res 20: 5 351–5 356

Authors' address: Dr. G. Darai, Institut für Medizinische Virologie, Universität Heidelberg, Im Neuenheimer Feld 324, D-69120 Heidelberg, Federal Republic of Germany.

Detection of virus or virus specific nucleic acid in foodstuff or bioproducts – hazards and risk assessment

M. Büttner, A. Oehmig, F. Weiland, H.-J. Rziha, and **E. Pfaff**

Federal Research Centre for Virus Diseases of Animals, Tübingen, Federal Republic of Germany

Summary. There are two possibilities for virus contamination of foodstuff and bioproducts of animal origin: i) the presence of endogenous virus as a result of an acute or subclinical infection of animal raw material used for food processing or ii) contamination of food in the course of processing or thereafter. The latter must be considered as the highest risk for human consumers since the viral contamination mostly is caused by virus shedding people and the transmitted viruses are obligate human pathogens. Food from animals consumed as raw material (e.g. oysters) is listed in a high risk category concerning viral contamination (e.g. hepatovirus). Virus contamination of bioproducts such as vaccines, blood products or biological material used in surgery and for transplantations also is more hazardous because the application of contaminating virus usually occurs by circumvention of the natural barrier systems of the body. Moreover, in many cases immunosuppressed people are treated with bioproducts. Due to an enclosing shield of high protein and lipid content in food and bioproducts viruses are well protected against physical and chemical influences, however most preparation procedures for food are destructive for viruses. The detection of pseudorabies virus and pestivirus in biological fluids was tested using polymerase chain reaction (PCR), reverse transcriptase (RT)-PCR and cell culture propagation. PCR is a powerful method to detect viral nucleic acid whereas the detection of infectious virus in cell cultures is more limited, e. g. due to protein and lipid destroying conditions. Virus contamination of bioproducts should be considered as a hazard no matter which method has been used for its detection. Examples are given about the contamination of cell lines and vaccines.

*

Control of animal health in livestock as well as meat inspection at slaughter are the main measures taken for the prevention of human infection with pathogenic microorganisms (zoonotic fungi, bacteria and their toxins, viruses). Since this paper contains the term „virus" in the title, it is easy to exclude BSE and related

Table 1. Human pathogenic viruses and zoonotic viruses detectable in foodstuff

Virus example	Frequent contamination	Proven etiolgy of disease
Hepatitis A	mussels, oysters	hepatitis
Hepatitis E	shellfish	thyphoid fever, hepatitis
Poliovirus	oysters, meat	meningitis, polyomyelitis
Norwalk virus	shellfish	epidemic gastroenteritis
	Zoonoses	
Phlebovirus	sheep/goat meat raw milk	Rifttal fever, hemorrhagia
Echovirus 4	meat	enteritis, meningitis

unconventional agents. Even if the prion theory should turn out to be wrong it would make no sense to continue the speculations about the presence of these agents in foodstuff or bioproducts as long as the pathogenicity for man and the replication strategies of those agents are still unclear [27]. Viruses are considered of low importance (2%) among the overall impact of food-borne microbial infections of man [15]. However, concerning secondary, exogenous contamination of food, primarily of raw consumed specialities such as oysters, viruses play an important role (Table 1). Despite of the situation with special food that is consumed in a raw condition, viral contamination of animal products in the food industry usually is of minor hazard because:

i) in contrast to other micro-organisms (fungi, bacteria) replication of viruses in contaminated foodstuff does not occur and therefore false treatment or incorrect storage of food is not problematic.
ii) preparation of food of animal origin under normal circumstances (cooking temperatures) is destructive for viral proteins and nucleic acids especially when the contamination is limited to the surface of foodstuff (secondary contamination).

Viruses that may be spread to internal layers of a food product prepared from animals usually are of animal origin (primary contamination or endogenous: e. g. dispersed during viremia). In this case primarily zoonotic viruses must be considered a hazard for man. Hazards for human health originating from endogenous or primary infections of animals used for the preparation of foodstuff were recognised in direct link to outbreaks of zoonoses such as a rift valley fever epidemic in Egypt 1977 [15, 17]. The existence of species-related viruses in meat and organs of animals at slaughter was seen in numerous cases world-wide during the preparation and follow up of primary cell cultures [15]. There is no doubt that the most dangerous viruses for man in food are the obligate man-pathogenic viruses. The highest risk for infection is associated with the consumption of food in a non-treated raw condition. The food specialities which are well known sources of virus infections of the

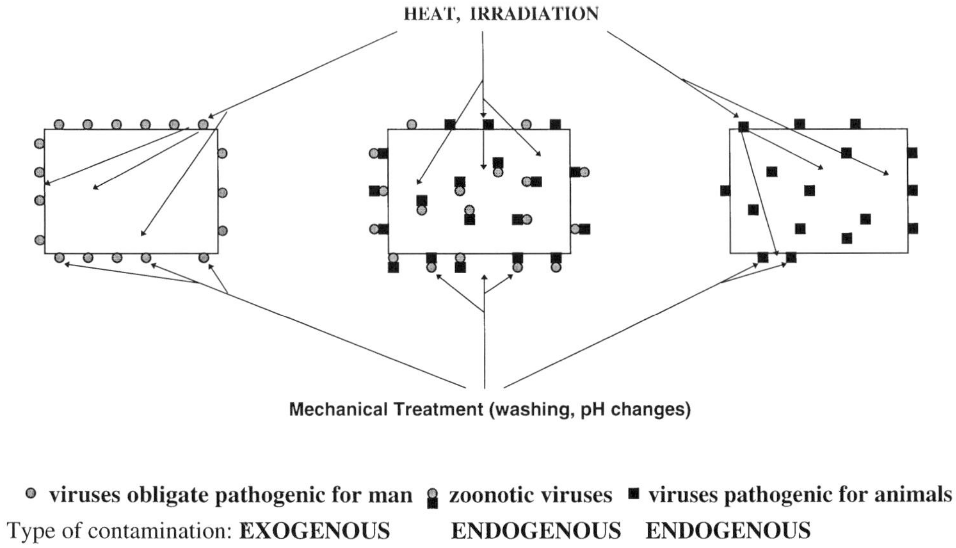

Fig. 1. Differences in virus contamination of foodstuff and influencing by physical treatment

consumers are also listed in Table 1. The second already mentioned dangerous viruses which can be transmitted by food are the zoonotic viruses. Besides the nature of the virus an important difference to the obligate human pathogens is the primary or endogenous existence of zoonotic viruses in the animal which can lead to a distribution of the virus in the whole body e.g. during viremia. Therefore in contrast to exogenous and mostly superficially contaminated food with human pathogens zoonotic viruses can be present also in deeper layers of tissues. This situation is important for the influence of various methods of food preparation on virus infectivity (Fig. 1). It is clear that the same can happen with all viruses that are classified as obligate animal pathogens and possibly are present at slaughter.

Besides the importance for effective prevention of human infections, it must be mentioned that contamination of animal fodder with viruses or unconventional agents (e.g. the scrapie pathogen) can have fatal effects for livestock production and even companion animals. For example in the past little care has been taken for surveillance of pig food resulting in severe epidemics of classical (CSF) as well as African swine fever (ASF) [21, 25].

The methods for the detection of viruses in foodstuff cannot be differentiated from those used in virology in general. However the substrate foodstuff, especially processed food and the mostly low concentrations of contaminating virus determine special diagnostic procedures to be applied. From food bacteriology it is known that the distribution of contaminating micro-organisms in a given substance contributes most to success or failure of recovery attempts (Table 2). It is interesting that not the quality of a test method is the most limiting factor in virus recovery but the heterogeneity of distribution

Table 2. Factors affecting recovery of micro-organisms contaminating foodstuff

List of factors involved	% Importance
Heterogeneity of distribution of target organisms	~ 38
Degree of clumping	~ 20
Goodness of the analytical method	~ 12
Effectiveness of resuscitation	~ 5

of a contaminant. It is clear that low amounts of a contaminant favour the heterogeneity of its distribution and have a negative effect on recovery. The efforts in molecular biology and the successful application of molecular biology in diagnosis of micro-organisms have made access to processed food as a diagnostic substrate. Therefore protection of consumers via analysis of foodstuff has gained more attention since contamination of food before or after processing with human viruses is well known [4–6, 13, 22]. Methods of molecular biology like nucleic acid hybridisation or polymerase chain reaction (PCR) are very sensitive tools for detection of viral contamination especially when only degraded material is available for diagnosis [8, 19]. Since contamination means that the virus is not present in amounts that allow the application of diagnostic methods of low sensitivity like electron microscopy, amplification of the contaminant usually must be tried. Moreover virus enclosing substances mostly do not allow the application of direct and easy to perform methods. The amplification method of choice in classical virology is virus propagation in cell cultures. However this method requires low toxicity and absence of bacteria and fungi in the substances to be screened for virus presence. With the exception of raw meat and organs taken immediately after slaughter and products frozen under sterile conditions foodstuff does not fulfil these requirements. This is one of the reasons why polymerase chain reaction (PCR) has become the dominating method for the detection of virus contamination of foodstuff and bioproducts [8]. Its high sensitivity and specificity are further well appreciated attributes. Taking into account the condition of the samples available for detection of virus contamination in food and bioproducts PCR must be recommended, however it does not allow any conclusion about the presence of infectious virus. Many bioproducts (Table 3) allow the performance of both methods since they are provided as a „sterile“ material (e.g. vaccines), however high specificity in a one step diagnostic procedure will only be ensured by PCR. Moreover an important advantage of PCR is saving of time until specific diagnosis can be accomplished, Virus contamination may not be detectable immediately in cell culture due to low or

Table 3. Effect of pH on the detection of herpesvirus or pestivirus by cell culture propagation or PCR

pH cell culture medium	Pseudorabiesvirus cell culture	PCR	CSFV cell culture	RT-PCR
<3.7	–	–	–	–
3.8–4.3	–	+	–	–
4.7–10	+	+	+	+
10.3–11	–	+	–	–
11–12.8	–	–	–	–

Table 4. Cell culture contamination with bovine pestivirus

Type of cell line			Method used for detection
Ruminants cells			
ATCC	CPA	bovine	immunofluorescence test IF
”	CPAE	”	polymerase chain reaction PCR
”	FBHE	”	indirect immunoperoxidase test IP
”	CPA47	”	IF, IP, PCR
”	SBAC	”	IF, IP, PCR
”	BL-3	”	IF, IP, PCR
”	EJG	”	IF, IP, PCR
”	MDBK	”	PCR, IF flow cytometry[a]
”	SCP	ovine	PCR, IF, IP
”	Ch 1Es	caprine	PCR, IF, IP
”	Fc 2Lu	*cat*	PCR, IF, IP
”	CRFK	*cat*	PCR, IF, flow cytometry[a]
”	RK-13	*rabbit*	PCR, IF, IP
Experimental infections of cell lines with pestivirus			
”	3 cell lines	*cat*	PCR, IP
”	3 cell lines	*rabbit*	PCR, IP
”	LLC-MK2	*monkey*	PCR, IP

ATCC American Type Culture Collection
[a] Experiments at BFAV, Tübingen

non-permissive condition of the cells or may cause trouble because of the presence of a non-cytopathogenic (non-cp) virus or due to loss of virus replication.

With a DNA- and a RNA-virus, namely pseudorabies virus, an alpha Herpesvirus of swine and classical swine fever virus (CSFV), a pestivirus within the *Flaviviridae* family the influence of pH was tested on the performance of PCR and cell culture detection of virus. Both viruses play an important role as contaminants of animal fodder [21, 28]. The range of detection of the DNA virus was clearly improved by using PCR and nested PCR (Table 3). In these experiments the virus containing fluid was cell culture medium and the pH was simply manipulated in vitro by the addition of hydrochloric acid or sodium hydroxide solution. The performance of reverse transcriptase PCR (RT-PCR) for the detection of pestivirus was limited to a pH range between 4.7 and 10. Using a pan-Pestivirus PCR and nested PCR which amplifies a sequence of the conserved non-translated region of the pestiviruses it was possible to detect classical swine fever virus (CSFV) RNA [24]. When RT-PCR was used for the detection of CSFV in the thawing juice of contaminated porcine muscle positive results could only be obtained with virus amounts of 10^4 $TCID_{50}$/ml and more. The use of pooled thawing juice as diagnostic material can be of advantage for the control of imported frozen meat for example instead of time consuming immuno-histology [12]. The pH of thawing juice of porcine meat is between 5.3 and 5.7. In such fluids the virus was not detectable at all in cell cultures. The same results were obtained with the juice from moulded meat. Besides the fact that reverse transcription is a critical step which is not necessary for the detection of DNA viruses, the PCR detection of virus in substances like thawing juice of meat may be disturbed by substances interfering with viral DNA or RNA extraction such as albumin [23]. Even if there are some limiting factors as for example extreme pH values and PCR interfering substances in certain products PCR is the most powerful tool for virus detection in foodstuff. We could show that even irradiation of the pseudorabies virus at 3.0 kGray for 54 h and treatment with the inactivating chemical β-Propiolactone up to a final concentration of 0.1% did not interfere with the PCR detection. The assessment of positive results in diagnostic food virology depends on the impact of the virus that has been detected. The following criteria should be considered:

1. The positive results in food virology obtained with ubiquitous animal viruses that are not pathogenic for man are problematic and primarily of interest in experimental models. They should be told the consumers with care.
2. Generally it can be said that the detection of obligate man pathogenic viruses such as hepatitis viruses and zoonotic viruses is a diagnostic result that should be considered a hazard independent of the method that has been applied for their detection.

The question whether a once detected virus in food is still infectious is not of primary importance as long as the source of contamination has not been traced back and has been clearly identified. Contamination of bioproducts or biologicals is much more important and dangerous as foodstuff contamination for two reasons:

1. Bioproducts are applied without prior treatment (for example by heat) and mostly they are applied by injection or by surgical manipulation.
2. A great part of the recipients of bioproducts are immuno-compromised people.

The term bioproducts or biologicals in a broad sense covers all medical products having a natural substance as source material [10, 11]. In Fig. 2 the most important bioproducts in medicine and veterinary medicine are listed. The

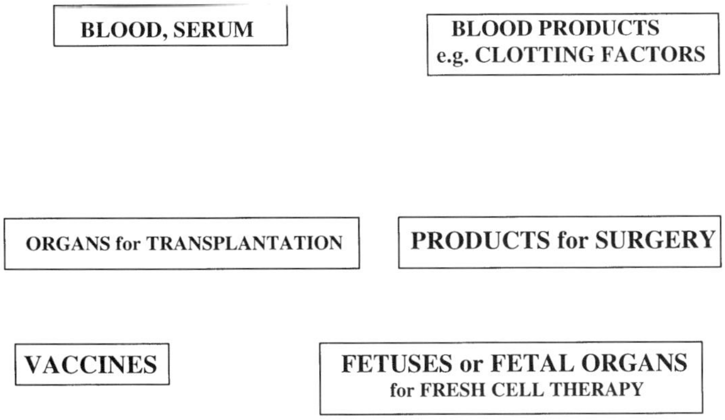

Fig. 2. Biological products with a high risk of virus contamination

production and maintenance of cell cultures is the basis for the production of many bioproducts. One of the persistent problems in the production of bioproducts is the contamination of cell cultures [10, 11, 18]. An example as a consequence thereof is the contamination of vaccines. Essentially the same categories of viral contaminants can be found in bioproducts as in foodstuff, however animal viruses are often underestimated as contaminants. The contamination of fetal bovine serum (FBS) which is used worldwide as a source of growth factors for nearly all kinds of cell cultures is a good example. Fetal bovine sera are distributed commercially and virus contamination can play a role in medicine as well as in veterinary medicine. Contaminants of FBS can be different viruses of bovine origin, however the most frequent contaminant is bovine viral diarrhea virus (BVDV), a pestivirus which exists in two biotypes, a cytopathogenic (cp) and a non-cytopathogenic (non-cp) one [18, 21]. The latter is the prominent biotype in nature which can infect the fetus via the placenta and therefore is present very often in fetal serum [18]. The non cytopathogenic behaviour supports not only survival of the virus but also helps to escape detection. Non cytopathogenic BVDV is present in a variety of ruminant cell lines and moreover even in cell lines that may be considered as not permissive [1, 2]. In Table 4 a summary of published results about the detection of bovine pestivirus in cell lines is listed.

Besides PCR a reliable method for testing of virus contamination of cell lines is immuno-fluorescence combined with flow cytometry [3]. This procedure can enhance the detection of infected cells by a factor up to 10^4 compared to evaluation by microscopy. However due to the unknown amounts of virus in batches of bovine fetal serum and the likelihood of high dilution and therefore enormous heterogeneity of distribution in different batches detection of BVD virus contamination is difficult. In vaccine production the contamination of producer cells with pestivirus [9] has caused two different forms of trouble:

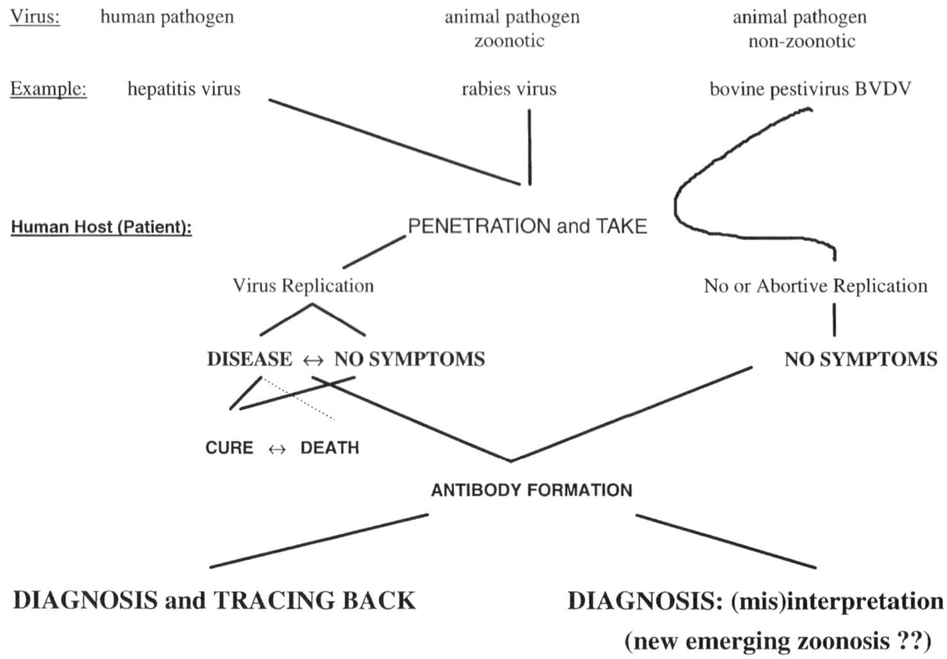

Fig. 3. Events after transmission of viral contamination in bioproducts

1. Contaminated vaccines for veterinary use caused disease and abortion in ruminants [14].
2. Contaminated vaccines used in human medicine induced antibody production and led to discussions about the existence of a man-specific pestivirus [7, 26].

What is likely to happen after treatment of an individual with a contaminated bioproduct for example after vaccination with a contaminated vaccine is displayed schematically in Fig. 3.

For bioproducts it should be demanded that no man-pathogenic adventitious virus ever has been present even if it is proven that it can not replicate at the time of detection. The same is valid for foodstuff contaminated by viruses that are obligate pathogens for man. Once a viral contamination has been diagnosed it should be specified and traced back to unravel the source of transmission primarily to stop further contamination and to learn more about the circumstances of the contamination.

A summary of considerations about the problematic of virus contamination of foodstuff and bioproducts as well as the interpretation of positive diagnostic results is given as follows:

From the virological point of view there is no need to demand for sterile food. However precautions should be taken for food to be consumed in a raw condition. It should be carefully examined for the presence of man-pathogenic viruses using the methods of highest sensitivity and fidelity.

Bioproducts possess an inherent risk of contamination. They must be considered as more dangerous for man and animals than virus contaminated food because:

i) bioproducts are applied without further treatment (e. g. heating) and mostly by circumvention of the barriers of the innate immune system.
ii) the majority of bioproducts are used for the treatment of immuno-suppressed or immuno-compromised people.

References

1. Bolin SR, Black JW, Frey ML, Katz JB, Ridpath JF, Roblin RO (1994) Detection of a cell line contaminated with hog cholera virus. JAVMA 205: 742–745
2. Bolin SR, Ridpath JF, Black J, Macy M, Roblin R (1994) Survey of cell lines in the American Type Culture Collection for bovine viral diarrhea virus. J Virol Methods 48: 211–221
3. Darzynkiewicz Z, Crissman H (1990) Flow cytometry. Academic Press, San Diego
4. Descenclos JC, Klontz KC, Wilder MH, Nainan OV, Margolis HS, Gunn RA (1991) A multistate outbreak of hepatitis A caused by the consumption of raw oysters. Am J Public Health 81: 1268–1272
5. Fang G, Araujo V, Guerrant RL (1991) Enteric infections associated with exposure to animals or animal products. Infect Dis Clin North Am 5: 681–701
6. Fleissner ML, Herrmann JE, Booth JW, Blacklow NR, Nowak NA(1989) Role of Norwalk virus in two foodborne outbreaks of gastroenteritis: definite virus association. Am J Epidemiol 129: 165–172
7. Giangaspero M, Vacirca G, Büttner M, Wolf G, Vanopdenbosch E, Muyldermans G (1993) Serological and antigenetical findings indicating pestivirus in man. Arch Virol [Suppl] 7: 53–62
8. Greiser-Wilke I, Fries R (1994) Methods for the detection of viral contamination in food of animal origin. Dtsch Tierarztl Wochenschr 101: 284–290
9. Harasawa R, Tomiyama T (1994) Evidence of pestivirus RNA in human virus vaccines. J Clin Microbiol 32: 1604–1605
10. Horaud F (1995) Viral iatrogenic contamination and safety of biologicals. Blood Coag Fibrinolyis 6: 5–9
11. Horaud F (1991) Viral safety of biologicals. Dev Biol Stand 75: 3–7
12. Krassnig R, Schuller W, Heinrich J, Werfring F, Kalaus P, Fruhwirth M (1995) Isolation of hog cholera virus (HCV) in imported frozen meat of wildboars. Dtsch Tierärztl Wochenschr 102: 56
13. Le Guyader F, Apaire-Marchais V, Brillet J, Billaudel S (1993) Use of genetic probes to detect hepatitis A virus and enterovirus RNAs in wild shellfish and relationship of viral contamination to bacterial contamination. Appl Environ Microbiol 59: 3963–3968
14. Løken T, Krogsrud J, Bjerkas I (1991) Outbreaks of border disease in goats induced by a pestivirus-contaminated Orf vaccine, with virus transmission to sheep and cattle. J Comp Pathol 104: 195–209
15. Mayr A (1980) New emerging viral zoonoses. Vet Rec 106: 503–506.
16. Minor P (1989) Adventitious viral agents in biological products. Dev Biol Stand 70: 173–179
17. Morse SS, Schluederberg A (1990) Emerging viruses: the evolution of viruses and viral diseases. J Infect Dis 162: 1–7

18. Nuttal PA, Luther PD, Scott EJ (1977) Viral contamination of bovine foetal serum and cell cultures. Nature 266: 835–837
19. Saiki RK, Gefland DH, Stoffel S, Scharf SJ, Hiduchi, Horn GT, Mullis KB, Enrlich HA (1988) Primer-directed enzymatic amplification of DNA with a thermostable DNA polymerase. Science 239: 487–491
20. Tagaki M, Ishikawa K, Nagai H, Sasaki T, Gotoh K, Koyama H (1995) Detection of contamination of vaccines with the reticuloendotheliosis virus by reverse transcriptase polymerase chain reaction (RT-PCR). Virus Res 40: 113–121.
21. Thiel H-J, Plagemann PGW, Moennig V (1996) Pestiviruses In: Fields BN, Knipe DM, Howley PM (eds) Virology vol 1, 3rd ed. Lippincott-Raven, Philadelphia New York, pp 1 059–1 073
22. Torne J, Miralles R, Tomas S, Saballs P (1988) Typhoid fever and acute non A, non-B hepatitis after shell-fish consumption. Eur J Clin Microbiol Infect Dis 7: 581–582
23. Vandenvelde C, Scheen R, Defoor M, Duys M, Dumon J, Van Beers D (1993) Suppression of the inhibitory effect of denatured albumin on the polymerase chain reaction by sodium octanoate: application to the routine clinical detection of hepatitis B virus at its infectivity threshold in serum. J Virol Methods 42: 51–64
24. Vilcek S, Herring AJ, Nettleton PF, Lowings JP, Paton DJ (1994) Pestiviruses isolated from pigs, cattle and sheep can be allocated into at least three genogroups using polymerase chain reaction and restriction endonuclease analysis. Arch Virol 136: 309–323
25. Wilkinson PJ (1989) African swine fever virus In: Pensaert MB (ed) Virus infections of porcines. Elsevier, Amsterdam Oxford New York Tokyo, pp 17–35 (Virus Infections of the Vertebrates, Vol 2)
26. Wilks CR, Abraham G, Blackmore DK (1989) Bovine pestivirus and human infection. Lancet i: 107
27. Will RG, Ironside JW, Zeidler M, Cousens SN, Estibeiro K, Alperovitch A, Poser S, Pocchiari M, Hofman A, Smith PG (1996) A new variant of Creutzfeldt-Jakob disease in the UK. Lancet 347: 921–925
28. Wittmann, G, Rziha, H.-J. (1989) Aujeszky's disease in pigs. In: Wittmann G (ed) Herpesvirus diseases of cattle, horses and pigs. Developments in veterinary virology. Kluwer, Boston Dordrecht London, pp 230–325

Authors' address: Dr. M. Büttner, Federal Research Centre for Virus Diseases of Animals, Paul-Ehrlich-Str. 28, D-72076 Tübingen, Federal Republic of Germany.

Rapid molecular detection of microbial pathogens: breakthroughs and challenges

S. D. Pillai

Environmental Science Program, Texas A & M University
Research Center, El Paso, Texas, U.S.A.

Summary. Microbiological contamination of foods and drinking water is a global problem, and a significant amount of expense is being incurred as a result of such contamination. The microorganisms associated with almost half of all disease outbreaks still go unidentified, primarily as a result of inadequate monitoring and surveillance. Though significant improvements have been made in refining molecular methods for detecting infectious agents, a majority of these methods are being employed only on clinical samples where pathogen densities are much higher than those found in environmental and food samples. Comparative evaluations of the various protocols in terms of cost, sensitivity, specificity, speed, and reproducibility need to be undertaken so that the true applicability of these methods can determined. In the future, molecular methods, especially gene amplifications and in situ hybridizations, will find increasing applications in the differentiation of viable and non-viable organisms, in predicting antimicrobial resistance, and in the identification and characterization of unculturable microorganisms. Though molecular detection methods will not totally replace conventional methods, they will significantly enhance our ability to detect microbial pathogens rapidly.

Introduction

The presence of microbial pathogens in foods and drinking water is a global public health problem. In addition to its health implications, it has significant financial ramifications as well. In the United States alone, food borne diseases cost the economy anywhere from eight to ten billion dollars annually [42]. The economic impacts can be even more severe in developing countries. World-wide, nearly five billion waterborne infections occur annually in Africa, Asia and Latin America. An estimated 25,000 people died every day in 1980 from contaminated drinking water [15]. Unfortunately, the microorganisms associated with almost half of all disease outbreaks still go unidentified primarily as a result of inadequate monitoring and surveillance. The incidence of virus caused disease is greatly underestimated due to a number of reasons, including the lack of efficient, timely, and sensitive methods for detecting foodborne viruses. Since the infective doses of viruses are much lower than of bacterial pathogens, current methods are often not sensitive or are too costly for routine purposes [20].

An effective pathogen monitoring program has to be based on pathogen detection methods that are rapid, reliable, easy to use, and cost-effective. Over the last five to ten years, there have been numerous reports on the development of molecular microbial detection technologies that claim to have widespread applicability for the rapid and effective detection of microbial pathogens. It should be pointed out that most of these molecular methods were originally developed for detecting human pathogens in clinical samples and were later modified to be applicable for detecting pathogens in food samples. This paper details some of the molecular tools that are currently available for detecting microbial pathogens and discusses challenges still confronting rapid pathogen detection.

Breakthroughs

Nucleic acid probes

Nucleic acid probes or "gene probes" are single strands of DNA or RNA which bind to their complementary sequences when present within a mixture of different nucleic acid molecules. The ability of complementary nucleic acid strands to bind like a "zipper" is the fundamental basis of gene probe technology. Due to the advances in gene sequencing and the availability of sequence databases, commercial nucleic acid probes specific to a variety of microbial pathogens are currently available (Table 1). Gene probes (varying in range from tens to thousands of bases long) can be labeled with enzymes, radioisotopes or chemiluminiscent moities so that the probe + target hybrid molecules can be detected in the presence of non specific nucleic acid sequences. Hybridizations are performed either on a solid matrix, in a liquid phase or directly within tissues (in situ hybridization). The advantages of using liquid phase (solution) hybridizations over the solid phase formats are the

Table 1. Examples of microbial pathogens for which commercial probe assays are currently available

Detection in clinical samples	Culture confirmation
Group A Streptococci	*Campylobacter* spp.
Trichomonas vaginalis	Enterococci
Neisseria gonorrhoeae	*Salmonella* spp.
Chlamydia trachomatis	*Mycobacterium tuberculosis* complex
Human papillomavirus	*Listeria monocytogenes*
Legionella pneumophila	Human papillomavirus
	Haemophilus influenzae
	Histoplasma capsulatum
	Neisseria gonorrhoeae

Even though PCR is a DNA driven amplification process, it has been exploited to detect RNA targets also by making use of the enzyme reverse transcriptase (RT) to convert RNA into cDNA molecules. These cDNA molecules thus end up as targets for PCR amplifications. This ability to amplify RNA has been used to detect RNA viruses in a variety of samples including clinical and water samples [9]. Very recently the role of PCR in identifying unculturable infectious agents has also been described [14]. Below are some examples of how PCR is being used in the identification and characterizations of viral pathogens:

Hepatitis A virus

Because typical methods of isolating and purifying viral nucleic acids from clinical samples are time consuming, Jansen et al. [21] developed an antigen capture PCR method for HAV. In this method, anti-HAV monoclonal antibodies coated to the surface of reaction tubes were used to concentrate HAV particles. Following capture the viral RNA was released and RT-PCR was performed to detect the pathogen. This method was shown to be successful in a clinical setting. Desenclos et al. [10] successfully used the antigen capture method and RT-PCR to test oyster homogenates and environmental samples for the investigation of an outbreak of HAV.

Rotaviruses

PCR assays have been described for the detection and typing of rotaviruses [18]. At first glance it may be difficult to envision that PCR would play a significant role in routine laboratory detection given the simplicity of virus detection by electron microscopy or by RNA PAGE profile. However, the sensitivity of rotavirus detection by traditional techniques requires more than 10^7 virions per gram of specimen compared to rotavirus detection by PCR which is reported to be 1000 times more sensitive, detecting about 10^3 to 10^4 copies [47]. Thus, PCR can be used to detect low level virus shedding and the potential role of environmental surfaces and toys in the person-person transmission of the virus.

Norwalk virus

This group of viruses is probably the leading cause of non-bacterial gastroenteritis in developed countries. Since identification by tissue culture is currently not feasible, detection by other approaches becomes the only alternative. Recently, RT-PCR based approaches for the detection of Norwalk virus in fecal samples have been reported [22]. The potential of PCR to play an important role in viral molecular epidemiology was recently reported by Clewley [9]. PCR sequencing of amplification products showed that Norwalk virus associated acute gastroenteritis outbreak strains in the UK were indeed related at the antigenic and genomic levels.

Unculturable and unknown infectious agents

A number of syndromes exist in which the infectious etiology is likely but the pathogen resists cultivation with standard microbiological techniques. For emerging infectious diseases, rapid identification and characterization of the responsible agents are crucial first steps for epidemic control. Representational difference analysis, RDA [28], consensus PCR [37], and cDNA screening [27] are some of the molecular methods available for the identification of unculturable infectious agents.

Representational difference analysis (RDA) (Fig. 1) is one of the more robust methods of identifying new infectious agents since it does not require prior knowledge of the agent's class. The technique relies on the selective enrichment of DNA fragments present in diseased tissue but absent in healthy tissues of the same individual. This method was used by Chang et al. [7] to suggest that a new human herpes virus was associated with Kaposi's sarcoma present in homosexual men diagnosed with AIDS.

The RDA method relies on initially digesting DNA from both healthy and diseased tissues by restriction enzymes. PCR primers are ligated to both sets of

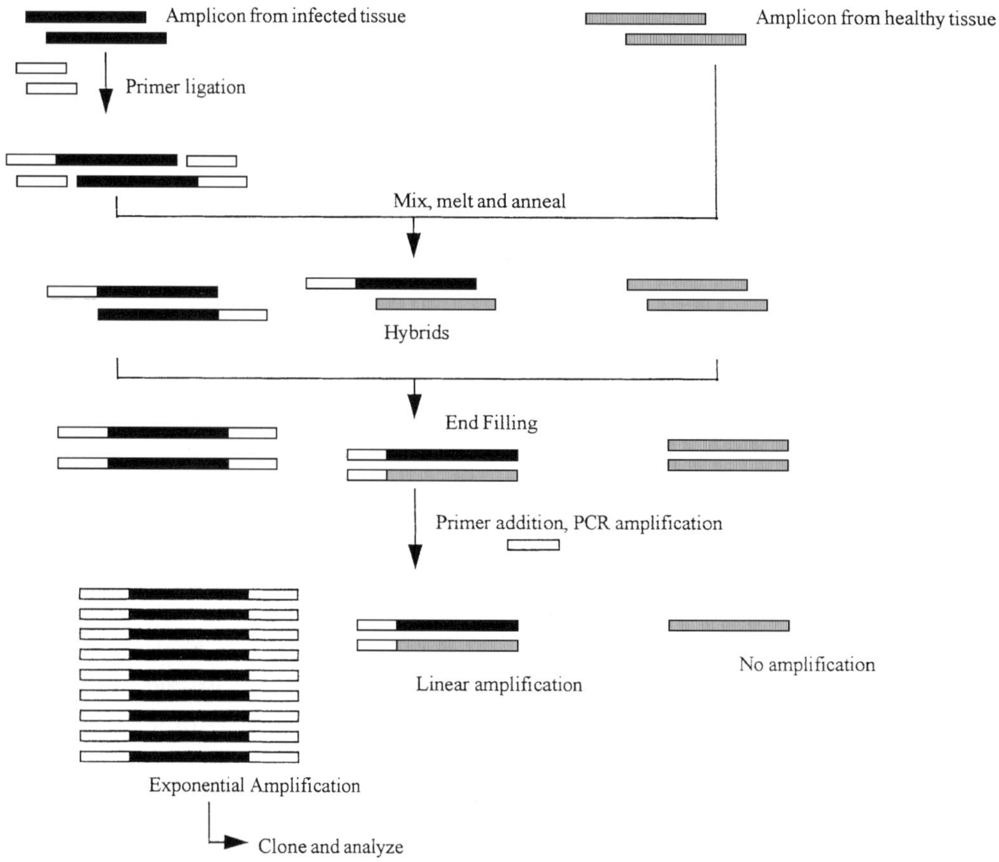

Fig. 1. Schematic representation of the Representational Difference Analysis (RDA) method of pathogen detection (adapted from [14])

DNA fragments and non specifically amplify the mixtures. Unique strands of DNA from diseased tissue representing restriction fragments of a suspected agent are isolated in a subtractive hybridization process coupled to PCR amplification. Initially, the priming sequences ligated on the DNA restriction fragments are removed. New primer sequences are ligated only to the diseased tissue fragments and are hybridized with an excess of healthy tissue fragments. Fragments common to both healthy and diseased tissues will anneal to each other and since the healthy tissue fragments are in excess, any given human fragment derived from the diseased tissue will reanneal to a complementary strand from the healthy tissue. Thus DNA fragments from the infectious agents will not have complementary strands and will therefore have priming sites and be capable of undergoing subsequent PCR amplification. Several rounds of representational difference analysis are performed which successively enrich the mixture for unique DNA sequences present only in the diseased tissue.

Consensus sequence-based PCR (Fig. 2) relies on the use of highly conserved DNA sequences such as the rRNA sequences to amplify DNA from related organisms not yet discovered. Unlike using RDA, using consensus sequences to amplify DNA requires some knowledge of the suspected agent's phylogenetic relationship to other organisms. Consensus sequences are likely to exist among many of the classes of viruses that eventually could be used in screening panels of diseased tissues when a viral cause is suspected.

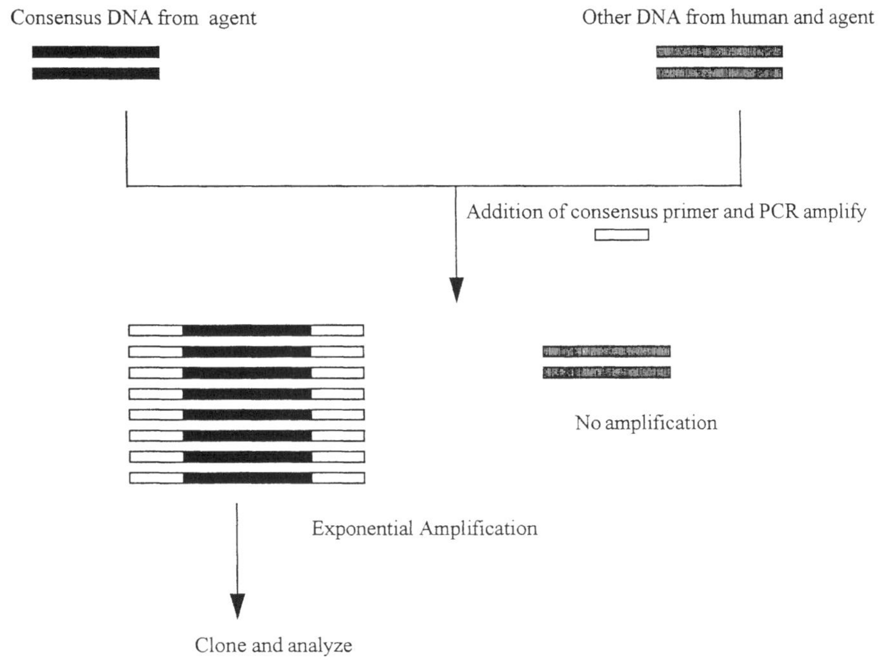

Fig. 2. Schematic representation of the consensus-PCR method of pathogen detection (adapted from [14])

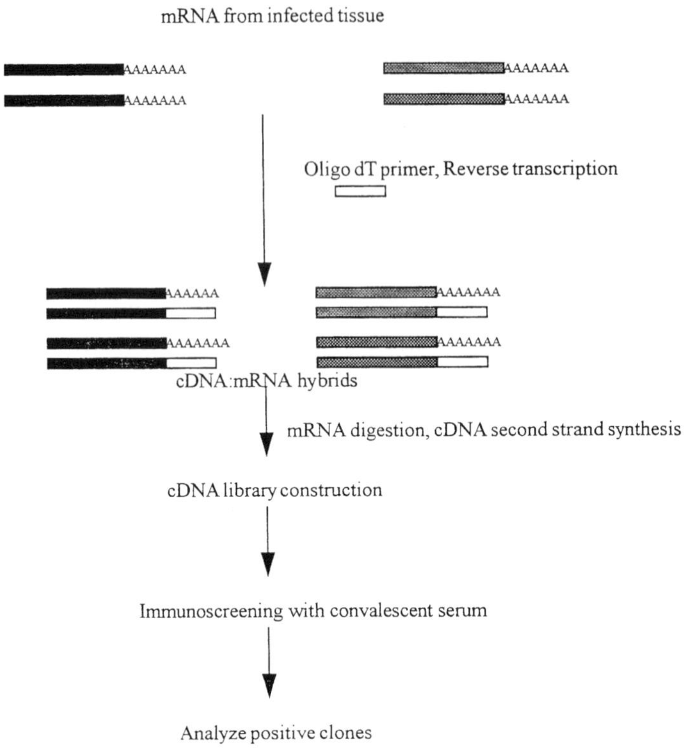

Fig. 3. Schematic representation of the cDNA method of pathogen detection (adapted from [14])

Consensus primers were used in the genetic identification of a hantavirus associated with an outbreak of acute respiratory illness [31].

cDNA screening was successfully used to identify the cause of most cases of non-A non-B hepatitis, Hepatitis C virus. Conventional techniques failed to identify the agent responsible for most cases of NANB hepatitis despite evidence that the disease was caused by a bloodborne virus. Choo et al. [8] initially made a cDNA expression library from RNA isolated from an infected chimpanzee. Hepatitis C virus cDNA was identified by immunologic detection of cDNA that was encoding the viral protein (Fig. 3).

Other amplification protocols

Kwoh et al. [26] developed one of the first non-PCR amplification protocols, termed the Transcription based Amplification System (TAS). It involves the sequential involvement of reverse transcriptase (formation of cDNA), followed by RNA polymerase which catalyzed the synthesis of multiple copies of the original RNA template. Using this methodology, they demonstrated that it was possible to detect as few as one HIV-1 infected cell in the midst of a background of 10^6 cells. Further improvements of the TAS protocol were made by Guatelli et al. [19] with the development of the 3SR (Self Sustaining Sequence Replication) amplification technology. The particular improvement of 3SR over TAS was the use of RNase H which replaced heat denaturation of the RNA-DNA duplexes. Since heat denaturation steps are typically not involved in 3SR, this amplification could be conducted under isothermal conditions. Currently, 3SR protocols can provide a 10^8 fold amplification in about 30 min [17]. There are a couple of 3SR methods that are available for detecting pathogens in clinical samples, but there are no published reports of 3SR methods being used to detect pathogens within a food matrix.

Strand displacement amplification (SDA) is one of the more recent amplification systems in use today which shows a lot of promise [45]. This process, like 3SR is isothermal and does not require temperature cycling. In SDA, single stranded nicks are made in the DNA-primer hybrids that are initially formed. DNA polymerases are then utilized to extend the strands which are subsequently nicked using restriction endonucleases. The nicked gaps are then filled in by the polymerases, which results in the displacement of the originally synthesized strand, and which are then available for reannealing to the primers to initiate a new amplification cycle. One of the drawbacks of SDA is that the target size can become a limiting factor due to the processivity of the DNA polymerase on double stranded templates. Though this method has been demonstrated on relatively "clean" templates, there is very little information as to how it would perform in the presence of extraneous matrices and background microbial populations like those found in foods.

Some of the methods that have been developed to amplify probe molecules hybridized to target sequences include the Qβ replicase for amplifying RNA probe molecules [29], the Ligase Chain Reaction [2], and branched probes. The basic principle behind the Qβ replicase system is to produce RNA probes that can anneal to the target of interest as well as serve as substrates for the Qβ replicase which can replicate the RNA probe molecule. The RNA probe is initially hybridized to the target sequence and all unbound probe molecules are washed off or removed by treatment with RNase. When the Qβ replicase is added, only those probe molecules hybridized to the target will be amplified many times over to facilitate detection. Wiedmann et al. [49] have demonstrated the use of Ligase Chain Reaction (LCR) to differentiate *Listeria monocytogenes* from other *Listeria* species. LCR is a highly specific and sensitive technique that can differentiate DNA target sequences that differ by as little as one base pair. In LCR, probe pairs are designed such that their 3′ end abut one another at the site on the target which is expected to show single base pair variabilities. If the base pair at that site matches the nucleotide at the 3′ end of the upstream primer, the adjoining probes will be joined by the enzyme ligase. Currently, a thermostable ligase is used in LCR amplifications to mimic an amplification analogous to PCR. One of the potential drawbacks are the generation of non-specific amplification products due to blunt end primer annealing, which, however, has been overcome for the most part by the use of thermostable ligases and higher stringency conditions. Wiedmann et al. [49] have demonstrated that as few as 10 CFU of *L. monocytogenes* could be detected (in pure culture studies) using a non isotopic PCR coupled LCR assay. In this method, PCR was initially used to obtain a *Listeria* specific amplification product. This product was then used as the target for LCR to specifically amplify *L. monocytogenes* using biotin and dUTP labeled primers for non isotopic product detection. The products were analyzed in a microtitre plate format using a semi-quantitative approach.

Detection of pathogens in foods of animal origin

A number of gene amplification protocols for detecting food borne microbial pathogens have been published. However, many of these studies have utilized spiked samples rather than naturally contaminated samples to demonstrate the efficacy of the particular method [34]. There is still only very limited amount of information on the efficacies of the various protocols for detecting pathogens especially toxigenic *E. coli, Salmonella* spp., *Campylobacter* spp., and *Listeria* spp. in naturally contaminated samples. Moreover, since a majority of the reports have focused on bacterial pathogens, there is also only limited information on methods to detect viruses in food samples.

Poultry products

Foods of animal origin are believed to be one of the major sources of food-borne salmonellosis. Mahon et al. [30] used primers specific to Salmonella's origin of DNA replication as well as those specific to the virulence plasmid in a multiplex-PCR format to detect *Salmonella* on chicken skin. They did, however, have to rely on an 18 h enrichment culture to achieve a 15% positive detection rate compared to only about a 7% detection rate by culture approaches. Soumet et al. [39] evaluated six different sample processing methods for detecting *Salmonella*-specific nucleic acid in poultry products and reported that a minimum of 10 h pre-enrichment is necessary for efficient amplification. Bej et al. [3] have recently reported that by using a combination of pre-enrichment, PCR, and hybridization methods (to detect PCR amplified products), low level detection of the pathogen is possible. Cano et al. [6] utilized a resin extraction method to demonstrate the detection of 10CFUs of *Salmonella* spp. from ground chicken within 6 h. Giesendorf et al. [16] report detecting 25CFU/g of *Campylobacter* spp. in chicken products after an enrichment step. Uyttendaele et al. [44] were one of the few who utilized a non-PCR based gene amplification method (NASBA) to detect <10 CFU of *Campylobacter jejuni* per 10**g** of chicken samples. Tsen and Chen [43], using proteinase K to detach bacterial cells from the matrix, reported detecting as few as 1–10 cells of spiked *S. aureus* in poultry samples. Wang et al. [46] were successful in detecting low numbers of *Listeria monocytogenes* in spiked meat samples.

Beef products

One of the early studies done to demonstrate the potential of PCR to detect microbial pathogens in meat products were by Wernars et al. [48] who reported specific detection of as few as 3 CFU/g toxin producing *E. coli* in ground beef. Gannon et al. [13] used a multiplex primer approach to detect 1 CFU/g to toxigenic *E. coli* in spiked ground beef. There have ben other reports of rapid pathogen detection in meat products but many of them have utilized either spiked samples to validate the methodology or have used time consuming enrichment steps. Niederhauser et al. [32] successfully detected *Listeria* spp., in spiked and naturally contaminated food samples including raw and cooked meat products. They have reported a detection sensitivity of about 10 bacteria/ 10 g sample within a total time span of about 56 h which included enrichment. Ferreira et al. [12] used *Clostridium botulinum* type A primers to demonstrate that clostridial vegetative cells could be detected in canned meat samples. Szabo et al. [40] have also reported detecting *Cl. botulinum* types A, B, and E in low levels in meat samples. Kapperud et al. [24] using spiked samples have reported detecting as few as between 10–30 CFU of *Yersinia enterocolitica* per gram of meat sample (without enrichment) and as low as 2CFU/g with enrichment steps. Uyttendaele et al. [44] utilized the NASBA method of gene amplification to detect low numbers of *Campylobacter* in meat samples. Tsen and Chen [43] have also reported low level detection of *S. aureus* in spiked meat samples.

Seafood products

Jones et al. [23] reported that by using a combination of PCR and DNA hybridization, <40 *Salmonella* cells could be detected in oysters. Using preenrichment, Bej et al. [4] reported detecting <10 cells in oyster samples. Koch et al. [25] have reported detecting as few as 1CFU/g of *Vibrio cholerae* from a variety of seafoods such as oysters, crabmeat, and shrimp. Their protocol did involve enrichment steps and the results were based on spiked studies. Reverse transcriptase based PCR was used on Norwalk virus particles seeded into oysters [1]. Interestingly, HAV could not be detected by RT-PCR when it was added to whole oysters but only when spiked into extracts of oysters suggesting that sample matrix inhibition is critical. A method to extract rotavirus and hepatitis viral RNA from oyster samples for hybridization analysis was reported earlier by Zhou et al. [51].

Challenges

Though there are a number of published protocols demonstrating that molecular methods can be used to detect microbial pathogens in foods especially those of animal origin, it is essential that their sensitivity values be comparable to those obtainable by conventional methods. The use of artificially inoculated food samples to mimic natural contamination can be fraught with serious limitations. Candrian [5] recently reported that the detection sensitivity of PCR based detection method dropped to approximately 15% when naturally contaminated samples were tested as compared to 100% when artificially spiked samples were used for validation studies. This discrepancy is primarily because the physical association of microbial cells to the food matrix is extremely complex and which cannot be simulated in spike studies. The use of naturally contaminated samples permit the comparative evaluations of the different protocols in terms of speed, detection sensitivity, cost and reproducibility.

Another challenge facing molecular detection of pathogens is in the question of sample size. What is the minimum sample size (that has to be analyzed) that can be termed *representative*? How do we obtain this *representative* sample? Considering the different types of matrices and the rather natural "patchy" distribution of microorganisms on surfaces, it is essential that standardized methods for obtaining statistically and ecologically relevant samples for molecular analyses be developed and verified.

Challenges facing the application of nucleic acid based detection tools is in the interpretation of results when detecting nucleic acids originating from a pathogenic microorganism. Do we have to confirm the presence of virulence genes in addition to, or rather than, species specific genes? What can we infer about the viability of the organism(s) from probe or amplification results? Can gene probe hybridizations and gene amplifications be made truely quantitative? Can the presence of nucleic acids solely be indicative of potential health risks? Considering that many of the molecular methods cannot differentiate between dead and viable microorganisms, does this mean that new methods need to be developed to verify in situ gene expression in the pathogens present in samples?

Future outlook

The spectrum of infectious diseases in changing worldwide. Due to population growth, urban crowding, and international travel the spread of infectious diseases from person to person, from animal to person, and from foods to person in an increasing global problem. Additionally, the emergence of antibiotic resistant pathogenic strains pose a formidable challenge to public health agencies around the world. In order to effectively address the spread of the old and the emerging infectious diseases, it is imperative that surveillance systems for the detection, investigation, and identification of factors influencing their emergence be developed. Undoubtedly, molecular tools for the detection of the organisms and for the typing and subtyping of emerging pathogens would be expected to become a cornerstone of this surveillance program. Specific molecular tools are already in use today in many surveillance programs around the world. However, many of these techniques still need perfecting especially with regards to their ability to predict antimicrobial drug resistance and the identification of environmental reservoirs of pathogens. For example, rapid screening for multidrug resistant *Mycobacterium tuberculosis, Salmonella typhi* and *S. aureus* will become extremely critical in the near future. Additionally, the availability of molecular tools to rapidly differentiate viable from non-viable pathogenic strains in environmental and clinical samples would become critical. So also has to be our ability to detect latent infections as expeditiously as possible.

The ability to control the emergence of infectious diseases depends to some degree on our ability to identify the natural reservoirs and routes of transmission of these organisms. This would imply that rapid molecular methods be available to detect the pathogens in a variety of environments including soils, water, aerosols, foods, and clinical samples. Many of the molecular methods that have been reported recently have tremendous potential for this purpose. There needs to be a concerted effort in conducting round robin investigations of selected protocols in multiple laboratories. The availability of such multi-laboratory tested protocols would be a boon to public health agencies and laboratories around the world. For example, international monitoring capability could be improved by the development of standardized testing and reporting protocols. We could anticipate further development in the methods to process the samples and for molecular typing of organisms. It must be mentioned, however, that molecular methods, however, rapid and sophisticated they become, will not totally replace traditional culture techniques. Molecular techniques will nevertheless significantly improve our ability to predict and prevent the onset of infectious diseases in the near future.

Acknowledgements

This work was supported by funds from the project TEX 8239 administered by the Texas Agricultural Experiment Station of the Texas A&M University System.

References

1. Atmar RI, Metcalf TG, Neill FH, Estes MK (1993) Detection of enteric viruses in oysters by using polymerase chain reaction. Appl Environ Microbiol 59: 631–635
2. Barany, F (1991) The ligase chain reaction (LCR) in a PCR world. PCR Methods Appl 1: 5–16
3. Bej AK, Southworth JP, Law R, Mahbubani MH, Jones DD (1996) Detection of *Salmonella* in chicken meat using PCR. Food Testing Anal 2: 17–20
4. Bej AK, Mahbubani MH. Boyce MJ. Atlas RM (1994) Detection of *Salmonella* spp. in oysters by PCR. Appl Environ Microbiol 60: 368–373
5. Candrian U (1995) Polymerase chain reaction in food microbiology. J Microbiol Methods 23: 89–103
6. Cano RJ, Rasmussen SR, Sanchez-Fraga G, Palmores JC (1993) Fluorescent detection-polymerase chain reaction (FD-PCR) assay on microwell plates as screening test for salmonellas in food. J Appl Bacteriol 75: 247–253
7. Chang Y, Cesarman E, Pessin MS, Lee F, Culpepper J, Knowles DM, Moore PS (1994) Identification of herpesvirus-like DNA sequences in AIDS-associated Kaposi's sarcoma. Science 266: 1865–1869

8. Choo QL, Kuo G, Weiner AJ, Overby LR, Bradley DW, Houghton M (1989) Isolation of a cDNA clone derived from a blood borne non-A, non-B viral hepatitis genome. Science 244: 359–362
9. Clewley JP (1995) The polymerase chain reaction (PCR) for human viral diagnosis, 1 ed. CRC Press, Boca Raton
10. Desenclos J-CA, Klontz KC, Wilder MH, Nainan OV, Margolis HS, Gunn RA (1991) A multistate outbreak of hepatitis A caused by the consumption of raw oysters. Am J Publ Health 81: 1268–1272
11. Echeverria P, Seriwatana J, Sethabutr O, Taylor DN (1985) DNA hybridization in the diagnosis of bacterial diarrhea. Clin Lab Med 5: 447–462
12. Ferreira JL, Baumstark BR, Hamdy MK, McCay SG (1993) Polymerase chain reaction for detection of type A *Clostridium botulinum* in foods. J Food Prot 56: 18–20
13. Gannon VPJ, King RK, Kim JY, Thomas EJG (1992) Rapid and sensitive method for detection of shiga-like toxin-producing *Escherichia coli* in ground beef using the polymerase chain reaction. Appl Environ Microbiol 58: 3809–3815
14. Gao S-J, Moore PS (1996) Molecular approaches to the identification of unculturable infectious agents. Emerg Infect Dis 2: 159–167
15. Gerba CP (1996) Pathogens in the environment. In: Pepper IL, Gerba CP, Brusseau ML (eds) Pollution science. Academic Press, New York, pp 279–299
16. Giesendrof BAJ, Quint WGV, Henkens MHC, Stegeman H, Huf FA, Niesters HGM (1992) Rapid and sensitive detection of *Campylobacter* ssp. in chicken products by using the polymerase chain reaction. Appl Environ Microbiol 58: 3804–3809
17. Gingeras TR, Kwoh DY (1992) In vitro nucleic acid target amplification techniques: issues and benifits. Praxis Biotechnol 4: 403–429
18. Gouvea V, Glass RI, Woods P, Taniguchi K, Clark HF, Forrester B, Fang Z-Y (1990) Polymerase chain reaction amplification and typing of rotavirus nucleic acid. J Clin Microbiol 28: 276–282
19. Guatelli JC, Gingeras TR, Richman DD (1989) Nucleic acid amplification in vitro: detection of sequences with low copy numbers and application to diagnosis of human immunodeficiency virus type 1 infection. Clin Microbiol Rev 2: 217–226
20. Hill WE (1996) The polymerase chain reaction: applications for the detection of foodborne pathogens. Crit Rev Food Sci Nutr 36: 123–173
21. Jansen RW, Siegl G, Lemon SM (1990) Molecular epidemiology of human hepatitis A virus defined by an antigen-capture polymerase chain reaction method. Proc Natl Acad Sci USA 7: 2867–2871
22. Jiang X, Wang J, Graham DY, Estes MK (1992) Detection of Norwalk virus in stool by polymerase chain reaction method. J Clin Microbiol 30: 2529–2534
23. Jones DD, Law R, Bej AK (1993) Detection of *Salmonella* spp. in oysters using polymerase chain reactions (PCR) and gene probes. J Food Sci 58: 1191–1196
24. Kapperud G, Vardund T, Skjerve E, Hornes E, Michaelsen TE (1993) Detection of pathogenic *Yersinia enterocolitica* in foods and water by immunomagnetic separation, nested polymerase chain reactions and colorimetric detection of amplified DNA. Appl Environ Microbiol 59: 2938–2944
25. Koch WH, Payne WL, Wertz BA, Cebula TA (1993) Rapid polymerase chain reaction method for detection of *Vibrio cholerae* in foods. Appl Environ Microbiol 59: 556–560
26. Kwoh DY, Davis GR, Whifield KM, Chappelle HI, DiMichele LJ, Gingeras TR (1989) Transcription-based amplification system and detection of amplified human immunodeficiency virus type 1 with a bead based sandwich hybridization format. Proc Natl Acad Sci USA 86: 1173–1177

27. Linnen JJ, Wages Jr J, Zhang-Keck ZY, Fry KE, Krawczynski KZ, Alter H, Koonin E, Gallagher M, Alter M, Hadziyannis S, Karayiannis P, Fung K, Nakatsuji Y, Shih JW-K, Young L, Piatak Jr M, Hoower C, Fernandez J, Chen S, Zou J-C, Morris T, Hyams KC, Ismay S, Lifson JD, Hess G (1996) Molecular cloning and disease association of hepatitis G virus: a transfusion-transmissible agent. Science 271: 505–508
28. Lisitsyn N, Lisitsyn N, Wigler M (1993) Cloning the differences between two complex genomes. Science 259: 946–951
29. Lizardi PM, Guerra CE, Lomeli H, Tussie-Luna I, Kramer FR (1988) Exponential amplification of recombinant-RNA hybridization probes. Bio/technology 6: 1197–1202
30. Mahon J, Murphy CK, Barrow PA (1994) Comparison of multiplex PCR and standard bacteriological methods of detecting *Salmonella* on chicken skin. Lett Appl Microbiol 19: 169–172
31. Nichol ST, Spiropoulou CF, Morzunov S (1993) Genetic identification of a hantavirus associated with an outbreak of acute respiratory illness. Science 262: 914–917
32. Niederhauser C, Candrain U, Hofelein C, Jermini M, Buhler H-P, Luthy J (1992) Use of polymerase chain reaction for detection of *Listeria monocytogenes* in food. Appl Environ Microbiol 58: 1564–1568
33. Persing DH (1993) In vitro nucleic acid amplification techniques. In: Persing DH, Smith TF, Tenovar FC, White JJ (eds) Diagnostic molecular microbiology: principles and applications. American Society of Microbiology, Washington DC, pp 3–25
34. Pillai SD, Ricke SC (1995) Strategies to accelerate the applicability of gene amplification protocols for pathogen detection in meat and meat products. Crit Rev Microbiol 21: 239–261
35. Pillai SD, Josephson KL, Pepper IL (1994) Comparison of three probe labelling methods to detect PCR amplification products. J Rapid Methods Autom Microbiol 2: 299–309
36. Reif TC, Johns M, Pillai SD, Carl M (1994) Identification of capsule forming *Bacillus anthracis* spores with the PCR and a novel dual probe hybridization format. Appl Environ Microbiol 60: 1622–1625
37. Relman DA (1993) The identification of uncultured microbiol pathogens. J Infect Dis 168: 1–8
38. Saiki RK, Gelfand DH, Stoffel S, Scharf SJ, Higuchi R, Horn GT, Mullis KB, Erlich HA (1988) Primer directed enzymatic amplification of DNA with a thermostable DNA polymerase. Science 239: 487–491
39. Soumet C, Ermel G, Fach P, Colin P (1994) Evaluation of different DNA extraction procedures for the detection of *Salmonella* from chicken products by polymerase chain reaction. Lett Appl Microbiol 19: 294–298
40. Szabo EA, Pemberton JM, Gibson AM, Desmarcheller PM (1994) Polymerase chain reaction for detection of *Clostridium botulinum* types A, B and E in food, soil and infant feces. J Appl Bacteriol 75: 539–544
41. Tenovar FC, Unger ER (1993) Nucleic acid probes for detection and identification of infectious agents. In: Persing DH, Smith TF, Tenovar FC, White JJ (eds) Diagnostic molecular microbiology: principles and applications. American Society of Microbiology, Washington DC, pp 3–25
42. Todd ECD (1989) Preliminary estimates of costs of food borne disease in the United States. J Food Prot 52: 595–601
43. Tsen H-Y, Chen T-R (1985) Use of the polymerase chain reaction for specific detection of type A, D and E enterotoxigenic *Staphylococcus aureus* in foods. Appl Microbiol Biotechnol 37: 685–690

44. Uyttendaele M, Schukkink R, van Gemen B, Debevere J (1995) Detection of *Campylobacter jejuni* added to foods by using a combined selective enrichment and nucleic acid sequence based amplification (NASBA). Appl Environ Microbiol 61: 1341–1347
45. Walker GT, Frasier MS, Schram JL, Little MC, Nadeau JG, Malinowski DP (1992) strand displacement amplification – an isothermal, in vitro DNA amplification technique. Nucleic Acids Res 20: 1691–1696
46. Wang R-F, Cao W-W, Johnson MG (1982) 16S rRNA based probes and polymerase chain reaction method to detect *Listeria monocytogenes* cells added to foods. Appl Environ Microbiol 58: 2827–2831
47. Ward RL, Berstein DI, Young EC, Sherwood JR, Knowlton DR, Schiff GM (1986) Human rotavirus studies in volunteers: determination of infectious dose and serological response to infection. J Infect Dis 154: 871–880
48. Wernars K, Delfgou E, Soentoro PS, Notermans S (1991) Successful approach for detection of low numbers of enterotoxigenic *Escherichia coli* in minced meat by using the polymerase chain reaction. Appl Environ Microbiol 57: 1914–1919
49. Wiedmann M, Barany F, Batt CA (1993) Detection of *Listeria monocytogenes* with a non isotopic polymerase chain reaction-coupled ligase chain reaction assay. Appl Environ Microbiol 59: 2743–2745
50. Zerbani M, Musiani M, Venturoli S, Gallinella G, Gibellini D, Gentilomi G, La Placa M (1990) Rapid screening for B19 parvovirus DNA in clinical specimens with a digoxigenin-labeled DNA hybridization probe. J Clin Microbiol 28: 2496–2499
51. Zhou Y-J, Estes MK, Jiang X, Metcalf TG (1991) Concentration and detection of hepatitis A virus and rotavirus form shellfish by hybridization tests. Appl Environ Microbiol 57: 2963–2968

Authors' address: Dr S. D. Pillai, Environmental Science Program, Texas A&M University Research Center, 1380 A&M Circle, El Paso, TX 79927, U.S.A.

Where do we stand with oral vaccination of foxes against rabies in Europe?

W. W. Müller

Dedicated to the memory of the late Dr. L. G. Schneider

WHO/OIE Collaboration Centre for Rabies Surveillance and Research
at the Federal Research Centre for Virus Diseases of Animals,
Tübingen, Germany

Summary. The oral vaccination of wild animals was first attempted in 1962 after the repeated failure of poisoning or trapping to control movement of the disease in these species. Foxes were chosen for research purposes because they are a problem animal species and are exquisitely susceptible to rabies. The first successful laboratory studies with attenuated vaccine came in 1971, and the first successful field trial was carried out in Switzerland beginning in 1978. In the 1980's several European countries joined the trials. In the following years many improvements were made: the chicken head was replaced by machine-made baits for easy mass production, the hand placement of vaccine baits was to a greater extent being replaced by aerial distribution (small aircraft or helicopter), and several new vaccines were developed. Additionally, the European Union supported the oral vaccination financially. There was a great impact on the rabies situation. When the second country, Germany, joined the field trial in 1983 the total of reported rabies cases in Europe amounted to 23 002, in 1995 a total of 8 134 cases was reported. In spite of the great improvement made in the past years, in the beginning of the 1990's several severe set-backs were experienced. The paper elaborates on reasons for these set-backs and suggests a strategy to overcome them.

Historical notes

The idea to develop an oral rabies vaccine for foxes was born in North America where the first research was carried out. Baer [1] mentions that especially two severe outbreaks, one in Arctic foxes in Canada and the other in foxes in central New York State and the failure of poisoning and trapping the main carrier species, the fox, to control the disease led to attempts to vaccinate these animals.

If a fox is to be vaccinated you need a bait, a vaccine and, alternatively, a device to inject the vaccine into the animal.

In 1970, there were four virus candidates considered worthwhile to be tested in trials:

CVS, LEP, ERA, HEP (Challenge Virus Standard, Low Egg Passage Flury, Evelyn-Rokitniki-Abelseth, and High Egg Passage Flury). In the end ERA (an SAD-Street Alabama Dufferin-derivative) was selected because it was innocuous for foxes after intra-muscular injection and because it was immunogenic in a variety of animals [1].

Numerous collaborative laboratory safety tests were carried out [1] until it was finally settled on a SAD-BHK vaccine to be tested in field trials in Europe in 1978, since in the mid-1970s Switzerland and Germany had actively joined in the development of oral rabies vaccines.

A live attenuated rabies vaccine delivered into the nature? To all the different species? What were the dangers for humans?

Wandeler et al. [27] summarized the desirable requirements for vaccine and bait as follows:

The live attenuated rabies vaccine to be used for the immunization of free-living animals:

(1) Should orally immunize target animals
(2) Should not be pathogenic for humans, for the target species, and for other species eating the bait
(3) Should not be excreted
(4) Should not easily revert to higher pathogenicity
(5) Should be free from pathogenic contamination
(6) Should be storable
(7) Should be stable at environmental temperatures for several days, but not for prolonged periods
(8) Should be easy and inexpensive to produce
(9) Should bear at least one genetic marker

The baits to be used as vehicles for a live attenuated rabies vaccine:

(1) Should be attractive to the target species
(2) Should be eaten without being hoarded
(3) Should be rejected by other species (including humans)
(4) Should reach a large proportion of the target population
(5) Should not inactivate the vaccine
(6) Should deliver the vaccine into the oral cavity
(7) Should be able to incorporate a biologic marker (e.g., tetracycline)
(8) Should be easily available and inexpensive

Next to the epidemiological, virological and serological work to be done, there was a great effort to investigate ecology and biology of the fox and the other animals living in the fox biotope. As an example of the many papers regarding this subject, the work of Wandeler et al. [26] is mentioned.

A vivid account of the contributions of scientists and institutions in the last 30 years of control of rabies in wildlife, and here especially in regard to oral vaccination, was given by Winkler and Bögel [30]. Still more technical details were presented in a paper by Winkler [29].

In all these research efforts one driving force to coordinate the work on oral vaccination needs to be mentioned: the Veterinary Public Health Unit of WHO Headquarters Geneva. It organized many meetings resulting in an exchange of data and experience and recommending priorities for the next steps to be taken.

The begin of field trials in Europe

In 1978 the first field trial of oral vaccination was carried out in the Canton of Valais in Switzerland [19]. The SAD strain adapted and grown in BHK 21 cells was used, later often referred to as SAD Bern. A minimum dose of 10^7 tissue culture infectious units was used in 1.8 ml of vaccine per vaccine bait and stabilized against thermal inactivation by adding 10% egg yolk. The vaccine was placed in a container and deposited under the skin of a slaughterhouse chicken head which was used as bait.

15 vaccine baits per km^2 were evenly distributed either in non-infected areas to prevent the introduction of rabies or in infected areas. The vaccine baits were hand-placed or distributed from helicopters in difficult terrain. Usually a spring and autumn campaign were carried out.

In certain areas the bait uptake was controlled to determine how many vaccine baits were taken and by which animals.

The bait uptake could be evaluated in a different way due to the fact that a marker – chlortetracycline – was injected into the chicken heads, which could later be found in bones or teeth of control animals from the vaccination areas which were shot and supplied by hunters.

Control animals were called in approximately four weeks after placing the vaccine baits. In the laboratory, antibodies were evaluated to determine the seroconversion rate.

The rabies isolates from the vaccination area were characterized with monoclonal antibodies to distinguish between virus field isolates and vaccine virus. This was a measure to determine the occurence of vaccine-induced rabies.

Following the same principles as in the Switzerland field trial five years later the second European field trial was started in Germany [15, 16, 28]. However, the SAD Bern strain was repeatedly cloned and was grown on cloned BHK cells. It was called SAD B19 and had changed to remarkable temperature stability, high titre, and low residual pathogenicity for small rodents [15].

A new form of bait preparation proved to be effective – the so-called "Gullet baiting system". It reduced the need for manpower [15].

Improvement of the method and financial support

The interest in oral vaccination of foxes was great due to the initial success. Nevertheless, the preparation of vaccine baits with chicken heads was limiting the production of large quantities for the field trial since mostly voluntary laboratory staff was used. Therefore, a decisive breakthrough was achieved by the development of a bait consisting of fat and fish meal that could be produced by machine. And more, when tested in the field, the then called Tuebingen fox bait achieved a higher seroconversion rate and consequently, a better protection of the fox population [17].

The switching from the chicken head bait to a machine manufactured bait allowed an extension of the field trials in Germany in the various Bundesländer (federal states) and an export of vaccine baits to other European countries [17]. The production of vaccine baits in Germany increased from 150 000 per year with chicken heads to 2.6 million Tübingen vaccine baits in 1987 [17].

In the 1980's several European countries joined the field trials (see Table 1). And during these years there was encouraging success. Finland was infected after many years, most likely from the former Soviet Union [11], and the Netherlands were infected from Belgium or Germany. Both countries were treated with oral vaccination, became rabies free, and are free today (Table 1).

A further improvement came about when aerial distribution of vaccine baits on a large scale by small aircraft as well as by helicopter was tested. Here, vaccine baits could be well placed even in difficult terrain and, the inefficiency which must occur when hundreds of hunters and helpers are involved could be avoided. Nevertheless, to a certain extent there is still today hand placing carried out, i.e. near human habitation where aerial distribution of vaccine baits is difficult.

Motivated by the fact that the SAD strain is a live virus and has residual pathogenicity to rodents, research has gone on regarding new vaccines for oral

Table 1. Rabies development in countries practicing oral vaccination of foxes in Europe

Countries	Cases 1977	Max. cases (year)	Cases 1994	Cases 1995	Start of oral vaccination
AUT	3 058	4044 (1978)	254	95	1986
BEL	68	842 (1989)	61	213	1986
CZH[a]	428	2106 (1984)	221	178	1989
DEU[a]	6 738	9162 (1983)	1376	856	1983 1989[b]
FIN	–	63 (1988)	–	–	1988
FRA[a]	1 668	4214 (1989)	99	38	1986
HUN	736	1466 (1987)	949	1134	1992
ITA	97	448 (1983)	36	11	1984
LUX	34	205 (1982)	1	15	1986
NET[a]	2	65 (1984)	–	–	1988
POL	1 287	3084 (1992)	2227	1973	1993
SVK	167	564 (1994)	564	266	1992
SVN	10	1766 (1981)	839	1084	1988
SWI[a]	1 041	1413 (1981)	225	23	1978

[a]Bat rabies cases not considered

[b]Start of oral vaccination in former German Democratic Republic

use. Stöhr and Meslin [20] mention recently that at least 6 vaccines are presently in use in Europe for large scale vaccination campaigns. Next to the above mentioned SAD Bern and SAD B19 vaccines a vaccinia-rabies-glycoprotein recombinant (VRG) vaccine was developed [5] and later used in the field in Belgium, France and Luxembourg [20].

In France a modified live vaccine was constructed from SAD Bern, the SAG-1, an escape mutant selected in the presence of an antiglycoprotein monoclonal antibody [8]. It was used in France and Switzerland in the field and is now being further developed to the SAG-2 as double avirulent derivative of SAD Bern [7].

A second modified live vaccine in Germany was produced from SAD Bern by Sinnecker et al. and called SAD P5/88 [18].

The strain SAD Bern constitutes the vaccine part of a vaccine bait, manufactured in the Czech Republic called Lysvulpen [25].

In the Slovak Republic a vaccine is used today with a vaccine deriving of an other SAD strain: Vnukovo [21].

Summarizing what is published until now on vaccine induced rabies, other than the pathogenicity in laboratory trials of rodents, there are reports only on one strain, the SAD Bern. Wandeler [24] has reported on 3 animals (a cat, a stone marten, a fox cub) during vaccination campaigns in Switzerland with their virus isolates identical in monoclonal antibody pattern to the vaccine virus. Two of the isolates were given orally to and injected into dogs and cats. The inoculated animals did not demonstrate any clinical signs of rabies, but developed antibodies. Bingham et al. [2] report of the death of 2 chacma baboons (*Papio ursinus*) out of four wild-caught animals which were given the vaccine orally in laboratory trial in Zimbabwe. A possible explanation of the incident is: the baboons which succumbed to the vaccine virus were immunosuppressed due to the stress of caging.

No vaccine induced rabies occurred in the following primate safety experiments: Rupprecht et al. [13] used the vaccinia rabies recombinant vaccine for squirrel monkeys (*Saimiri sciurius*) and chimpanzees (*Pan troglodytes*), Schneider (pers. comm.) used the SAD B19 vaccine on chimpanzees (*Pan troglodytes*) and Neubert (pers. comm.) the SAD P5/88 vaccine on baboons (*Papio hamadryas*).

Looking at the development of rabies in the countries where oral vaccination is practiced there is everywhere an obvious improvement, no matter which of the above mentioned vaccine baits were used [25]. In France, where more than one type of vaccine bait was used, an evaluation of efficacy was attempted. However, the many parameters involved and the different conditions of the biotope where the vaccine baits were placed make it difficult to come to easy conclusions. For example, some parameters like the fox population density, which is so important for oral vaccination, can only be estimated. Masson et al. [10] comparing three vaccines go as far as singling one out as being the favourite, but without referring to parameters of the biotope, without figures on fox population density, without reliable statistics and without any comparison of seroconversion rates. It may be added that exactly these three vaccines tested in France by Masson et al. suffered serious set-backs when used in Belgium, Germany and Switzerland [25].

The WHO Rabies Reference Centres in Tübingen and Malzèville and WHO Headquarters Geneva have organized meetings in West and East Europe for countries practising oral vaccination. This has helped to establish a policy to enlarge the vaccinated areas along their common borders and to exchange knowledge and experience.

A great impact in improving the rabies situation in Europe resulted from the financial support to the oral vaccination by the EU to its member countries as well as to limited areas of certain countries bordering the EU.

To summarize the development of rabies two facts may be quoted: when the second country, Germany, joined the field trial in 1983 the total of reported rabies cases in Europe amounted to 23 002, in 1995 a total of 8 134 cases was reported with a diminishing tendency in 1996 [25]. In most of the countries of western and central Europe, there are only scattered cases left (see Fig. 1). However, an increased application of vaccine baits would still have to be applied for some time for final eradication (see under chapter corrected strategy).

Set-backs and criticism

Soon after the beginning of the oral vaccination there was the experience that rabies could be successfully and in a short time eradicated when oral vaccination was applied after a heavy epidemic. Obviously, the epidemic reduced the fox population decisively and thereafter the remaining population was immunized effectively. On the other hand, any deviation from the recommended strategy – two vaccinations during the year with 15 to 20 vaccine baits per km^2 – mostly because of shortage of money, slowed down the process of improving the rabies situation. In the beginning of the 1990's several countries regularly vaccinating all their infected areas and their neighbours as well noticed an increase of rabies cases. Until 1994 the countries Switzerland, Germany and Belgium made this experience [25]. The reasons for such a development are now known: there were reinfections of formerly rabies-free areas (or expected to be free) and outbreaks from residual foci, both, as a rule, in combination with high fox populations.

What happens with fox populations in areas which had become rabies-free or nearly free after the application of oral vaccination was investigated by our institute in Bavaria and Baden-Württemberg [4, 22]. Both studies could connect the improved rabies situation or the absence of rabies to an increase of the fox population density.

Regarding the mechanism of how fox population density, rabies incidence and oral vaccination relate, Schenzle [14] developed a mathematical model. The gist of the paper can be summarized as follows: because oral vaccination causes an increase of the fox population size, the number of vaccine baits per fox decreases. A critical point of population immunity must be reached, otherwise residual foci would lead to failure in eradicating the disease.

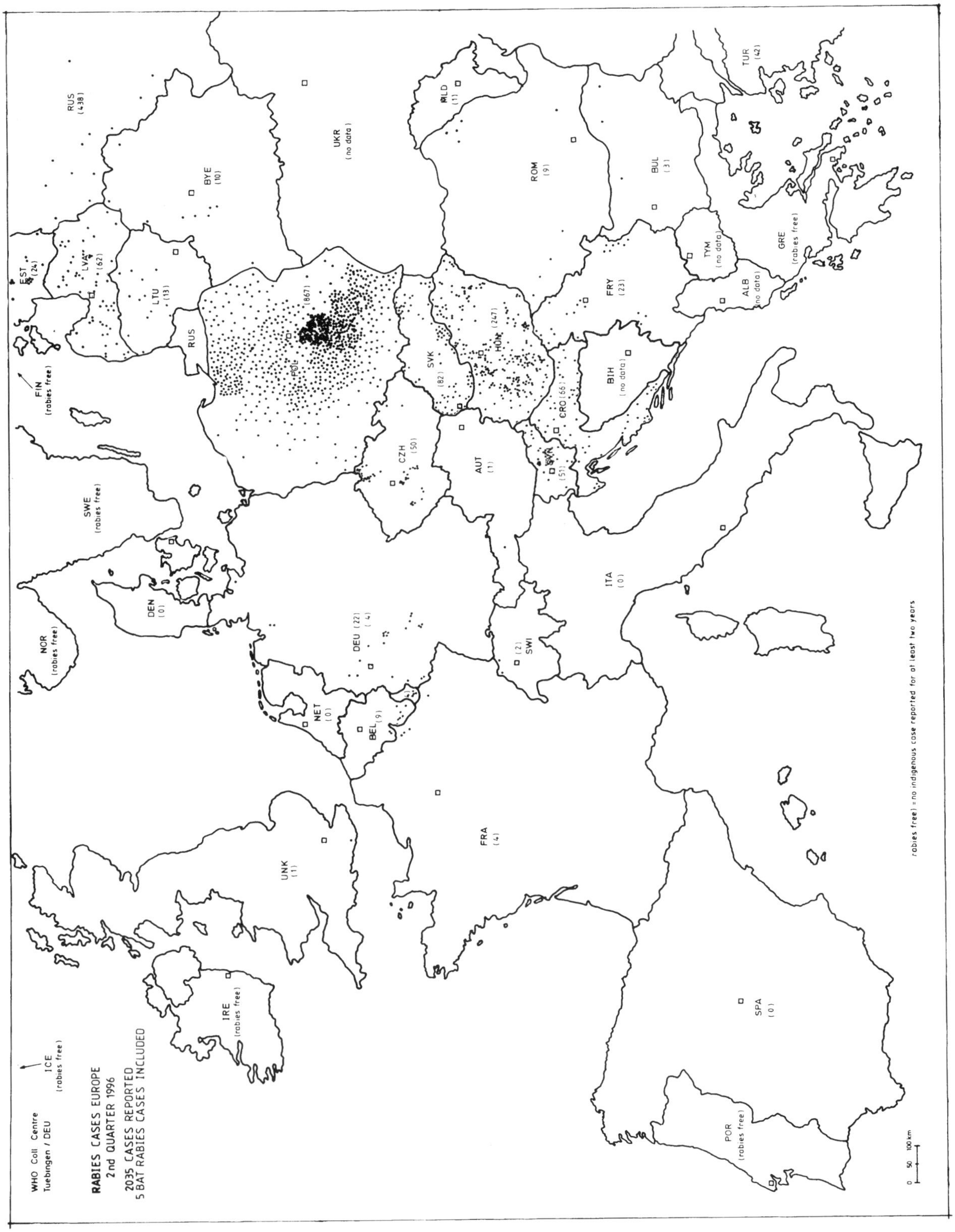

Fig. 1. Distribution of rabies cases in Europe during the second quarter of 1996 [25]

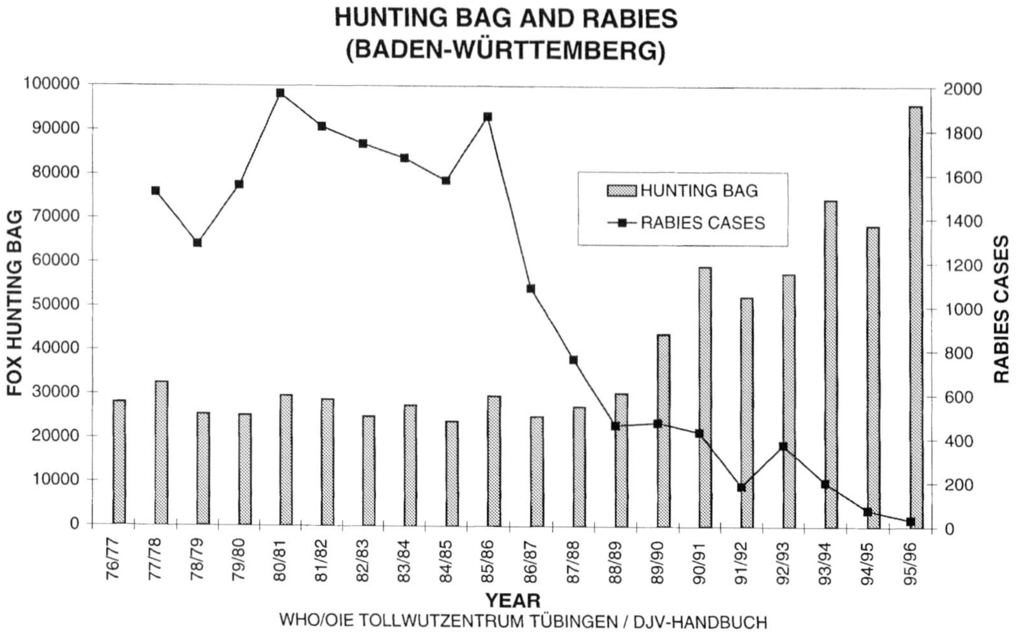

Fig. 2. Course of annual rabies cases and annual hunting bag in Baden Württemberg, Germany. Begin of oral vaccination in 1983

The fox population density is commonly measured by the changes in the annual hunting bag. Figure 2 shows clearly for the state of Baden-Württemberg, Germany: when rabies decreased or disappeared in certain areas, the hunting bag of foxes rose. Taken into consideration that biotopes can be very different, having a different carrying capacity for a fox population, the increase of the hunting bag can be even more pronounced as shown in the figure. The answer to this development is being dealt with in the following chapter, Corrected strategy.

Beside the set-back of the rabies development in different countries and the worry how to cope with it, there has been criticism by biologists and hunting researchers on other effects of the increase of the fox population.

Rühe [12] presented a thesis that due to the disappearence of rabies and the increase of the fox population the fox tapeworm (echinococcosis) would be distributed, a zoonosis much worse than the rabies, and he demanded the stop of oral vaccination of foxes against rabies. Echinococcosis is a disease which has found greater attention more recently but the threat to humans is still to be evaluated. Walter and Kimmig [23], for example, mentioned a survey of 437 foresters, hunters and forest keepers in an endemic fox tapeworm area resulting in no signs of infection and no proven antibody activities. Vos [22] who evaluated a rabies-free area for 7 years after oral vaccination found no increase of echinococcosis in foxes. And no other disease, such as mange, often incriminated as well, has taken the place of rabies.

In regard to echinococcosis, research is under way to treat fox populations using the baits in the same way as for the oral rabies vaccination, but mixing into the bait an anthelmintic [6]. Still, there are many discrepancies in regard to spread, transmission and control of the disease.

More problematical are reports that endangered prey animals of the fox could be extinguished due to the increased fox population. Here ground nesting birds like curley (*Numerius arquata*), pewit (*Vanellus vanellus* L.) and bustard (*Otis tarda*) are often mentioned. This is a problem of the balance in nature we are running into so often today. I think that only well designed research projects can help these animals as hunting to control foxes is obviously not an effective remedy.

Corrected strategy

Once the dense fox populations could be related to the set-backs of the oral vaccination the technical realization of an increased vaccine bait application had to be considered.

Two vaccinations in one year with 15 vaccine baits per km^2 each time were not enough thus, an increase of vaccine baits up to 30 per km^2 has been tried. Another way of increasing to population immunity was a third campaign in summer. The autumn vaccination then improves the percentage of animals immunized in summer.

Breitenmoser et al. [3] suggest an additional vaccination in early summer (May to June) to enhance the immunization of young foxes before their dispersal.

In Baden-Württemberg, Germany, a double vaccination has been tested within two to four weeks which can be used for the spring, summer or autumn campaign according to the riskiness of the situation. Serological results in 9 areas of double vaccination of different campaigns have shown in 1995 between 69 and 95% fox population immunity. All of these areas were trouble spots, but, in 1996 the rabies situation turned to nought cases (results to be published).

In connection to the latter field trials there are two important points to be raised: when is there an indication for a double vaccination and the costs.

I would name four indications:

- areas which do not respond to the biannual campaigns,
- residual cases in good fox biotopes,
- areas to be protected from fierce moving epizootics (cordons),
- areas 2 years after the last rabies cases in good fox biotopes, to take care of residual foci not discovered by surveillance.

In regard to costs it must be said that the earlier one changes to alternative methods (more vaccine baits) the more money can be saved overall. However, in regard to the indication given above there is also the face that fox populations cannot exactly be determined.

In regard to the chance to cope with set-backs, it is encouraging to note that all previously mentioned countries have again improved their rabies situation with an increase of vaccine baits per annum [25].

What remains is the question of increased fox populations. As a veterinarian I feel not really competent on the issue. Therefore I quote a biologist who has long followed the situation on foxes and rabies. Macdonald [9] said 1987 in one of his papers, I quote – "This has led some to voice a further qualm that eradication of rabies is undesirable in so far as it will cause a population explosion in foxes, and thus a crash in the abundance of game. The former proposition is improbable on biological grounds (to the extent that animal populations are fundamentally limited by their food). The second proposition is improper on moral and economic grounds as an argument against eradicating rabies, although predation by foxes may reduce game bags, and sometimes even breeding stocks."

I may say that the larger part of Europe with fox biotopes has no rabies. And, the episode of fox-mediated rabies in Central Europe is rather short – it is thought to have started at the Polish/Russian border around 1939 and reached Germany in 1947, and France in 1968.

Synopsis of present situation

1. The success so far rectifies that oral vaccination of foxes against rabies is to be continued until the disease is eradicated.

   The method though is, due to high fox populations, to be varied in as much as vaccine baits per annum and area (km^2) are to be increased compared to the method initially practiced.
2. High fox populations must not be controlled with a dangereous disease like rabies, as often asked for by hunters and biologists. Fox-mediated rabies only came to western Europe in the 1960's. And, even today there are large areas of fox biotopes in Europe without rabies.

   It is not to be expected that the fox populations will greatly expand since they are limited by their food supply.

   Control should be attempted by hunting or other methods still to be decided on.
3. Wherever animals like ground nesting birds are in danger of being strongly reduced in number or eradicated by foxes, research projects should find means to protect them.
4. Rabies and echinococcosis are separate and distinct diseases which must be dealt with separately. Scientific research on the fox tape worm (echinococcosis) especially in regard to spread, transmission and control is extremely necessary. Figures on the importance of the tape worm as zoonosis need to be collected. Fortunately, initial progress on the control of the fox tape worm has already being made in as much as foxes can be treated with the same bait used in oral vaccination against rabies but combined with an anthelmintic instead of the vaccine.

References

1. Baer GM (1988) Oral rabies vaccination: an overview. Rev Infect Dis 10: S644–S648
2. Bingham J, Foggin CM, Gerber H, Hill FWG, Kappeler A, King AA, Perry BD, Wandeler AI (1992) Pathogenicity of SAD rabies vaccine given orally in chacma baboons (Papio ursinus). Vet Rec 131: 55–56
3. Breitenmoser U, Kaphegyi T, Kappeler A, Zanoni R (1995) Significance of young foxes for the persistance of rabies in northwestern Switzerland. In: Immunobiology of viral infections. Proc 3rd Congress Europ Soc Vet Virol: 391–396
4. Illig M (1995) Epidemiologische Verlaufsformen der Tollwut in Baden-Württemberg unter Berücksichtigung der Impfeinsätze. Inaugural-Dissertation der Justus-Liebig-Universität Gießen
5. Kieny MP, Lathe R, Drillien R, Spelmer D, Skory S, Schmitt D, Wiktor T, Koprowski H, Lecocq JP (1984) Expression of rabies virus glycoprotein from a recombinant vaccinia virus. Nature 312: 163–166
6. Kimmig P, Schelling U (1991) Aktuelle Probleme der Echinokokkose (Echinococcus multilocularis). Öff Gesundheitswesen 53: 596–599
7. Lafay F, Bénéjean J, Tuffereau C, Flamand A, Coulon P (1984) Vaccination against rabies: construction and characterization of SAG2, a double avirulent derivative of SAD Bern. Vaccine 12: 317–320
8. Leblois H, Tuffereau C, Blancou J, Artois M, Aubert A and Flamand A (1990) Oral immunization of foxes with avirulent rabies virus mutants. Vet Microbiol 23: 259–266
9. Macdonal DW (1988) Rabies and foxes: The social life of a solitary carnivore. In: Agriculture. Vaccination to control rabies in foxes. Commission of the European Communities, Luxembourg, pp 5–13
10. Masson E, Aubert MFA, Barrat J, Vuillaume P (1996) Comparison of the efficacy of the antirabies vaccines used for foxes in France. Vet Res 27: 255–266
11. Nyberg M, Kulonen K, Neuvonen E, Ek-Kommonen C, Nuorgam M and Westerling B (1992) An epidemic of sylvatic rabies in Finland. Acta Vet Scand 33: 43–57
12. Rühe F (1992) Die Tollwutimpfung von Füchsen – gefährlich für den Menschen? Die Pirsch 13: 3–5
13. Rupprecht CE, Hanlon CA, Cummins LB, Koprowski H (1992) Primate responses to a vaccinia–rabies glycoprotein recombinant virus vaccine. Vaccine 10: 368–374
14. Schenzle D (1995) Zur Frage der weiteren Tollwutbekämpfung in Deutschland. Dtsch Tierärztl Wochenschr 102: 421–424
15. Schneider LG, Cox JH (1983) Ein Feldversuch zur oralen Immunisierung von Füchsen gegen die Tollwut in der Bundesrepublik Deutschland. I. Unschädlichkeit, Wirksamkeit und Stabilität der Vakzine SAD B19. Tierärztl Umsch 38: 315–324
16. Schneider LG, Wachendörfer G, Schmittdiel E, Cox JH (1983) Ein Feldversuch zur oralen Immunisierung von Füchsen gegen die Tollwut in der Bundesrepublik Deutschland. II. Planung, Durchführung und Auswertung des Feldversuchs. Tierärztl Umschau 38: 476–480
17. Schneider LG, Cox JH, Müller WW, Hohnsbeen KP (1988) Current oral rabies vaccination in Europe: an interim balance. Rev Infect Dis 10: S654–S659
18. Sinnecker H, Apitsch L, Berndt D, Schrader C, Gogolin J, Egert J (1990) Die Entwicklung des Tollwutlebendimpfvirus SAD/Potsdam 5/88 zur oralen Fuchsimpfung sowie seine Charakterisierung am Mausmodell. Mh Veterinärmed 45: 77
19. Steck F, Wandeler A, Bichsel P, Capt S, Schneider LG (1982) Oral immunisation of foxes against rabies. A field study. Zbl Vet Med B 29: 372–396
20. Stöhr K, Meslin FM (1996) Progress and setbacks in the oral immunisation of foxes against rabies in Europe. Vet Rec 139: 32–35

21. Svrcek S, Durove A, Ondrejka R, Zavadova J, Süliova J, Benisek Z, Vrtiak OJ, Feketeova J, Madar M (1995) Immunogenic and antigenic activity of an experimental oral rabies vaccine prepared from the strain Vnukovo-32/107. Vet Med Czech 40: 87–96
22. Vos A (1993) Aspekte der Dynamik einer Fuchspopulation nach dem Verschwinden der Tollwut. Inaugural-Dissertation der Ludwig-Maximilians-Universität München
23. Walter K, Kimmig P (1995) Serologische Untersuchungen auf FSME, Borreliose, Hantavirus- und Fuchsbandwurminfektionen bei Förstern, Jägern und Waldarbeitern im Ostalbkreis/Baden-Württemberg. Ärzteblatt Baden-Württemberg 8: 340–342
24. Wandeler AI (1988) Control of wildlife rabies: Europe. In: Campbell JB, Charlton KM (eds) Rabies. Kluwer Academic Publishers, Boston, pp 375–376
25. WHO Collaborating Centre for Rabies Surveillance and Research Tübingen: Rabies Bulletin Europe. Quarterly issues available 1/77 to 2/96
26. Wandeler A, Müller J, Wachendörfer G, Schale W, Förster U, Steck F (1974) Ecology and biology of the fox in relation to control operations. Zbl Vet Med B 21: 735–773
27. Wandeler AI, Capt S, Kappeler A, Hauser R (1988) Oral Immunization of wildlife against rabies: concept and first field experiments. Rev Infect Dis 10: 649–653
28. Wilhelm U, Schneider LG (1990) Oral immunization of foxes against rabies: practical experiences of a field trial in the Federal Republic of Germany. Bull World Health Organ 68: 87–92
29. Winkler WG (1992) A review of the development of the oral vaccination technique for immunizing wildlife against rabies. In: Bögel K, Meslin F-X, Kaplan M (eds) Wildlife rabies control. Wells Medical, pp 82–96
30. Winkler WG, Bögel K (1992) Control of rabies in wildlife: a revolutionary approach now being tested in Europe and North America is the offspring of a research effort that began modestly in an Atlanta laboratory some 30 years ago. Sci Am 266: 86–92

Author's address: Dr. W.W. Müller, WHO/OIE Collaboration Centre for Rabies Surveillance and Research at the Federal Research Centre for Virus Diseases of Animals, Paul-Ehrlich-Strasse 28, D-72076 Tübingen, Germany.

Foot-and-mouth disease as zoonosis

K. Bauer

Landesuntersuchungsamt für das Gesundheitswesen Nordbayern,
Nürnberg, Federal Republic of Germany

Summary. Man's susceptibility to the virus of foot-and-mouth disease (FMD) was debated for many years. Today the virus has been isolated and typed (type O, followed by type C and rarely A) in more than 40 human cases. So no doubt remains that FMD is a zoonosis. Considering the high incidence of the disease (in animals) in the past and in some areas up to date, occurrence in man is quite rare.

In the past when FMD was endemic in Central Europe many cases of diseases in man showing vesicles in the mouth or on the hands and feet were called FMD. The first suggestion of a human infection with FMD was reported in 1695 by Valentini in Germany [7]. All reports before 1897, the year of the discovery of the virus of FMD by Loeffler and Frosch [2], were not of course confirmed either by isolation of the virus or by identification of immunoglobulins after infection.

Nevertheless the successful self-infection reported by Hertwig in 1834 most likely seems to have been FMD in man: each of three veterinarians drank 250 ml of milk from infected cows on four consecutive days. The three men developed clincial manifestations.

The diseases most often confused with FMD are infections with several viruses of the Coxsackie A group (this infection is referred to as "hand and mouth disease"), herpes simplex and sometimes vesicular stomatitis.

Beginning in 1921 up to 1969 at least 38 papers were published, which described clinically manifest FMD in man in more than 40 proven cases. One further report described an asymptomatic infection with FMD in man [10].

Criteria for establishing a diagnosis of FMD in man are the isolation of the virus from the patient and/or identification of specific antibodies after infection. Laboratory tests for diagnosis of human FMD are the same as for animals.

Proven cases of FMD in man have occurred in several countries in Europe, Africa and South America. The type of virus most frequently isolated man is type O followed by type C and rarely A. The incubation period in man, although somewhat variable, has not been found to be less than two days and rarely more than six days.

The disease in man is much like that seen in susceptible animals: vesicles develop on the hands, mostly on the fingers, occasionally on the feet and in the region of the mouth, especially on the tongue and palate. A fever of short duration occurs as the vesicles appear, after which there is rather rapid healing of the lesions.

The onset is characterized by mild headache, malaise with fever that reaches as high as 39.5 °C. A tingling, burning sensation of the fingers and palms precedes the development of vesicles between fingers at the lateral sites of the hand and volar surfaces of terminal phalanges. Similar sensations are felt in the feet if they are involved. Sometimes the only blisters occur in the oral cavity. Here the greatest discomfort arises. The pain involved in eating, drinking and talking is intense. Excessive salivation adds to the distress.

The aphthae may be as small as a pin head or as large as 2 cm in diameter. In this area the spinose layer of the epidermis experiences a colliquative necrosis. Initially the blister fluid is clear and yellowish but soon becomes inspissated. Blisters dry up within two to three days with skin being sloughed, showing the red basal layer of the epidermis. With care the areas heal quickly by first intention. Secondary blisters may appear up to five days after the primary ones have developed. Recovery is usually well advanced within a week after appearance of the last blister.

Virus content of the blister fluid may be up to $10^{6.5}$ ID_{50} and $10^{3.3}$ ID_{50} in the sloughed epidermis.

Type specific antibodies rise to 1:256 ND_{50} after three weeks and disappear after 16 weeks [3].

Reports of FMD in man ended when mass vaccination eradicated the disease in animals in Europe or reduced the number of outbreaks world wide. Though there are scientifically authenticated cases of human FMD infection, human beings are considered to be relatively non-susceptible to infection by FMD virus. Large numbers of people working in FMD laboratories in several countries have been essentially in daily contact with the virus and only rarely infection has been reported in such persons. In one laboratory only two cases occurred within more than fifty years. Three cases were reported from a large vaccine production unit [3]. Laboratory workers involved in FMD virus manipulation have been found to have antibody titers against the virus without having had any clinical expression of the disease. The percentage of these asymptomatic infections reached from 15% to 54% as published from three laboratories [4, 8, 10].

One paper reported the transfer of FMD virus in the nose of man from infected pigs to susceptible cattle. FMD virus was retained in the nose up to more than 24 hours [6]. None developed any signs of the disease nor was there any rise in antibody titer [5].

Summarizing one may say that no doubt remains that FMD is a zoonosis although of very low incidence. The latter is confirmed by unknown numbers of children in USA, Norway and Rumania who have been inadvertently exposed by being vaccinated with smallpox vaccines contaminated with FMD virus: no cases of the disease were seen in these persons [9].

References

1. Hertwig CA (1834) Übertragung tierischer Ansteckungsstoffe auf den Menschen. Med Vet Z 48
2. Loeffler F, Frosch P (1897) Summarischer Bericht über die Ergebnisse der Untersuchungen der Kommission zur Erforschung der Maul- und Klauenseuche bei dem Institut für Infektionskrankheiten in Berlin. Zbl Bakt I Orig 22: 257–259
3. Pilz W, Garbe HG (1965) Weitere Fälle von Maul- und Klauenseuche – MKS-Infektionen beim Menschen. Zbl Bakt I Orig 198: 154–157
4. Popa M, Iftimovici R, Dohotaru V, Lupascu L, Floren A, Anastasiu J (1963) Frecventa anticorpilor neutralizanti spezifici in single unor persoane au venit in contact cu virusul febrei aftoase. Lucr Stiintifice ale Inst de Ser si Vac Pasteur Bucresti 7: 79
5. Sellers RF, Donaldson AI, Herniman KAJ (1970) Inhalation, persistence and dispersal of foot-and-mouth disease virus by man. J Hyg Camb 68: 565–573
6. Sellers RF, Herniman KAJ, Mann JA (1971) Transfer of foot-and-mouth disease virus in the nose of man from infected to non-infected animals. Vet Rec 86: 447–449
7. Valentini MB (1695) Maul- und Klauenseuche bei Menschen, mehrere Erkrankungen in Hessen. Ephem Sat curios Cent I and II: 156 [cited by Bussenius and Siegel Z Klin Med 32: 147–187 (1897)]
8. Vetterlein W (1955) Der Nachweis von Antikörpern gegen Maul- und Klauenseuche im Serum gesunder, besonders infektionsexponierter Menschen. Zbl Bakt I Orig 162: 4–12
9. Vetterlein W (1963) Die Maul- und Klauenseuche beim Menschen. Vortrag Ges f. Hygiene Abt Seuchenschutz, Leipzig
10. Wisniewski J, Janjowska J (1969) Asymptomatic infection with foot-and-mouth disease in man. Przgl Epidem 23: 263–267 [cited in Excerpta Medica Sect IV 23: 131 (1970)]

Authors' address: Dr. K. Bauer, Rilkestr. 27, D-90419 Nürnberg, Federal Republic of Germany.

Molecular epidemiology of influenza

C. Scholtissek

Institute of Virology, Justus-Liebig-Universitat Giessen, Giessen, Germany

Summary. The genome of the influenza A viruses comprises eight single-stranded RNA segments, and this property makes genetic reassortment after double infection of a host with two different influenza A strains possible. Nature takes advantage of genetic reassortment during antigenic shift creating new pandemic strains. After concurrent infection of a host with both avian and human strains, the hemagglutinin gene of the human virus may be replaced by the allelic gene of the avian virus. This reassortment leads to a human virus strain that has avian hemagglutinin molecules on its surface, significant because the human population lacks neutralizing antibodies to this new glycoprotein. The Hong Kong pandemic of 1968 resulted from just such an event.

*

Influenza A viruses are found in many species, including humans, pigs, and horses; waterfowl are an especially large reservoir [2]. However, avian influenza viruses usually do not spread in the human population, nor do human strains circulate in birds. This strong interspecies barrier prompts a search for a species in which reassortment can and does occur. Both avian and human influenza viruses must be able to infect this species, and considerable data support swine as this host. Under natural conditions pigs can be infected relatively easily by both human and avian influenza viruses. Most of the pandemics in this century started in Southeast Asia presumably because of the high exposure of pigs to both humans and waterfowl. In Asia, pigs often live under the same roof as their owners. In addition, these animals forage in the rice fields after the harvest, where they can come in contact with feral ducks infected with influenza viruses. These practices encourage the concurrent infection of pigs with both human and avian influenza strains, leading to reassortment [8].

Although avian influenza viruses cross into species other than pigs only rarely, these events can have dramatic consequences. According to phylogenetic analysis, an avian influenza A virus crossed into the human population about 100 years ago without reassortment, possibly first circulating in pigs before establishing a new stable lineage in humans [9, 12]. Shortly thereafter, about 20 million people died of influenza during the Spanish flu pandemic of 1918–1919.

Thus, mechanisms other than reassortment can facilitate the entry of influenza A strains into new hosts. For example, the severe swine influenza outbreaks that occurred in Northern Europe during the winter of 1979–1980 were caused by an intact, unreassorted avian influenza A virus [10]. In light of the potentially considerable repercussions, we need to understand the underlying mechanisms by which avian influenza viruses enter new species.

To this end, we studied the rates at which several genes of influenza strains from pigs in Northern Europe mutated and evolved. After comparing these rates to those of reference human strains, we found that most of the swine virus genes mutated more rapidly than did the allelic genes of human isolates (Table 1). Only the HA1 region evolved more rapidly in viruses from humans than from swine, possibly because of stronger selection pressure from the human immune system. In light of its potential as a mechanism facilitating entry into new hosts, we further characterized the relative genetic instability of strains isolated from pigs in Northern Europe.

While attempting to plaque-purify the A/swine/Germany/2/81 (H1N1) isolate, we observed a widely heterogeneous plaque morphology. These plaques were neutralized to varying degrees by a monoclonal antibody (MAB11G4) raised against the hemagglutinin of the Sw/Ger/81 strain. Regardless of the size

Table 1. Rates of change in North European swine influenza A viruses and H1N1 human reference strains

Sequence analyzed	Rate of change[a]				No. of strains	Years of isolation
M	gene		protein			
			M1	M2		
Swine	1.3		2.2	1.7	5	1981–1993
Human	0.8		0.1	0.6	7	1950–1979
NS	gene		protein			
			NS1	NS2		
Swine	2.5		3.6	3.1	4	1981–1993
Human	1.5		2.4	0.2	14	1942–1988
NP	gene		protein			
Swine	1.9			2.0	5	1985–1993
Human	1.4			0.9	16	1940–1983
HA	gene		protein			
	HA1	HA2	HA1	HA2		
Swine	3.6	1.4	3.5	3.4	4	1981–1992
Human	3.7	2.8	5.9	1.8	7(HA1)	1979–1987
					3(HA2)	1934–1979

Data were obtained by direct sequence comparison
M Matrix; *NS* non-structural; *NP* nucleoprotein; *HA* hemagglutinin

$$^{a}\text{Rate of change} = \frac{\text{No. of changes in sequence} \times 10^3}{\text{No. of years} \times \text{no. of positions}}$$

For example, a rate of 1.4=14 changes among 500 positions in 2 years

Table 2. Heterogeneity of plaque size and neutralization after individual passage of A/sw/Germany/2/81 (H1N1) plaques

Plaque isolates	Antibody treatment	Plaque forming units	
		Small plaques	Large plaques
S/S/S/S	–	1.5×10^7	9×10^7
	+	1×10^6	8×10^5
S/S/S/L	–	1×10^7	1.5×10^8
	+	1×10^5	6.5×10^5
S/S/S/S/S	–	2×10^8	$< 10^5$
	+	2×10^5	5×10^4
S/S/S/L/L	–	1×10^7	2×10^8
	+	3×10^5	3×10^6
S/S/S/L mab[a]/L	–	5×10^6	1.5×10^8
	+	8×10^6	2×10^8

Small (*S*, about 1 mm) and large (*L*, about 5 mm) plaques were picked and propagated in embryonated chicken eggs. We harvested the allantoic fluids for the next round of plating and isolation. For example, S/S/S/S means that we picked a small plaque after each of four rounds of plating, then harvested the allantoic fluids, treated them with the MAB 11 G4 antibody as appropriate, and titered the resulting virus stock

[a]We picked a large plaque after treating the stock with MAB 11G4

of the isolated plaque, we obtained a heterogeneous population of plaques after its propagation. Again, neutralization by MAB11G4 of these plaques varied (Table 2). As a rule, an exceptionally high number of these plaques escaped antibody neutralization, further supporting the high genetic variability of this strain [4].

We obtained comparable results with the H1N1 European swine isolates A/swine/Arnsberg/6554/79, A/swine/France/Olid/80, and A/swine/Netherlands/12/85. In contrast, two later H1N1 strains, A/swine/Germany/8 533/91 and A/swine/Schleswig-Holstein/1/93, exhibited a homogeneous plaque morphology; all plaques were about 1 mm in diameter. Both of these strains could be completely neutralized with MAB11G4; the titer decreased at least by a factor of 10 000 after antibody treatment.

Mutation of one of the proteins in the polymerase complex of the earliest North European swine virus may account for the genetic instability we observed among H1N1 swine viruses isolated between 1979 and 1985. Such "mutator mutations," which can occur in influenza viruses [11], increase the likelihood that errors will be introduced during transcription and lead to a large number of variants. Leading to production of numerous variants with decreased infectivity, such a mutation might not be advantageous to viruses that continue to spread in their native host. However, mutator mutations may be necessary to establish stable lineages in new species. Perhaps only avian viruses that contain mutator mutations can cross into a different species (e.g., pigs). The presence of such a mutation may lead to production of a large number of variants in the new host, and those strains best adapted to the new species are selected and maintained. As long as the mutator mutation is present, the avian-derived swine virus (now adapted to a mammalian host) can cross into yet another species (e.g., humans). The events leading to the 1919–1920 Spanish flu pandemic seem to support this theory. The genetic stability of the swine strains isolated in 1991 and 1993 may have resulted from reversion of the mutator mutation. Once a stable lineage is established, the mutator mutation no longer confers a selective advantage to the virus.

Particularly in light of the devastating outcome of the Spanish flu outbreak, we must be alert to the potential impact of new European swine viruses. The recent swine isolates from Northern Germany (Oldenburg and Schleswig-Holstein) are no longer genetically unstable (presumably because they have lost the mutator mutation) and are therefore unlikely to spread to new hosts. However, H1N1 swine viruses from other parts of Europe may still be highly unstable and capable of cross-species spread to humans.

Presumably, antigenic shift (creation of a pandemic strain by reassortment) also requires a mutator mutation. Our attempts to obtain reassortants by double infection of tissue cultures led to the observation that hemagglutinin (HA) and matrix (M) genes segregated well when both of the parental strains were of avian origin, but not when we crossed avian and human strains. Further, the avian-human seggregants (which were obtained only after strong selection pressure) had low titers [5, 6]. Therefore, it seems that reassortment of an avian HA with a human M1 protein (as resulted in the 1968 Hong Kong pandemic) required multiple mutations before optimal HA-M protein cooperation is achieved. Because of their higher rates of mutation and evolution, viruses containing mutator mutations presumably would achieve this result sooner than would strains lacking such mutations.

In 1992, influenza viruses that were reassortants between North European swine isolates and human viruses were isolated from sick children in the Netherlands [1]. Fortunately, these viruses failed to further spread in the human population, perhaps because they lacked a mutator mutation. Because pigs seem to play such an important role in the generation of human pandemic strains, future surveillance programs should address both swine viruses and the genetic stability of any avian-human reassortants. Only those reassortants that are genetically unstable warrant considerable concern.

Acknowledgements

This work was supported by the Deutsche Forschungsgemeinschaft, SFB 272, and by the Fonds der Chemischen Industrie.

References

1. Claas ECJ, Kawaoka Y, De Jong JC, Masurel N, Webster RG (1994) Infection of children with avian-human reassortant influenza virus from pigs in Europe. Virology 204: 433–457

2. Hinshaw VS, Webster RG (1982) The natural history of influenza A viruses. In: Bear AS (ed) Basic and applied influenza research. CRC Press, Boca Raton, pp 79–104
3. Kawaoka Y, Krauss S, Webster RG (1989) Avian-to-human transmission of the PB1 gene of influenza A virus in the 1957 and 1968 pandemics. J Virol 63: 4603–4608
4. Ludwig S, Stitz L, Planz O, Van H, Fitch WM, Scholtissek C (1995) European swine virus as a possible source for the next influenza pandemic? Virology 212: 555–561
5. Rott R, Orlich M, Scholtissek C (1979) Correlation of pathogenicity and gene constellation of influenza A viruses. III Non-pathogenic recombinants derived from highly pathogenic parent strains. J Gen Virol 44: 471–477
6. Scholtissek C, Rott R, Orlich M, Harms E, Rohde W (1977) Correlation of pathogenicity and gene constellation of an influenza A virus (fowl plague). I: Exchange of a single gene. Virology 81: 74–80
7. Scholtissek C, Rohde W, von Hoyningen V, Rott R (1978) On the origin of the human influenza virus subtypes H2N2 and H3N2. Virology 87: 13–20
8. Scholtissek C, Naylor E (1988) Fish farming and influenza pandemics. Nature 331: 215
9. Scholtissek C, Ludwig S, Fitch WM (1993) Analysis of influenza A virus nucleoproteins for the assessment of molecular genetic mechanisms leading to new phylogenetic virus lineages. Arch Virol 131: 237–250
10. Schultz U, Fitch WM, Ludwig S, Mandler J, Scholtissek C (1991) Evolution of pig influenza viruses. Virology 183: 61–73
11. Suárez P, Valcácel J, Ortín J (1992) Heterogeneity of the mutation rates of influenza A viruses. Isolation of mutator mutants. J Virol 66: 2491–2494
12. Webster RG, Bean WJ, Gorman TO, Chambers TM, Kawaoka Y (1992) Evolution and ecology of influenza A viruses. Microbiol Rev 56: 152–179

Authors' address: Dr. C. Scholtissek, Institute of Virology, Justus-Liebig-Universitat Giessen, Frankfurter Str. 107, D-35392 Giessen, Federal Republic of Germany.

Influenza virus: transmission between species and relevance to emergence of the next human pandemic*

R. G. Webster

Department of Virology and Molecular Biology, St. Jude Children's Research Hospital, Memphis, Tennessee, U.S.A.

Summary. Although influenza viruses are not spread from human to human through the conventional food chain, this is not necessarily the case for the transmission of the precursors of the human pandemic influenza viruses. Aquatic birds of the world are the reservoirs for all influenza A viruses; the virus is spread by fecal-oral transmission in untreated water. Influenza A viruses are frequently transmitted to domestic poultry and two of the 15 subtypes H5 and H7 can become highly pathogenic and have the capacity to decimate commercial poultry flocks. Less frequently, avian influenza viruses are transmitted between species-to pigs, horses and sea mammals. This transmission involves mutational, reassortant or recombinational events and can occur through fecal contamination of unprocessed avian protein or through the water. The transmission of avian influenza viruses or virus genes to humans is postulated to occur through pigs that act as the intermediate host. This involves either multiple mutational or reassortant events and is believed to occur by airborne transmission. Once avian influenza viruses are established in mammals, they are transmitted from animal to animal by the respiratory airborne route. The transmission of avian influenza virus from their reservoir in wild aquatic birds to domestic poultry and to mammalian species including humans can be prevented by treatment of the water supply and of avian protein sources with disinfectants or by heating. Agricultural authorities have recommended the separation of wild aquatic and domestic poultry and of pig and poultry farming. It is theoretically possible to reduce the possibility of the next pandemic of influenza in humans by changes in agricultural practices so that ducks are separated from pigs and people.

*This paper is dedicated to the late Dr. Peter A. Bachmann who isolated and characterized the first avian H1N1 influenza virus from ducks in the 'Englischer Garten' Munich. This virus is a reference strain and possible ancestor of the influenza virus that transmitted to pigs in Europe in the late 1970s. His contributions live on.

Introduction

Since the central theme of this book is the transmission of disease agents through the food chain, we will examine influenza viruses from this perspective. Influenza A viruses are found in humans, pigs, horses, sea mammals, and also in wild aquatic and domestic birds [23]. How are they spread? Between mammals, influenza is an airborne infection, but between birds, influenza can be either an airborne or waterborne infection. Influenza viruses of aquatic birds periodically transmit to domestic birds sometimes with catastrophic effects; this transmission can be either airborne or waterborne. Less frequently, avian influenza viruses transmit to mammals and three to four times in the past century this transmission has initiated a pandemic of influenza in humans. The method of spread of avian influenza viruses to mammals remains unresolved, but could be either airborne or waterborne. In this report we will consider recent examples of interspecies transmission of influenza A viruses and the possible prevention of emergence of the next human pandemic which is considered imminent.

The reservoirs of influenza A viruses

The available evidence indicates that aquatic birds are the reservoirs of all 15 subtypes of influenza A viruses. We will first consider the replication of influenza A in aquatic birds. In wild ducks, influenza viruses replicate preferentially in the cells lining the intestinal tract, cause no disease signs, and are excreted in high concentrations in the feces (up to $10^{8.7}$ 50% egg infectious doses per g) [24]. Avian influenza viruses have been isolated from freshly deposited fecal material and from unconcentrated lake water. This information indicates that waterfowl have a very efficient way to transmit viruses; i.e., via fecal material in the water supply. If one considers that a large number of susceptible young ducks are hatched each year throughout the world, it is understandable that many birds are infected by virus shed into water. This would explain the high incidence of virus infection in Canadian ducks, particularly juveniles when up to 30% can be shedding virus prior to fall migration. Transmission by feces also provides a way for wild ducks, as they migrate through an area, to spread their viruses to other domestic and feral birds [10].

The avirulent nature of avian influenza infection in ducks may be the result of virus adaptation to this host over many centuries, creating a reservoir that ensures perpetuation of the virus. This speculation strongly suggests that ducks occupy a unique and very important position in the natural history of influenza viruses. Influenza viruses of avian origin have been implicated in outbreaks of influenza in mammals, such as seals [6], whales [11] and pigs [21], as well as in domestic poultry [13].

Wild avian to domestic avian transmission of influenza viruses

A recent example of the emergence of an influenza virus from the aquatic bird reservoir was the appearance of a highly pathogenic H5N2 influenza virus in domestic chickens in Mexico [13]. In October of 1993, there was decreased egg production and increased mortality among Mexican chickens, in association with serologic evidence of an H5N2 influenza virus. First isolated from chickens in May of 1994, after spreading widely in the country, the virus caused only a mild respiratory syndrome in specific-pathogen-free chickens. Because eradication of the virus by destruction of infected birds posed major obstacles to the poultry industry in Mexico, we were able to conduct a "field experiment" to determine the fate of an avirulent virus after repeated cycles of replication in millions of chickens. By the end of 1994, the virus had mutated to contain a highly cleavable hemagglutinin (HA), but remained only mildly pathogenic in chickens. Within months, however, it had become lethal in poultry. Nucleotide sequence analysis of the HA cleavage site of the original avirulent strain revealed R-E-T-R, typical of avirulent viruses and unlike the K-K-K-R sequence characterizing viruses responsible for the 1983 outbreak in U.S. poultry. Both mildly and highly pathogenic isolates contained insertions and a substitution of basic residues in the HA connecting peptide, R-K-R-K-T-R, which made the HA highly cleavable in trypsin-free chicken embryo fibroblasts. Phylogenetic analysis of the HA of H5 avian influenza viruses, including the Mexican isolates, indicated that the epidemic virus had originated from the introduction of a single virus of the North American lineage into Mexican chickens (Fig. 1). This H5N2 influenza virus was initially spread between chickens by the airborne route, but as the virus became more

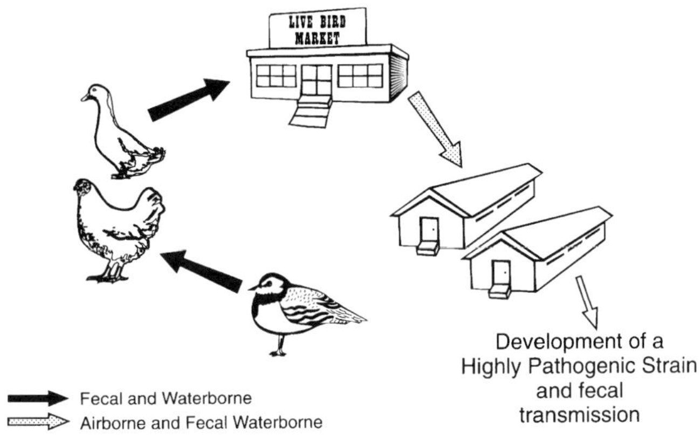

Fig. 1. Emergence and spread of an H5N2 influenza virus in domestic poultry. Diagram illustrating the spread of H5N2 influenza viruses from wild aquatic migrating birds (Ruddy turnstones) to free-ranging "backyard" poultry (ducks, chickens). These birds may be marketed through live-bird markets where commercial birds are also marketed. Non-sanitized crates can carry avian influenza viruses from live-bird markets to commercial chicken farms. In Mexico, the initial spread of virus on commercial farms was mainly airborne and the practice of selling live birds at open markets and returning unsold live birds to commercial farms may have facilitated the spread. The virus mutated during multiple cycles of replication and became highly pathogenic; these variants were mainly transmitted through fecal contamination

pathogenic it spread predominantly through the feces. Thus, waterborne transmission supplemented airborne transmission offering the possibility of greater spread through water and fecal contamination. It must be kept in mind that domestic poultry has become a major protein source for humans in many countries so influenza viruses do have the capacity to disrupt a major food source. Influenza A viruses transmit readily from wild avian species to domestic species because there is no host barrier. Thus, influenza viruses of all subtypes transmit to turkeys frequently, especially to range reared birds [10].

It became apparent during the Chicken/Pennsylvania (H5N2) avian influenza outbreak that live-bird markets in the major cities of the USA are an important "mixing pot" between between "backyard" raised domestic poultry and the heavily populated commercial poultry industry. These markets are now being monitored on a regular basis. It was at live-bird markets in New York City that the precursor H5N2 virus of the current Mexico virus was first detected in the chicken population [17]. This virus did spread to commercial farms in the USA, but was eradicated before it became established.

The human pandemics of the 20th century

The available evidence indicates that the pandemics of human influenza of this century originated from the Eurasian avian lineage. Epidemiological evidence supports the proposition that the Asian/57 (H2N2), Hong Kong/68 (H3N2) and the Russian/77 (H1N1) viruses all originated from China. Phylogenetic studies suggest that the catastrophic "Spanish" influenza of 1918 may also be of Eurasian origin [7] and raises the question of whether the precursors of the disastrous "Spanish influenza" outbreak of 1918 were of American or of Asian origin. Since the high mortality associated with the catastrophic 1918 epidemic occurred in the second year of spread of this virus [4], it is possible that this virus originated in Asia and spread to North America before it acquired the property of high lethality in humans.

Both the Asian/57 (H2N2) and Hong Kong/68 (H3N2) pandemics originated by reassortment. In 1957 the Asian pandemic virus acquired three genes (PB1, HA and NA) from the avian influenza gene pool in wild ducks by genetic reassortment and kept five other genes from a circulating human strain [14]. After the Asian strain appeared, the H1N1 strains disappeared from humans. In 1968 the Hong Kong pandemic virus acquired two genes (PB1 and HA) from the duck reservoir by reassortment and kept six genes from the virus circulating in humans.

A surprising discovery from phylogenetic analyses of amino acid changes was that avian influenza viruses, unlike mammalian strains, show low evolutionary rates [8]. In fact, influenza viruses in aquatic birds appear to be in evolutionary stasis, with no evidence of net evolution over the past 60 years. Nucleotide changes have continued to occur at a similar rate in avian and mammalian influenza viruses, but these changes do not result in amino acid changes in the avian viruses, while all eight mammalian influenza gene segments continue to accumulate changes in amino acids. The high level of genetic conservation suggests that avian viruses are approaching or have reached an adaptive optimum, wherein nucleotide changes provide no selective advantage. It also means that the source of genes for pandemic influenza viruses exists phenotypically unchanged in the aquatic bird reservoir.

Overall, the most important implication of phylogenetic studies is that the ancestral viruses which caused the "Spanish" influenza in 1918, as well as the viruses that provided gene segments for the Asian/1957 and Hong Kong/1968 pandemics, are still circulating in wild birds, with few or no mutational changes.

Avian to mammalian transmission of influenza A viruses

The primordial source of all influenza A viruses in mammals is from the aquatic bird reservoirs in the world where each of the known subtypes are perpetuated [23]. These avian viruses can be divided into two different populations, one in Eurasia and one in The Americas.

The transmission of avian influenza viruses to mammalian species including pigs, horses, sea mammals and humans occurs much less frequently for there are host range barriers that must be breached. These include receptor specificities to gain entry to mammalian cells and acquisition of mutations and/or recombination events that permit viruses to replicate. In the paper by Scholtissek in this book [19], he deals with the transmission of avian influenza viruses to pigs in Europe in 1979 and the establishment of the virus in the pig population of Europe. A mutator mutation is postulated to account for the high rate of genetic variation in the avian-like influenza viruses in European pigs [15]. The pig is also considered to be the "intermediate host" for the transmission of avian influenza viruses to humans [15, 22].

Humans do not possess the receptors necessary for infection with avian influenza viruses (α2–3 galactose linkage to sialic acid) while pigs have both α2–3 and α2–6 gal receptors in their respiratory tract (Y. Kawaoka, pers. comm.). It is postulated that the avian H1N1 influenza currently circulating in European pigs will be the precursor of the next human pandemic virus [15]. Avian-human reassortants have been detected in pigs in Italy [2] and counterparts of these viruses have been isolated from children with influenza in The Netherlands [3].

A second example of transmission of an avian influenza virus to mammals occurred in horses in China when an avian H3N8 influenza virus transmitted to horses and caused a serious epidemic of influenza. In 1989, a new H3N8 influenza virus appeared in horses in Northern China that derived all of its gene segments from an avian influenza virus [9]; this is the most recent example of interspecies transmission where an entire avian influenza virus transferred to mammalian species. This virus initially caused severe respiratory disease in herds with 20% mortality, but in subsequent years the virus has caused typical equine influenza, and it is not certain whether this virus has continued to circulate. It is clear, however, that influenza viruses continue to transmit between species sometimes with fatal results.

How are these viruses spread?

Avian influenza viruses in wild aquatic birds are spread by fecal-oral transmission through the water supply [15] and it is probable that initial transmission of avian influenza viruses to mammals including pigs and horses also occurs by fecal contamination of water (Fig. 2). Scholtissek has postulated that the utilization of fecal material from ducks for fish farming in Asia may contribute to transmission of avian influenza viruses to pigs [20]. Another direct method of transfer is by feeding pigs untreated garbage or the carcases of dead avian species. Raising pigs under chicken houses and feeding them dead avian carcases has been observed on rare occasions in the US; and H5N2 influenza virus was isolated from pigs living under chicken houses in Pennsylvania during the outbreak in 1982. It is not uncommon to raise both pigs and poultry on the same commercial farms and from the perspective of the control of inter-species transmission of influenza, this is undesirable for it may facilitate interspecies transmission of influenza viruses. After transmission to pigs, horses or humans, the method of spread of influenza is mainly by the respiratory route.

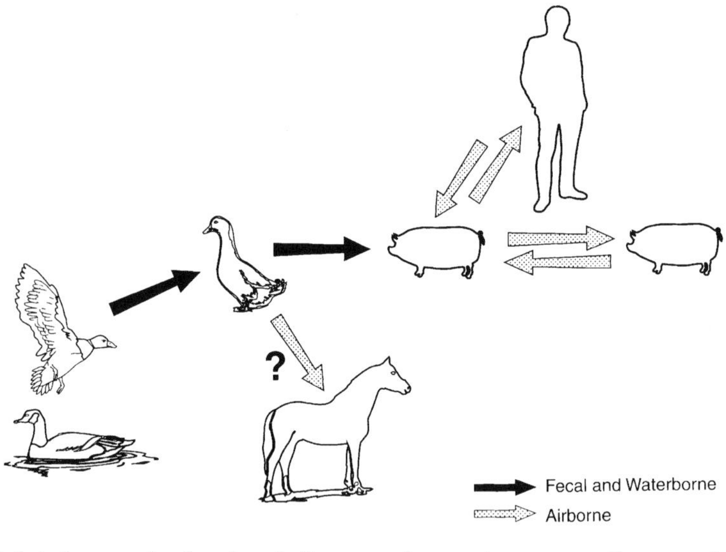

Fig. 2. Postulated spread of avian influenza viruses to mammalian species. Influenza viruses in the aquatic bird reservoirs of the world are transmitted primarily by fecal contamination of water to their offspring which during migration spread virus to range reared domestic avian species. The method of transmission to pigs, horses or sea mammals may be by fecal contamination of untreated water or avian protein fed to these animals. Although avian influenza viruses are relatively frequently transferred to pigs, they infrequently become infected and even less frequently transmit the virus from animal to animal. The spread from mammal to mammal is airborne and transmission from pigs to humans can occur relatively frequently, but the establishment of avian influenza viruses in humans occurs very rarely

Can the emergence of pandemic strains be prevented?

With the realization that there is a reservoir of all known influenza A virus subtypes in aquatic wild birds in nature, recommendations have been made by agricultural authorities to prevent direct or indirect contact between domestic poultry and wild birds. One of the classic mistakes made by chicken and turkey farmers is to raise a few domestic ducks on a pond near their poultry barns, for these birds attract wild ducks. Each of the highly pathogenic outbreaks of avian infuenza that have occurred in recent years – H5N2 in chickens and turkeys in Pennsylvania and surrounding states in 1983 to 1984 [1], H5N8 in turkeys in Ireland in 1983 [16], and H7N7 in chickens in Victoria, Australia, in 1985 [5], H5N2 in Mexico in 1993 [13], could probably have been prevented if the domestic poultry had been raised in ecologically controlled houses that maintained a high standard of security and limited access.

Turkeys raised on open range in Minnesota are frequently infected with influenza viruses. In a 2-year period, 97 flocks had virological or serological evidence of influenza virus infection [10]. Eight different serotypes were isolated from the turkeys, and antigenically similar viruses were isolated from sentinel ducks. Straightforward preventative measures have been recommended to minimize contact between the reservoir of influenza viruses in nature and domestic poultry. These are all aimed at minimizing contact between domestic poultry and wild birds, feces, and contaminated water and include the following: (i) do not walk directly from outside environments into poultry houses without washing boots and (ii) do not use untreated pond water for watering poultry. The frequency of influenza virus infection in turkeys raised in "closed" houses is much lower than among turkeys raised on the range.

If we assume that people, pigs, and aquatic birds are the principal variables associated with the emergence of new human pandemic strains, it may be possible to influence the occurrence of human pandemics of influenza. The principles applied to preventing outbreaks of influenza in domestic animals should be equally applicable here. We know that pandemic strains of human influenza emerge only rarely; however, the available information indicates that interspecies transmission of influenza viruses may not be so rare, for up to 10% of persons with occupational exposure to pigs develop antibodies to swine influenza virus [18]. We know that the majority of transfers of influenza viruses from pigs to humans are dead-end transfers in that they do not spread efficiently from human to human. As indicated above we do not know the frequency of virus transfer between the suspect species in Southern China. If there is an epicenter for pandemic influenza, and if there is a detectable frequency of transfer between people, pigs, and ducks, and if we understand the ecological and agricultural features involved in the transfer, pandemics may be preventable. If pigs are the major mixing vessel for influenza viruses, changes in the agricultural practices that separate pigs from people and ducks could conceivably prevent future pandemics.

Acknowledgements

This work was supported by Public Health Service grants AI-29680 and AI-08831 from the National Institute of Allergy and Infectious Diseases, by Cancer Center Support (CORE) grant CA-21765, and by the American Lebanese Syrian Associated Charities. We thank Dayna Baker for typing the manuscript.

References

1. Bean WJ, Kawaoka Y, Wood JM, Pearson JE, Webster RG (1985) Characterization of virulent and avirulent A/Chicken/Pennsylvania/83 influenza A viruses: potential role of defective interfering RNAs in nature. J Virol 54: 151–160
2. Castrucci MR, Donatelli I, Sidoli L, Barigazzi G, Kawaoka Y, Webster RG (1993) Genetic Reassortment between avian and human influenza A viruses in Italian pigs. Virology 193: 503–506
3. Claas ECJ, Kawaoka Y, De Jong JC, Masurel N, Webster RG (1994) Infection of children with avian-human reassortant influenza virus from pigs in Europe. Virology 204: 453–457
4. Crosby AW (1989) America's forgotten pandemic: The Influenza of 1918. Cambridge University Press, Cambridge
5. Cross GM (1986) The status of avian influenza in Australia. In: Proceedings of the 2nd International Symposium on Avian Influenza. U. S. Animal Health Association, Richmond, VA, pp 96–103
6. Geraci JR, St. Aubin DJ, Barker IK, Webster RG, Hinshaw VS, Bean WJ, Ruhnke HL, Prescott JH, Early G, Baker AS, Madoff S, Schooley RT (1982) Mass mortality of harbor seals: pneumonia associated with influenza A virus. Science 215: 1129–1131
7. Gorman OT, Bean WJ, Kawaoka Y, Donatelli I, Guo Y, Webster RG (1991) Evolution of influenza A virus nucleoprotein genes: implications for the origin of H1N1 human and classical swine viruses. J Virol 65: 3704–3714
8. Gorman OT, Bean WJ, Kawaoka Y, Webster RG (1990) Evolution of the nucleoprotein gene of influenza A virus. J Virol 64: 1487–1497
9. Guo Y, Wang M, Kawaoka Y, Gorman O, Ito T, Saito T, Webster RG (1992) Characterization of a new avian-like influenza A virus from horses in China. Virology 188: 245–255
10. Halvorson D, Karunakaran D, Senne D, Kelleher C, Bailey C, Abraham A, Hinshaw VS, Newman J (1983) Epizootiology of avian influenza-simultaneous monitoring of sentinel ducks and turkeys in Minnesota. Avian Dis 27: 77–85
11. Hinshaw VS, Bean WJ, Geraci JR, Fiorelli P, Early G, Webster RG (1986) Characterization of two influenza A viruses from a pilot whale. J Virol 58: 655–656
12. Hinshaw VS, Webster RG (1982) The natural history of influenza A viruses. In: Beare AS (ed) Basic and applied influenza research. CRC Press, Boca Raton, pp 79–104
13. Horimoto T, Rivera E, Pearson J, Senne D, Krauss S, Kawaoka Y, Webster RG (1995) Origin and molecular changes associated with emergence of a highly pathogenic H5N2 influenza virus in Mexico. Virology 213: 223–230
14. Kawaoka Y, Krauss S, Webster RG (1989) Avian-to-human transmission of the PBI gene of influenza A virus in the 1957 and 1968 pandemics. J Virol 63: 4603–4608
15. Ludwig S, Stitz L, Planz O, Van H, Fitch WM, Scholtissek C (1995) European swine virus as a possible source for the next influenza pandemic? Virology 212: 555–561
16. Murphy TM (1986) The control and epidemiology of an influenza A outbreak in Ireland, p 23. In: Recent advances in acute virus disease of poultry. Commission of the European Communities, Martinus Nijhoff, Dordrecht, The Netherlands

17. Saito T, Horimoto T, Kawaoka Y, Senne DA, Webster RG (1994) Emergence of a potentially pathogenic H5N2 influenza virus in chickens. Virology 201: 277–284
18. Schnurrenberger PR, Woods GT, Martin RJ (1970) Serologic evidence of human infection with swine influenza virus. Am Rev Respir Dis 102: 356–361
19. Scholtissek C (1997) Molecular epidemiology of influenza. In: Kaaden OR, Czerny CP, Eichhorn W (eds) Viral zoonoses and food of animal origin. A re-evaluation of possible hazards for human health. Springer, Wien New York, pp 99–103 (Arch Virol [Suppl] 13)
20. Scholtissek C, Naylor E (1988) Fish farming and influenza pandemics. Nature 331: 215
21. Scholtissek C, Burger H, Bachmann PA, Hannoun C (1983) Genetic relatedness of hemagglutinins of the H1 subtype of influenza A viruses isolated from swine and birds. Virology 129: 521–523
22. Schultz U, Fitch WM, Ludwig S, Mandler J, Scholtissek C (1991) Evolution of pig influenza viruses. Virology 183: 61–73
23. Webster RG, Bean WJ, Gorman OT, Chambers TM, Kawaoka Y (1992) Evolution and ecology of influenza A viruses. Microbiol Rev 56: 152–179
24. Webster RG, Yakhno MA, Hinshaw VS, Bean WJ, Murti KG (1978) Intestinal influenza: replication and characterization of influenza viruses in ducks. Virology 84: 268–278

Authors' address: Dr. R. G. Webster, Department of Virology and Molecular Biology, St. Jude Children's Research Hospital, 332 N Lauderdale, P. O. Box 318, Memphis, TN 38101, U.S.A.

Functional chimeric HN glycoproteins derived from Newcastle disease virus and human parainfluenza virus-3

R. Deng[1], **A. M. Mirza**[1], **P. J. Mahon**[2], and **R. M. Iorio**[1]

[1] Department of Molecular Genetics and Microbiology, University of Massachusetts Medical School, Worcester, Massachusetts, U.S.A.
[2] Department of Biology, Assumption College, Worcester, Massachusetts, U.S.A.

Summary. Newcastle disease virus (NDV) is primarily a respiratory tract pathogen of birds, particularly chickens, but it occasionally produces infection in man. Human parainfluenza virus type 3 (hPIV3) is a common respiratory pathogen, particularly in young children. These two viruses gain entry to host cells via direct fusion between the viral envelope and the cell membrane, mediated by the two surface glycoproteins: the hemagglutinin-neuraminidase (HN) and fusion (F) proteins. Promotion of fusion by HN and F requires that they are derived from homologous viruses. We have constructed chimeric proteins composed of domains from heterologous HN proteins. Their ability to bind cellular receptors and to complement the F protein of each virus in the promotion of fusion were evaluated in a transient expression system. The fusion specificity was found to segregate with a segment extending from the middle of the transmembrane anchor to the top of the putative stalk region of the ectodomain. All of the chimeras, in which the globular domain is derived from the NDV HN and various lengths of the stalk region are derived from the hPIV3 HN maintain receptor binding activity, but some have markedly reduced neuraminidase (NA) activity. Decrease in the NA activity of the chimeras correlates with alteration in the antigenic structure of the globular domain. This suggests that the stalk region of the HN spike is important for maintenance of the structure and function of the globular domain of the HN protein spike.

Introduction

Paramyxoviruses are a group of enveloped negative-stranded RNA viruses which includes Newcastle disease virus (NDV) and human parainfluenza virus 3 (hPIV3). NDV infects poultry and a variety of other avian species causing fowl pest, a disease of considerable agricultural and economic significance. Occasionally, it produces infection in humans, especially in poultry workers and laboratory personnel, causing conjunctivitis without corneal involvement

[1, 12]. HPIV3 is an important etiologic agent of respiratory tract infection in infants. Both NDV and hPIV3 are thought to bind to sialic acid-containing cellular receptors and gain entry into the cell by direct fusion between the viral envelope and the cellular membrane, mediated by the two viral surface glycoproteins: the hemagglutinin-neuraminidase (HN) and fusion (F) protein [4].

HN is a homotetrameric, type II integral membrane glycoprotein, existing as a spike consisting of a short hydrophilic cytoplasmic tail, a hydrophobic membrane span region and a large ectodomain [36, 41]. Although the structure of the ectodomain of the HN spike has not been solved, electron micrographs suggest that it consists of a large terminal globular domain, or "head", supported by a long stalk extending out from the membrane [41]. Based on the conservation of neuraminidase (NA) active site residues between the HN of paramyxoviruses and NA protein of influenza virus, it has been predicted that the globular head of the HN spike has a similar folding pattern to that of the NA protein: six beta-sheets, each comprised of four anti-parallel strands [7]. Little has been known about the structure of the stalk of the HN spike, except that a cysteine residue present in some strains of NDV is involved in inter-molecule disulfide-linkage to stabilize dimerization of the protein [25].

At least three functions have been found associated with HN. First, it mediates the attachment of the virion to sialic acid-containing receptors on the target cell surface to initiate infection [4]. Most monoclonal antibodies to NDV's HN have been shown to neutralize virus infection by blocking its attachment function [19]. The second is its neuraminidase activity which is thought to be required for the virus life cycle by enhancing virion mobility and releasing budding virions from the infected cells [8, 16, 21, 22, 29]. Both attachment function and NA activity of the HN protein have been localized to the globular domain of the protein [25, 41].

The third function of the HN protein has been referred to as its fusion specificity [2, 3, 10, 13–15, 42]. There is a great deal of evidence to suggest that HN provides more for the fusion process than attachment function [20, 28]. It has been shown that promotion of fusion requires that HN and F proteins are derived from the homotypic virus [2, 10, 13, 15, 42]. For NDV and hPIV3, fusion occurs only when a homologous pair of HN and F are co-expressed. HN protein from one virus can not complement the heterologous F protein in the promotion of fusion [10, 13]. This suggests that homologous HN and F proteins may specifically communicate with each other during the promotion of fusion. We have previously reported that a segment consisting of the transmembrane anchor and most of the putative stalk region of the HN protein is responsible for the fusion specificity of the HN proteins from hPIV3 and NDV [10]. Similar fusion specific domains on HN proteins of hPIV2 and Sendai virus have recently been reported [39, 42].

Extending our previous study [10], we have constructed a number of additional HN chimeras consisting of sequences from hPIV3 and NDV. By evaluating the structure and function of the chimeras, first, we have mapped a minimal domain that determines the fusion specificity of the hPIV3 HN to a region extending from the middle of the transmembrane anchor to the top of the putative stalk of the protein, and second, we demonstrate that the structure of the stalk of the HN protein influences the NA activity and, to a lesser extent, the antigenic structure in the globular domain of the HN spike.

Materials and methods

Recombinant plasmid vectors

The insertion of the HN and F genes of the Australia-Victoria isolate of NDV into pBluescript (SK+) (pBSK) (Stratagene Cloning Systems, La Jolla, CA) to generate pBSK-NDV-HN and pBSK-NDV-F plasmids has been described previously [26]. The hPIV3 HN and F genes were inserted into the same plasmid vector to generate pBSK-hPIV3-HN and pBSK-hPIV3-F [10].

Construction of chimeric HN proteins

Construction of chimeras CH1, CH1(−13), CH1(−10), CH1(−8), CH1(−6), CH4, CH5, CH6 and CH7 has been described previously [10]. CH5 was used as template to construct CH5(−7), using a chimeric oligonucleotide primer, NDV (40–78)-hPIV3 (94–117). The latter was used to generate CH5(−14) by using a primer of NDV (62–99)-hPIV3 (115–130). CH5(−17), CH5(−19) and CH5(−20) were constructed by using CH5(−14) as template and chimeric oligonucleotide primers as follows, respectively, NDV (80–118)-hPIV3 (133–154), NDV (81–108)-hPIV3 (125–182) and NDV (95–114)-hPIV3 (130–142).

Chimera CH1(+15) was constructed by using CH1 as template and chimeric primer hPIV3 (359–384)-NDV (331–393). Then CH1 (+15) was used as template to construct CH1(+30), using chimeric primer hPIV3 (286–309)-NDV (286–352). The latter was used as template to generate CH1(+46), using primer hPIV3 (241–261)-NDV (248–307).

Transient expression system

The chimeric HN, wt HN and F proteins were expressed in BHK-21 cells using the T7 RNA polymerase expression system [11]. The protocols for maintenance of the cells, infection with recombinant vaccinia virus, and transfection have been described previously [26].

FACS analysis

The transfection efficiency and cell surface expression were quantitated and the antigenic structure of the HN protein was evaluated by fluorescence-activated cell sorter (FACS) analysis, as described previously [9, 26]. A mixture of monoclonal antibodies (MAb) specific to seven overlapping antigenic sites (named 1, 2, 3, 4, 12, 14 and 23) previously identified on the globular head of NDV HN [18, 19] was used to quantitate the cell surface expression level of the chimeras. To probe the structural alteration of the globular domain of the chimeras, a MAb specific to a highly conformation-dependent epitope (site 23) and a MAb that recognizes a conformation-independent linear epitope (site 14) were used in the FACS analysis.

Hemadsorption and fusion assays

The receptor binding activity of the chimeric HN proteins expressed on the cell surface was determined by their ability to adsorb guinea pig erythrocytes [9]. The ability of the HN chimeras to complement the F protein from NDV or hPIV3 in the promotion of fusion was investigated by co-expression of the chimeric HNs with the appropriate F protein in BHK cells. The protocol for determining the fusion activity has been described previously [10].

Neuraminidase activity assay

The neuraminidase activity of the chimeras expressed at the cell surface was determined as described previously [25]. The assay was performed at pH 6 using neuraminlactose as substrate. After subtraction of the background absorbance obtained with vector-transfected cells, the data were corrected for differences in transfection efficiency and cell surface expression level of each chimera and expressed as the percentage of that obtained with wt NDV HN.

Results

A chimera with 25% of the N-terminal sequence from hPIV3 HN promotes fusion with hPIV3 but not NDV F

The first chimeric HN protein (CH1) was constructed by replacing the N-terminal 125 amino acid residues of NDV HN, encompassing its entire cytoplasmic tail, its transmembrane and most of its putative stalk region, with the corresponding N-terminal 141 residues from hPIV3 HN. This chimera is efficiently transported to the cell surface and retains its receptor binding activity as evidenced by its hemadsorption activity (Fig. 1A). It complements the F protein from hPIV3 quite efficiently in the promotion of fusion, having nearly two-thirds of the fusogenic activity of wt hPIV3 HN (Fig. 1A). However, CH1 fails to promote fusion when co-expressed with NDV F protein. These results suggest that the globular domain of the HN protein spike is not involved in the fusion specificity and the domain determining the fusion specificity can be localized to the N-terminal 25% of the HN molecule, consisting of the cytoplasmic tail, transmembrane anchor and most of the stalk region.

Mapping of the C-terminus of the fusion specific domain

As a first step in mapping the C-terminus of the fusion specific domain in hPIV3 HN, CH1 was used as parent template to construct chimera CH1 (−13), in which the 13 hPIV3-specific residues (from amino acid residue 129 to 141) at the membrane-distal end of the stalk of CH1 were replaced with the corresponding residues (from aa 111 to 125) in NDV HN. Although cell surface expression and receptor binding activity remain unaffected, the chimera completely loses its ability to complement hPIV3 F protein in the promotion of fusion (Fig. 1A and 1B). This result indicates that the replaced 13 amino acid residues are absolutely required for the fusion specificity and the C-terminal

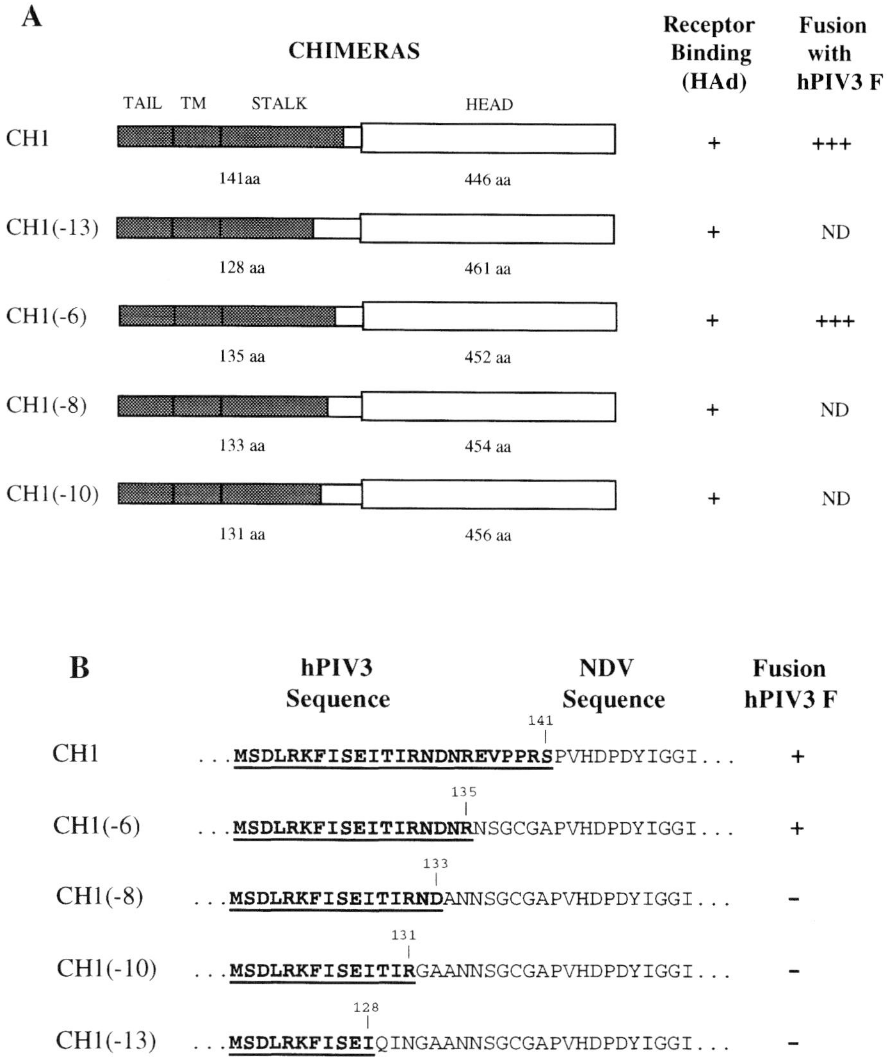

Fig. 1. Mapping the C-terminus of the fusion specific domain on hPIV3 HN. **A** In the chimeras, hPIV3 HN-derived sequences are the stippled areas, while the NDV HN-derived sequences are clear. The amino acid residues derived from both viruses are numbered under each chimera. CH1 was constructed by introduction of a *Bst EII* restriction site into each parent HN gene, followed by a ligation of the two complementary segments from the two HN gene. Using CH1 as template, a series of secondary chimeras were constructed in which the length of the hPIV3 HN-derived sequences were decreased as indicated by (−). The receptor binding activity was verified by hemadsorption of guinea pig erythrocytes. Fusion activity with hPIV3 F was assayed in cell monolayers co-expressing chimeric HNs with wt hPIV3 and expressed as + + ++ (>80% of wt), + + + (60–79% of wt), ++ (40–59% of wt), + (15–39% of wt), and ND-none detectable (< 15% of wt). **B** The amino acid sequences at the junctions between hPIV3 and NDV HN-derived parts in the chimeras are given. The hPIV3 HN-derived sequences are bolded and underlined. The position of the last hPIV3 HN-derived amino acid residue in each chimera is indicated above the residue

end of the fusion specific domain of hPIV3 HN is located within the sequence from residue 129 to 141.

To more finely map the C-terminal end of the domain, a series of chimeras was constructed. In chimeras CH1(−6), CH1(−8) and CH1 (−10), the 6, 8 and 10 hPIV-specific amino acid residues at the top of the stalk in CH1, respectively, were converted to the corresponding NDV-specific sequence (Fig. 1A). Co-expression of CH1(−6) with hPIV3 F protein results in extensive syncytium formation in the cell monolayers (Fig. 1A). However, neither CH1 (−8) nor CH1 (−10) can promote fusion with hPIV3 F protein, despite the fact that both chimeras are expressed at the cell surface in amounts comparable to both wt hPIV3 HN and CH1 (−6) and retain receptor activity (Fig. 1A). The only amino acid differences between the fusogenic CH1 (−6) and the nonfusogenic CH1 (−8) are at position 134 and 135 (Fig. 1B). The N 134 and R135 in CH1 (−6) have been changed to alanine and asparagine, respectively, in CH1 (−8) (Fig. 1B). Thus the C-terminal end of the fusion specific domain of hPIV3 HN is mapped to either residue 134 or 135. They are the 81st and 82nd amino acid residues, respectively, from the membrane and close to the top of the putative stalk region of the HN protein spike.

The transmembrane region but not the cytoplasmic tail is associated with the fusion specificity of hPIV3 HN

To determine whether the cytoplasmic tail and the transmembrane region are required for fusion specificity, chimeras CH5 and CH7 were constructed. CH5 has the cytoplasmic tail derived from NDV HN and the rest from hPIV3 HN (amino acid residue 25 to 572). In CH7, the cytoplasmic tail and the transmembrane region are derived from the NDV HN and the ectodomain from hPIV3 HN (amino acid residue 64 to 572). When co-expressed with hPIV3 F, CH5 promotes fusion very efficiently, 90% of wt level (Fig. 2A). However, CH7 fails to complement hPIV3 F in the promotion of fusion, despite the fact that it is expressed on the cell surface in amounts comparable to the wt and CH5 and retains its receptor binding activity (Fig. 2A). These data prove that the transmembrane region but not the cytoplasmic tail of the protein contributes to its fusion specificity.

Mapping of the N-terminus of the fusion specific domain

The results obtained with CH5 and CH7 indicate that the N-terminus of the fusion specific domain of hPIV3 HN resides within the sequence from residue 24 to 64. To further define the N-terminal end of the fusion specific domain, a series of chimeras was constructed using CH5 as the original parent template (Fig. 2A). In chimeras CH5 (−7), CH5 (−14) and CH5 (−20), 7, 14 and 20 amino acid residues, respectively, of hPIV3 HN sequence were replaced by the corresponding sequence from NDV HN. All three chimeras maintain their ability to bind cellular receptors (Fig. 2A). When co-expressed with hPIV3 F,

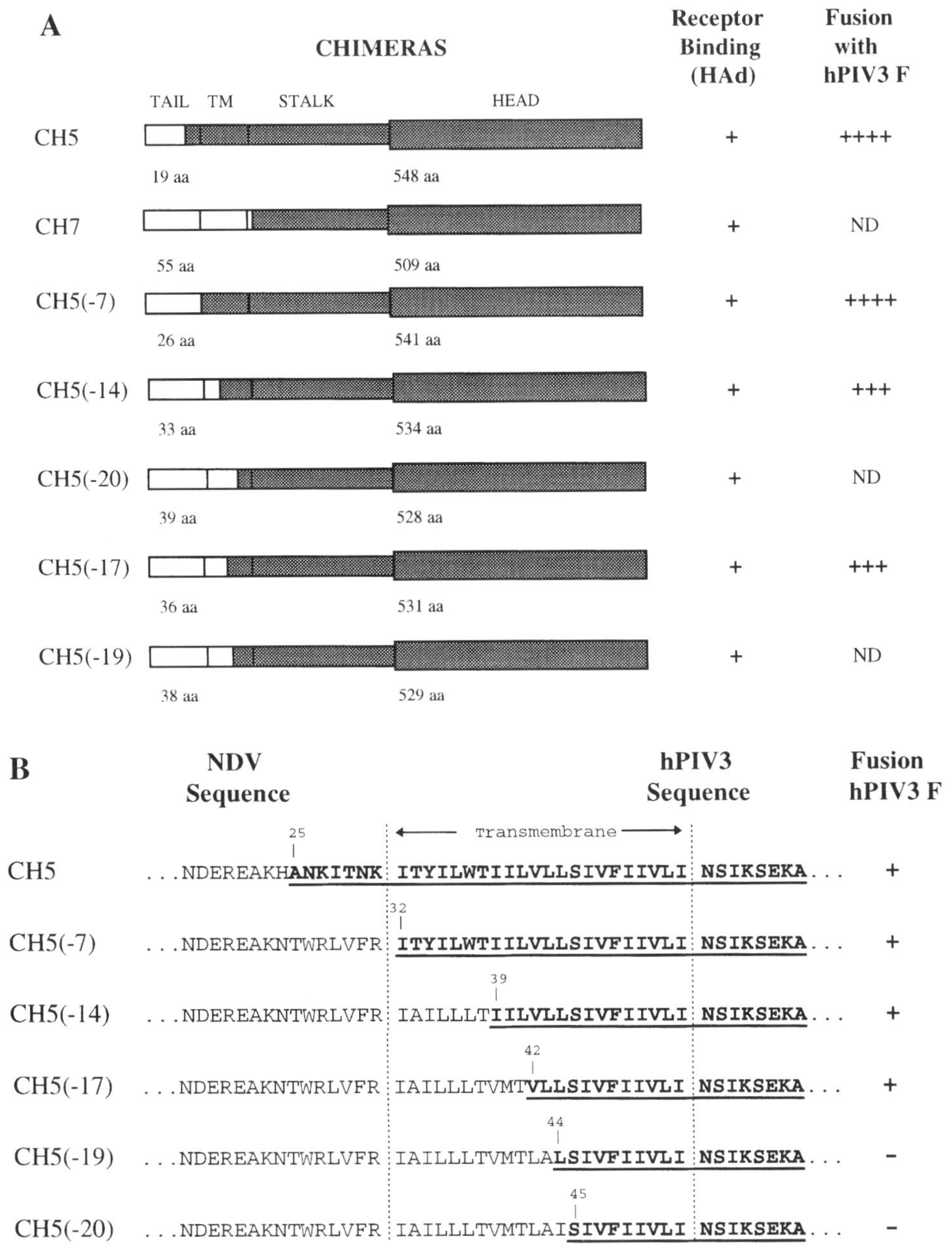

Fig. 2. Mapping the N-terminus of the fusion specific domain on hPIV3 HN. **A** In the chimeras, hPIV3 HN-derived sequences are the stippled areas, while the NDV HN-derived sequences are clear. The amino acid residues derived from both viruses were numbered under each chimera. CH5 and CH7 were constructed by introduction of a *Sph I* or *Sac I* restriction site, respectively, into each parent HN gene, followed by a ligation of the two complementary segments from the two HN genes. Using CH5 as template, a series of secondary chimeras were constructed in which the length of the hPIV3 HN-derived sequences were decreased as indicated by (−). The receptor binding activity was verified by hemadsorption of guinea pig erythrocytes. Fusion activity with hPIV3 F was assayed in cell monolayers co-expressing chimeric HNs with wt hPIV3 F and expressed as + + ++ (> 80% of wt), + + + (60–79% of wt), ++ (40–59% of wt), + (15–39% of wt), and ND-none detectable (< 15% of wt)). **B** The amino acid sequences at the junctions between hPIV3 and NDV HN-derived parts in the chimeras are given. The transmembrane anchor of the chimeras is indicated by two dotted lines. The hPIV3 HN-derived sequences are bolded and underlined. The position of the first hPIV3 HN-derived amino acid residue in each chimera is indicated above the residue

both chimeras CH5 (−7) and CH5 (−14) promote fusion efficiently in the cell monolayers. However, no cell fusion was detected in monolayers co-expressing CH5 (−20) and hPIV3 F. These results indicate that the N-terminus of the fusion specific domain of hPIV3 HN is located within the 6 amino acids from residue 39 to 45.

Next, two additional chimeras were constructed to more accurately map the N-terminal end of the domain. In CH5 (−17) and CH5 (−19), three and one amino acid residues of NDV sequence were converted back to the corresponding hPIV3 HN sequence, respectively. Despite the fact that both CH5 (−17) and CH5 (−19) are expressed at the cell surface and retain HAd activity, only CH5 (−17) maintains the ability to promote fusion with hPIV3 F (Fig. 2A). The only difference between the fusogenic CH5 (−17) and nonfusogenic CH5 (−19) is the two amino acids at position 42 and 43. The change of V42 and L43 in CH5 (−17) to leucine and alanine, respectively, in CH5 (−19) results in loss of its ability to complement hPIV3 F in the promotion of fusion (Fig. 2B). This defines the N-terminal end of the fusion specific domain of hPIV3 HN to either of these two residues. They lie in the middle of the transmembrane region of the HN protein (Fig. 2B).

The minimal domain of hPIV3 HN that determines the fusion specificity is from residue 42 to 135

Thus we have mapped both ends of the fusion specific domain of hPIV3 HN. The minimal domain of hPIV3 HN that can transfer its fusion specificity is from residue 42 to 135 (Fig. 3). It consists of 94 amino acid residues, 16.4% of the HN molecule. It extends from the middle of the transmembrane region up to the top of the putative stalk of the HN spike (Fig. 3).

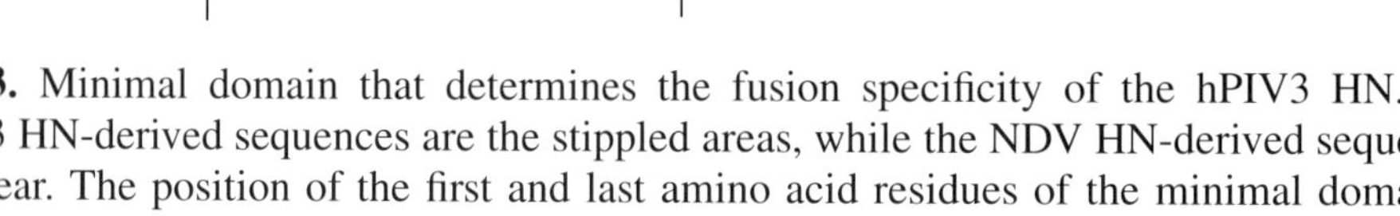

Fig. 3. Minimal domain that determines the fusion specificity of the hPIV3 HN. The hPIV3 HN-derived sequences are the stippled areas, while the NDV HN-derived sequences are clear. The position of the first and last amino acid residues of the minimal domain is indicated. The domain consists of 16.5% of the amino acid sequence of the hPIV3 HN molecule

A chimeric HN with N-terminal 25% of sequence from hPIV3 HN and globular head from NDV HN has reduced NA activity

One of the functions of the HN protein is its NA activity which cleaves sialic acid from receptor molecules. Although both NDV HN and hPIV3 HN exhibit this activity, the level of the activity is dramatically different in our expression system. Whereas NDV's HN expressed at the cell surface exhibits very high NA activity, we can not detected any NA activity in cells expressing hPIV3 HN (indicated in Fig. 4 as less than 1% of the NDV HN's level). A chimeric HN protein CH1, in which the N-terminal 25% of its sequence is derived from hPIV3 HN and the intact globular domain from NDV HN, has a markedly reduced neuraminidase activity, less than 10% of that of wt NDV HN (Fig. 4). This suggests that the NA activity defined by the globular domain can be influenced by the structure of the N-terminal region encompassing the cytoplasmic tail, transmembrane anchor and stalk of HN protein spike.

The structure of the NDV HN stalk influences its NA activity in the globular domain

The results obtained with CH1 indicate the importance of the N-terminal region for NA activity. To investigate this further, we have constructed a series of chimeric HN proteins with increasing lengths of NDV-derived sequence at the membrane-distal end of the stalk. Cell surface expression of the chimeras was determined by FACS analysis using a mixture of seven monoclonal antibodies specific to the globular head of NDV HN [18]. As shown in Table 1, all the chimeras were expressed at the transfected BHK cell surface. Chimeras CH1, CH1 (+15), CH6 and CH4 have expression levels comparable to that of wt NDV HN, ranging from 85.0% to 108.0% (Table 1). A significant decrease in the expression level was observed in CH1 (+30) and CH1 (+46), having 24.2% and 38.4% expression as wt HN, respectively (Table 1). This indicates that the

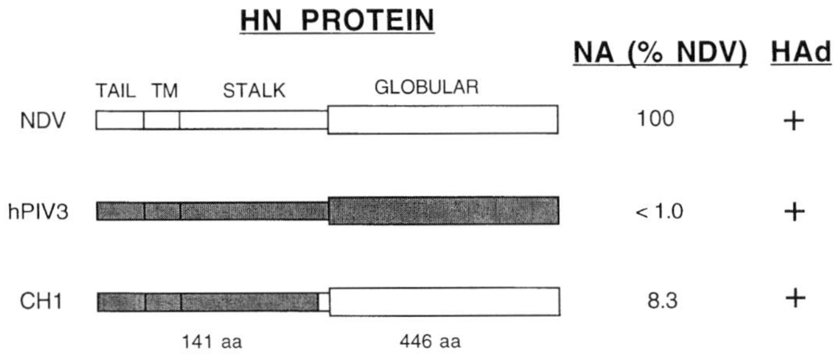

Fig. 4. Chimera CH1 in which the N-terminal 141 amino acid sequence is derived from hPIV3 HN and the globular domain from NDV HN has markedly reduced neuraminidase activity. The NA activity is measured at pH 6.0 using neuraminlactose as substrate, corrected for difference in cell surface expression, and expressed as percentage of wt NDV HN level. The receptor binding activity was verified by hemadsorption of guinea pig erythrocytes on the transfected cell monolayers

Table 1. Cell surface expression, receptor binding activity and antigenic structure of the chimeras

WT NDV HN or chimeras	NDV HN sequence	Cell surface[a] (% wt)	Attachment[b] (HAd)	Antigenic site 23[c]
WT NDV HN	1–571	100	+	100
CH1	126–571	85.0	+	55.8
CH1 (+15)	111–571	88.9	+	70.8
CH1 (+30)	96–571	24.2	+	57.4
CH1 (+46)	80–571	38.4	+	80.2
CH6	57–571	108.0	+	82.8
CH4	20–571	103.1	+	89.9

[a] Cell surface expression of the chimeras was determined by FACS analysis using a mixture of seven monoclonal antibodies (MAb) specific to the globular head of NDV HN and expressed as percentage of wt NDV HN

[b] Receptor binding activity of the chimeras was verified as hemadsorption of guinea pig erythrocytes on the transfected cell monolayers

[c] Alteration of the antigenic site 23 in the chimeras was detected by FACS analysis using a MAb to a conformational antigenic site 23 and a MAb to a linear epitope, antigenic 14 site. The data were presented as the percentage of the mean fluorescence obtained with the MAb to antigenic site 23 over that obtained with MAb to the antigenic site 14

amino acid residues from position 80 to 96 are important for normal surface expression of the HN protein. Receptor binding activity of the chimeras was assayed by hemadsorption. Similar to the parent chimera CH1, all the chimeras maintain receptor binding activity (Table 1), regardless of their difference in cell surface expression.

The neuraminidase activity of the chimeric proteins expressed on the cell surface is measured at pH 6.0 using neuraminlactose as substrate and is corrected for difference in expression level. The NDV-derived sequence in the parent CH1 begins at 78 residues from the membrane. When the NDV sequence is extended down to the 63rd residue from the membrane in CH1 (+15), NA activity is still only minimal, about 5% of wt level (Fig. 5). As the NDV-specific sequence approaches the membrane, NA activity is gradually increased, from 5.4% in CH1 (+15) to 88.6% in CH6 in which the NDV-specific sequence starts at the 9th residue from the membrane. Extension of the NDV-specific sequence to include the transmembrane anchor in chimera CH4 results in only a slight increase in the NA activity (Fig. 5). The greatest increment of NA activity results from the addition of residues 9 to 31 from the membrane, more than a 50% increase. These results suggest that the structure of the stalk region, particularly from residue 9 to 31 from the membrane, dramatically influences NA activity in the globular domain of the spike. Given the fact that all the chimeras maintain their receptor binding activity, it appears that the NA activity of NDV HN is more sensitive to the structure of the stalk than is its receptor binding activity.

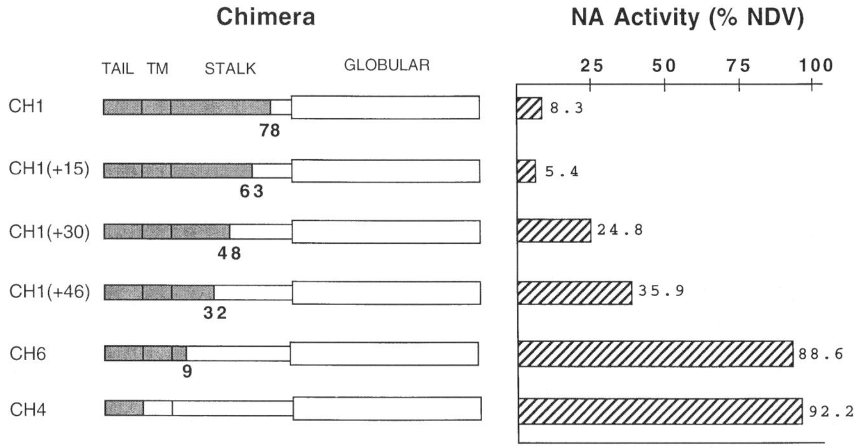

Fig. 5. Neuraminidase activity of chimeras with increasing lengths of NDV-derived sequence at the membrane-distal end of the stalk. The left panel shows the construction of the chimeras. Using CH1 as original template and chimeric oligonucleotides as primers, secondary chimeras CH1 (+15), CH1 (+30) and CH1 (+46) were constructed. CH6 and CH4 were generated by introduced of a *Sph I* or *Sac I* restriction site, respectively, into each parent HN genes, followed by a ligation of the two complementary segments from the two HN genes. The position of the first NDV-derived amino acid residue from the membrane is indicated under each construct. The bar graph on the right panel represents the NA activity of each chimera. The NA activity is measured at pH 6.0 using neuraminlactose as substrate, corrected for difference in cell surface expression, and expressed as percentage of wt NDV HN level

Decrease in NA activity of the chimeras correlates with alteration of their antigenic structure in the globular domain

To probe the effect of the stalk region on the antigenic structure of the globular domain of the HN spike, the ability of a monoclonal antibody to a highly conformation-dependent antigenic site (called site 23) on the globular head of the HN spike to recognize the chimeras was determined by FACS analysis. Meanwhile a MAb that recognizes a linear epitope (called antigenic site 14) defined by NDV HN residues 341 to 355 was used to standardize the cell surface expression of the chimeras. For each chimera, the data shown in Table 1 are the percentages of the mean fluorescence obtained with the MAb to antigenic site 23 to that obtained with MAb to the linear antigenic site 14.

Most of the chimeras exhibit a decrease, to different extents, in their ability to be recognized by the conformationally-sensitive MAb, ranging from 55.8% in CH1 to 89.9% in CH4 (Table 1). The most significant antigenic alteration was detected in chimeras CH1, CH1 (+15) and CH1 (+30) (Table 1). This suggests that the region from amino acid residue 80 to 125 affects the antigenic structure of the globular domain the most. This result correlates with the observation that the most significant decrease in NA activity is in these three chimeras (Fig. 5), all having less than 25% of the NA activity of wt NDV HN. This strongly suggests that the structure of the stalk influences the NA activity of the protein by altering the structure of the globular domain of the HN spike.

Discussion

Infection of cells by paramyxoviruses requires the binding of the virus to sialic acid containing receptors on the cell surface and the fusion of the envelope with a cellular membrane, mediated by the two viral surface glycoproteins, the HN and F proteins. The contribution HN makes to the fusion process is clearly much more complex than mere receptor binding. The second function provided by HN protein for the process is its fusion specificity. By constructing chimeric HN proteins, a minimal domain from amino acid residue 42 to 135 of hPIV3 HN that can transfer the fusion specificity for its homologous F protein has been identified. The domain extends from the middle of the transmembrane anchor to the top of the putative stalk region. A comparable domain that can transfer the fusion specificity of NDV HN was also identified [10]. Using a similar approach, the fusion specific domains on HN proteins from hPIV2 and Sendai virus have been identified [39, 42]. It appears that the involvement of the transmembrane region in fusion specificity is not exactly the same in different viruses. This may reflect another aspect of the virus-specific nature of the domain.

In a model of HN-F mediated fusion, it has been suggested that HN's binding to receptors triggers its own conformational change which in turn triggers a conformational change in F protein to release the fusion peptide [20]. Given the location on the HN spike of the domains responsible for receptor-binding and F protein specificity, one can readily envision a chain of events initiated by attachment of HN to cellular receptors mediated by the receptor binding site in its terminal globular domain. The information that receptor binding has taken place might then be transmitted down the spike structure through a conformational change to the fusion specific domain in the stalk and transmembrane region. This domain then communicates directly or indirectly with a complementary domain on the F protein spike, which in turn triggers a conformational change in F protein spike to its fusion-active form that actually mediates penetration of the target membrane.

The mechanism of fusion specificity has not been elucidated. One explanation has been proposed that the molecular architecture of the F protein requires a very precise distance to be bridged by the homotypic HN protein [14]. However, this is inconsistent with the observation that two conservative amino acid substitutions in the middle of the transmembrane region result in complete loss of the ability of the chimera to complement the hPIV3 F protein in the promotion of fusion (Fig. 2A). A direct virus-specific interaction between the homologous HN and F spikes has been proposed to explain the fusion specificity [10, 15, 34]. Using cross-linking, it has been shown that HN and F of measles virus [24] are closely associated with each other. A physical association between the NDV HN and F is supported by co-immunoprecipitation of the two proteins in the transient expression system (data not shown). This is consistent with the notion that the identified fusion specific domain on the HN spike identified interacts directly with corresponding domain(s) on the F spike, which triggers the conformational change of the F protein to release the fusion peptide. In a recent study, it was reported that hPIV3 F can down-regulate heterologous HN protein [40]. The authors propose that down-regulation of HN surface expression by heterologous F protein may be one of the determinants for the observed fusion specificity. It will be interesting to see whether the ability of chimeras to complement hPIV3 F in the promotion of fusion correlates with their cell surface expression in the presence of the hPIV3 F protein.

In contrast to influenza virus which has one glycoprotein (NA protein) with neuraminidase activity and the other (HA protein)) with receptor binding activity, paramyxoviruses have one glycoprotein (HN) with the two opposite functions. The regions of HN molecule responsible for receptor binding and NA activity have not yet been located on the amino acid sequence. There is some evidence that a single site in the molecule is responsible for both activities [33].

Consistent with this is the observation that MAbs to HN protein have been shown to inhibit both receptor binding and NA activities [5, 6, 18]. Also a single amino acid substitution in the NDV HN protein results in loss of the two functions in the mutated protein [35]. However, a great number of reports have suggested that receptor binding and NA activity are associated with independent sites [17, 18, 30–32, 37, 38]. MAbs have been shown to inhibit receptor binding activity independent of NA [18]. Escape mutants selected by these Mabs exhibit a decrease in only one of the two functions. In this study, we have demonstrated that the chimeras, with a stalk derived from a heterologous HN protein, maintain receptor binding activity but have markedly reduced NA activity. This is consistent with the idea that NA activity and receptor recognition may not be localized to a single domain in the globular head of the HN protein. Alternatively, NA activity may be more conformation-dependent than receptor recognition.

It has been thought that the stalk of the HN protein serves merely as a structural connection between the membrane and globular domain in which the multiple functions of the protein reside [27]. HN's amino-terminal 131 amino acid residues, encompassing the cytoplasmic tail, transmembrane and most of the stalk region, have been reported to be not necessary for the biological and antigenic capacities of the HN protein [41]. Consistent with this is that none of the neutralization mutants isolated so far maps within this region. However, our data demonstrate that the N-terminal sequence, including the transmembrane anchor and most of the stalk region of the HN protein, not only associates with its biological function-fusion specificity, but also influences the NA activity and antigenic structure in the globular domain. It is consistent with the observation that the membrane-anchor-less form of the NDV HN protein (CT-HN) exhibits markedly reduced NA activity, approximately 10% of the wt HN level [25] and loses its NA cooperativity which is observed in the wt NDV AV HN [23].

Acknowledgements

We thank M. Galinski for providing pTF7-hPIV3-HN and pTF7-hPIV3-F clones, R. Lamb for pSV103 NDV F clone, T. Morrison for the NDV HN clone, and B. Moss for the recombinant vaccinia virus. This work was supported by a grant awarded to R. I. from the National Institute of Health (Al-12467). R. D. was supported in part by a National Institute of Health Training Program in Viral Immunology (Al-07272).

References

1. Alexander DJ (1988) Newcastle disease: methods of spread. In: Alexander DJ (ed) Newcastle disease. Kluwer, Norwell, pp 262–272
2. Bousse T, Takimoto T, Gorman WL, Takahashi T, Portner A (1994) Regions on the hemagglutinin-neuraminidase proteins of human parainfluenza virus type-1 and Sendai virus important for membrane fusion. Virology 204: 506–514
3. Cattaneo R, Rose JK (1993) Cell fusion by the envelope glycoproteins of persistent measles viruses which caused lethal human brain disease. J Virol 67: 1493–1502
4. Choppin PW, Scheid A (1980) The role of viral glycoproteins in adsorption penetration and pathogenicity of viruses. Rev Infect Dis 2: 40–61
5. Coelingh KJ, Winter CC, Murphy BR, Rice JM, Kimball PC, Olmsted RA, Collins PL (1986) Conserved epitopes on the hemagglutinin-neuraminidase proteins of human and bovine parainfluenza type 3 viruses: nucleotide sequence analysis of variants selected with monoclonal antibodies. J Virol 60: 90–96
6. Coelingh KV, Winter CC, Jorgenson ED, Murphy BR (1987) Antigenic and structural properties of the hemagglutinin-neuraminidase glycoprotein of human parainfluenza type 3: sequence analysis of variants selected with monoclonal antibodies which inhibit infectivity, hemagglutination and neuraminidase activities. J Virol 61: 1473–1477
7. Colman PM, Hoyne PA, Lawrence MC (1993) Sequence and structure alignment of paramyxovirus hemagglutinin-neuraminidase with influenza virus neuraminidase. J Virol 67: 2972–2980
8. Compans RW, Dimmock NJ, Meier-Ewert H (1969) Effect of antibody to neuraminidase on the maturation and hemagglutinating activity of an influenza A2 virus. J Virol 4: 528–534
9. Deng R, Wang Z, Glickman RL, Iorio RM (1994) Glycosylation within an antigenic site on the HN glycoprotein of Newcastle disease virus interferes with its role in the promotion of membrane fusion. Virology 204: 17–26
10. Deng R, Wang Z, Mirza AM, Iorio RM (1995) Localization of a domain on the paramyxovirus attachment protein required for the promotion of cellular fusion by its homologous fusion protein spike. Virology 209: 457–469
11. Fuerst TR, Niles EG, Studier FW, Moss B (1986) Eucaryotic transient expression system based on recombinant vaccinia virus that synthesizes bacteriophage T7 RNA polymerase, Proc Natl Acad Sci USA 83: 8122–8126
12. Ginsberg HG (1988) Paramyxoviruses. In: Dulbecco R, Ginsberg HS (eds) Virology, 2nd ed. JB Lippincott, Philadelphia, pp 239–259
13. Heminway BH, Yu Y, Galinski MS (1994) Both surface glycoproteins are necessary for human parainfluenza type 3 mediated cell fusion. Virus Res 31: 1–16

14. Horvath CM, Paterson RG, Shaughnessy MA, Wood R, Lamb RA (1992) Biological activity of paramyxovirus fusion proteins: factors influencing formation of syncytia. J Virol 66: 4564–4569
15. Hu X, Ray R, Compans RW (1992) Functional interactions between the fusion protein and hemagglutinin-neuraminidase of human parainfluenza viruses. J Virol 66: 1528–1534
16. Huberman K, Peluso RW, Moscona A (1995) Hemagglutinin-neuraminidase of human parainfluenza 3: role of the neuraminidase in the viral life cycle. Virology 214: 294–300
17. Iorio RM, Bratt MA (1984) Monoclonal antibodies as functional probes of the HN glycoprotein of Newcastle disease virus: antigenic separation of the hemagglutinating and neuraminidase sites. J Immunol 133: 2215–2219
18. Iorio RM, Glickman RL, Riel AM, Sheehan JP, Bratt MA (1989) Functional and neutralization profile of seven overlapping antigenic sites on the HN glycoprotein of Newcastle disease virus: monoclonal antibodies to some sites prevent viral attachment. J Virus Res 13: 245–262
19. Iorio RM, Syddall RJ, Sheehan JP, Bratt MA, Glickman RL, Riel AM (1991) Neutralization map of the HN glycoprotein of Newcastle disease virus: domains recognized by monoclonal antibodies that prevent receptor recognition. J Virol 65: 4999–5006
20. Lamb RA (1993) Paramyxovirus fusion: a hypothesis for changes. Virology 197: 1–11
21. Lui C, Air GM (1993) Selection and characterization of a neuraminidase-minus mutant of influenza virus and its rescue by cloned neuraminidase gene. Virology 194: 403–407
22. Lui C, Eichelberger MC, Compans RW, Air GM (1995) Influenza type A virus neuraminidase does not play a role in viral entry, replication, assembly, or budding. J Virol 69: 1099–1106
23. Mahon PJ, Deng R, Mirza AM, Iorio RM (1995) Cooperative neuraminidase activity in a paramyxovirus. Virology 213: 241–244
24. Malvoisin E, Wild TF (1993) Measles virus glycoproteins: studies on the structure and interaction of the haemagglutinin and fusion proteins. J Gen Virol 74: 2365–2372
25. Mirza AM, Sheehan JP, Hardy LW, Glickman RL, Iorio RM (1993) Structure and function of a membrane anchor-less form of the hemagglutinin-neuraminidase glycoprotein of Newcastle disease virus. J Biol Chem 268: 21425–21431
26. Mirza AM, Deng R, Iorio RM (1994) Site-directed mutagenesis of a conserved hexapeptide in the paramyxovirus hemagglutinin-neuraminidase glycoprotein: effects on antigenic structure and function. J Virol 68: 5093–5099
27. Morrison T, Portner A (1991) Structure, function, and intracellular processing of the glycoproteins of paramyxoviridae. In: Kingsbury DW (ed) The paramyxoviruses. Plenum, New York, pp 347–382
28. Moscona A, Peluso RW (1991) Fusion properties of cells infected with human parainfluenza virus type 3: receptor requirements for viral spread and virus-mediated membrane fusion. J Virol 66: 6280–6287
29. Palese P, Compans RW (1976) Inhibition of influenza virus replication in tissue culture by 2-deoxy-2,3-dehydro-N-trifluoroacetylneuraminic acid (FANA): mechanism of action. J Gen Virol 33: 159–163
30. Portner A (1981) Evidence for two different sites in the HN glycoprotein involved in neuraminidase and hemagglutinating activities. In: Bishop DHL, Compans RW (eds) The replication of negative-strand viruses. Elsevier/North-Holland, New York, pp 465–470
31. Portner A, Scroggs RA, Metzger DW (1987) Distinct functions of antigenic sites of the HN glycoprotein of Sendai virus. Virology 158: 61–68

32. Ray R, Compans RW (1986) Monoclonal antibodies reveal extensive antigenic differences between the hemagglutinin-neuraminidase glycoproteins of human and bovine parainfluenza 3 viruses. Virology 148: 232–236
33. Scheid A, Choppin PW (1974) Identification and biological activities of paramyxovirus glycoproteins. Activation of cell fusion, hemolysis and infectivity by proteolytic cleavage of an inactive precursor protein of Sendai virus. Virology 57: 375–490
34. Sergel T, McGinnes LW, Peeples MA, Morrison TG (1993) The attachment function of the Newcastle disease virus hemagglutinin-neuraminidase protein can be separated from fusion promotion by mutation. Virology 193: 717–726
35. Sheehan JP, Iorio RM (1992) A single amino acid substitution in the hemagglutinin-neuraminidase of Newcastle disease virus results in a protein deficient in both functions. Virology 189: 778–781
36. Schuy W, Garten W, Linder D, Klenk HD (1984) The carboxyterminus of the hemagglutinin-neuraminidase of Newcastle disease virus is exposed at the surface of the viral envelope. Virus Res 1: 415–426
37. Smith GW, Hightower LE (1980) Uncoupling of the hemagglutinating and neuraminidase activities of Newcastle disease virus (NDV). In: Fields B, Jaenish R, Fox CF (eds) Animal virus genetics. Academic Press, New York, pp 623–632
38. Smith GW, Hightower LE (1982) Revertant analysis of a temperature-sensitive mutant of Newcastle disease virus with defective glycoprotein: Implications of the fusion glycoprotein in cell killing and isolation of a neuraminidase-deficient hemagglutinating virus. J Virol 42: 659–668
39. Tababayashi K, Compans RW (1996) Functional interaction of paramyxovirus glycoproteins: identification of a domain in Sendai virus HN which promotes cell fusion. J Virol 70: 6 112–6 118
40. Tanaka Y, Heminway BR, Galinski MS (1996) Down-regulation of paramyxovirus hemagglutinin-neuraminidase glycoprotein surface expression by a mutant fusion protein containing a retention signal for the endoplasmic reticulum. J Virol 71: 5 005–5 015
41. Thompson SD, Laver WG, Murti KG, Portner A (1988) Isolation of a biologically active soluble form of the hemagglutinin-neuraminidase protein of Sendai virus. J Virol 62: 4 653–4 660
42. Tsurodome M, Kawano M, Yuasa T, Tabata N, Nishio M, Komada H, Ito Y (1995) Identification of regions on the hemagglutinin-neuraminidase protein of human parainfluenza virus type 2 important for promoting cell fusion. Virology 213: 190–203

Authors' address: Dr. R. Deng, Molecular and Cellular Virology, Animal Health Biological Discovery, Central Research Division, Pfizer Inc., Groton, CT 06340, U.S.A.

Viral factors determining rotavirus pathogenicity

U. Desselberger

Clinical Microbiology and Public Health Laboratory,
Addenbrooke's Hospital, Cambridge, U.K.

Summary. The pathogenicity of rotaviruses depends on multiple viral and host factors. Evidence is presented for the involvement of a number of viral genes (coding for structural and non-structural proteins) in the ability of the virus to cause diarrhoea. Different genes are important in different rotavirus–host systems suggesting that there is no single viral pathogenicity factor.

*

Rotaviruses are the major viral cause of acute gastroenteritis in infants and young children and in the young of a wide variety of mammalian species and are responsible for at least 850 000 deaths per year, mainly in developing countries [27].

The virus has a genome consisting of 11 segments of double-stranded (ds) RNA. The genomes of several rotavirus strains have been sequenced completely and firm gene-protein assignments been made [10, 13, 30]. There are six structural proteins: VP1, VP2 and VP3 (coded for by RNAs 1–3) constituting the core, VP6 (coded for by RNA6) forming the inner capsid, and the 2 outer capsid proteins, VP7 (coded for by RNA 7, 8 or 9, depending on strain) and VP4 (coded for by RNA4). The remaining five RNAs (RNA 5, two out of three RNAs 7, 8 or 9, RNA 10 and RNA 11) code for nonstructural proteins (Fig. 1). The detailed morphology has been elucidated [39, 56].

At least 6 rotavirus groups (A to F) have been distinguished. The vast majority of rotavirus infections of humans involve group A rotaviruses, of which there are several subgroups. Group and subgroup specificities are determined by epitopes on the inner capsid protein VP6. At least 14 different VP7-specific and over 20 different VP4-specific serotypes or genotypes have been distinguished among group A rotaviruses [12, 22]. VP7-specific types are termed G types (VP7 being a *g*lycoprotein), and VP4-specific P types (VP4 being a *p*rotease-sensitive protein). VP7 and VP4 both elicit specific neutralizing antibodies. As VP7 and VP4 are coded for by different RNA segments, both surface phenotypes can segregate independently by reassortment, and various combinations of P and G types have been found, both in man

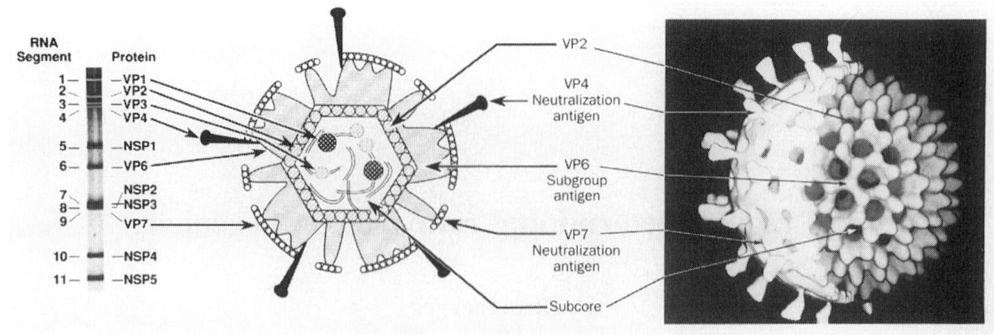

Fig. 1. RNA profile, protein products (gene protein assignment), and particle structure (diagram, reconstruction from cryo-electron micrograph) of rotaviruses (from [30]; with permission of authors and publisher)

and animals [12, 22]. Various rotavirus serotypes cocirculate in different geographical locations at any one time [27].

In the Anglo-Saxon literature, the terms pathogenicity and virulence are often used as synonyms [31, 46]. Here we define pathogenicity in rotaviruses as the ability to cause disease (diarrhoea). Virulence describes the capacity of a virus to cause disease, in relation to other closely-related viruses [52]. Virulence and pathogenicity depend on the number of viral and host factors. The viral factors relate to dose, capacity to spread between hosts, capacity to spread in specific cell populations, capacity to cause damage; host factors are the presence or absence of viral receptors, age, immune status, nutrition status and genetic background. In the following, the contribution on different viral factors to pathogenicity will be reviewed.

Rotaviruses infect mature enterocytes in the villous epithelium of the host's small intestine (Fig. 2). Epithelial cells are killed and sloughed off, resulting in

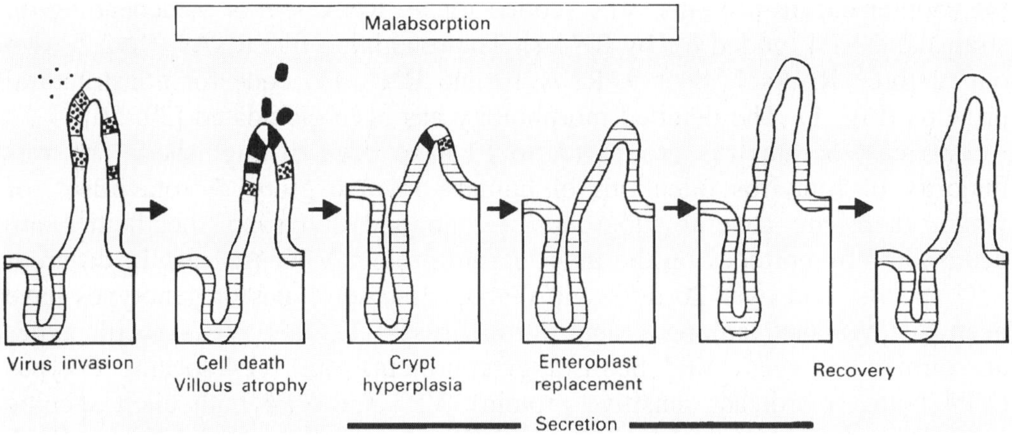

Fig. 2. Rotavirus pathogenesis. Development of damage to gut mucosa and ensuing diarrhoea (from [38]; with permission of author and publisher)

stunting of villi. The ensuing malabsorption leads to diarrhoea, the main symptom of symptomatic rotavirus infection. Epithelial loss is followed by compensatory crypt cell hyperplasia from where lost villous epithelium is replaced. This is accompanied by hypersecretion which contributes to the diarrhoea. It has also been postulated that changes in microcirculation associated with ischaemia play a role in rotavirus pathogenesis [20]. The alteration of the Na^+, K^+ and Ca^{++} homeostasis is known to occur in rotavirus-infected cell culture and may contribute to the symptoms observed during rotavirus disease. It has been shown that baculovirus-expressed NSP4 (NS28) protein of rotaviruses (the product of RNA segment 10) elevates the intracellular Ca^{++} level by release of Ca^{++} from the endoplasmic reticulum [50, 51].

Subclinical and inapparent rotavirus infections are frequent. It is believed that many inapparent infections are due to the presence of host factors such as immunologic memory from prior exposure to rotaviruses [29]. In addition, age-dependent resistance to disease has been reported [2]. It has been shown repeatedly that strains which are apathogenic can have full capacity to replicate to high titres in the small intestine [3, 4]. Recently it was shown in calves that low pathogenicity strains preferential colonise the proximal small intestine whereas high pathogenicity strains infect the whole small intestine [21].

Molecular investigations of naturally-occurring strains isolated from asymptomatic human neonates ("nursery strains") have shown that they possessed one of two distinct alleles of VP4 [9, 15, 17]. Initially, it was suggested that these VP4 alleles are responsible for apathogenicity. However, other studies have reported the presence of rotavirus strains with VP4 similar to that of the putative apathogenic human nursery strains (e.g. M37) in both symptomatic and asymptomatic children [42, 45]. Lack of clinical symptoms is possibly due to the presence of maternal antibodies.

VP4 has been implicated in pathogenicity since 1988, but its central position in virus replication and spread has been recognised for some time. VP4, one of the two outer capsid proteins, is the viral haemagglutinin [25], is posttranslationally cleaved by a trypsin-like protease and then enhances infectivity [14], is the determinant for plaque formation [25], binds to cellular receptors [41], and may be involved in host range restriction [26].

Using reassortants between two rotaviruses differing in virulence, virulence in mice segregated with VP4 [35]. Bridger et al. [3] tested the pathogenicity of two porcine rotavirus variants, called 4F and 4S, in gnotobiotic piglets. The two rotavirus variants have almost identical VP5 (NSP1), VP6 and VP7 genes but have very different VP4 genes showing only 67% nucleotide and 71% predicted amino acid homology [8]. During serial passage in piglets, variant 4F became highly pathogenic whereas the 4S variant remained apathogenic. Sequence analysis of the VP4 gene of different passages of variant 4F showed the appearance of an amino acid change in position 469 of the VP4 protein. The relative prevalence of viruses carrying the mutated VP4 gene correlated closely with the degree of emerging pathogenicity [6]. Recently, a single gene reassortant which carries the mutated VP4 gene of the pathogenic 4F strain in the genetic background of the apathogenic 4S variant (called S–F4) has been shown to be pathogenic in piglets, strongly suggesting that in this virus-host combination VP4 is the major determinant of pathogenicity [48]. The VP4 gene also segregated with plaquing efficiency and in vitro growth kinetics in several reassortants [49].

In sequence comparison dendrograms of VP4 genes, segregation into broadly species specific groups was found [8, 28], initially supporting the hypothesis of Kantharidis et al. [26] and Lopez et al. [28] that gene 4 may be involved in rotavirus host restriction. However, interspecies infections have been observed in nature [15, 18, 32, 33]. Differences in pathogenicity may occur by interspecies transmission of rotaviruses, as a consequence of their growth in the now heterologous host to much lower titres and with less cell damage. It has been shown repeatedly that natural rotavirus isolates are likely to be reassortants which picked up some of the genes from an animal or other human source [34, 37]. Interspecies transmission and gene reassortment amongst rotavirus has been implicated in the generation of novel strains which have produced pathogenicity. The VP4s of strains isolated from asymptomatic human neonates are most closely related to the VP4 of the porcine Gottfried strain [18] or to the VP4 of the bovine B223 strain [9, 15].

Pathogenicity and virulence are complex characteristics, and strong evidence has recently emerged that for rotaviruses they are also influenced by genes other than that coding for VP4. In reassortants made by co-infection of mice with a virulent mouse strain and simian virus (of low virulence in mice), the structural proteins were found not to be determinants of the ability to cause disease at low doses, but pathogenicity segregated with VP5 (NSP1, NS53) and VP8 (NSP2, NS35) [5]. This data is in some contrast to those of Offit et al. [35]. One possible explanation of the complete lack of association of pathogenicity with structural genes could be that the block to the growth of certain reassortants in mice occurs during replication and assembly of the progeny virus rather than at the stage of cell binding and entry.

Non-random segregation of the NSP1 gene in other previous reassortment studies has also implicated NSP1 in the determination of growth characteristics. An in vitro analysis of reassortants between the simian strains SA11 and RRV generated in vitro and in vivo (mice) demonstrated selection in favour of the SA11 NSP1 gene over that of the RRV strain [16]. Non-random segregation of the NSP1 gene associated with some gene linkage has also been observed in other reassortant studies carried out in vitro [19, 55]. It has been suggested that the rotavirus strains isolated from asymptomatic human neonates in India which have a number of bovine genes may have been aided in their spread through the human population by the NSP1 genes [55]. The NSP1 genes, which are the most variable rotavirus gene, seem to cluster loosely on the basis of species of origin in sequence dendrograms [11, 24]. However, a recent extensive phylogenetic analysis of NSP1 gene sequences suggested that this relationship between NSP1 type and species of origin is not absolute [55]. A report by Palombo and Bishop [36] showed that at least three distinct alleles of NSP1 occur amongst strains isolated from asymptomatic neonates.

Hoshino et al. [23] made a major effort to determine the contribution of different genes to pathogenicity in gnotobiotic piglets. Single gene reassortants were produced between a porcine virus which is pathogenic in piglets (SB–1A) and a human strain which is not (DS–1). It was shown that monoreassortants of SB–1A in which the genes coding for VP3, VP4, VP7 or NSP4 (NS28) were replaced by the corresponding gene of the DS–1 strain had lost their pathogenicity. Inversely, however, pathogenicity could be conferred on the DS–1 strain only by the co-assortment of all four of these SB–1A genes. Reassortants carrying only one, two or three or these SB–1A genes were not pathogenic.

Recently it has been shown that free NSP4 and peptides derived from it (spanning amino acids 114 to 135) are able to induce dose related diarrhoea when inoculated intraperitoneally or intraduodenally into six- to ten-day-old mice and rats. This effect could be blocked by mixing NSP4 with an NSP4-specific antibody prior to inoculation [1]. It could be demonstrated that the NSP4 peptide potentiated chloride secretion by a Ca^{++} dependent signalling pathway [50, 51], and suggested that NSP4 acted like a viral enterotoxin.

From the data reviewed here, a number of conclusions can be drawn. The genes coding for NSP1 and NSP2 which were found to be associated with pathogenicity in the mouse model [5] were not important for pathogenicity in gnotobiotic piglets [23]. The VP7 gene, implicated by Hoshino et al. [23], was not important in determining the pathogenicity of the porcine variants 4F and 4S as they are virtually identical in those strains [8]. This is not necessarily in contradiction to data of Hoshino et al. [23], since in their studies reassortants between homologous and heterologous strains were compared for pathogenicity allowing the suggestion that more than one gene from the homologous strain would be required to produce pathogenicity.

No single gene determines rotavirus pathogenicity under all circumstances. It appears that the gene or genes responsible vary from strain to strain and host to host. The overall genetic constellation of a rotavirus determines the growth

Table 1. Rotavirus genes associated with pathogenicity in different hosts

Gene segment	Gene product	Host	Reference
3	VP3	Pig	[23]
4	VP4	Mouse	[35]
		Man	[17]
		Pig	[3]
			[23]
			[6, 8]
5	NSP1 (NS53)	Mouse	[5]
7	NSP2 (NS35)	Mouse	[5]
9	VP7	Pig	[23]
10	NSP4 (NS28)	Pig	[23]
		Mouse, rat	[1]

properties and pathogenicity in a particular host. This is similar to the situation found in other virus families, for example reoviruses [52, 53] and influenza viruses [40, 43, 44]. It is conceivable that any of the 11 rotavirus genes could be a determinant of pathogenicity of a particular strain in a particular host. To date the genes coding for VP3, VP4, VP7 and NSP1, NSP2 and NSP4 have been associated with pathogenicity (Table 1). This could suggest that the functions performed by the products of these genes are more directly related to the pathogenic effects of all the virus than those of other genes.

There are still many questions open in rotavirus pathogenicity. For instance, it is not well understood why some strains which replicate very well and to titres comparable to pathogenic strains do not cause diarrhoea. Further studies in animal hosts are required to answer those questions. (A more detailed review on rotavirus pathogenicity is found in [7].)

References

1. Ball JM, Tian P, Zeng C, Morris AP, Estes MK (1996) Age-dependent diarrhoea induced by a rotaviral nonstructural glycoprotein. Science 272: 101–104
2. Bridger JC (1994) A definition of bovine rotavirus virulence. J Gen Virol 75: 2807–2812
3. Bridger JC, Burke B, Beards GM, Desselberger U (1992) The pathogenicity of two porcine rotaviruses differing in their in vitro growth characteristics and genes 4. J Gen Virol 73: 3011–3015
4. Bridger JC, Pocock DH (1986) Variation in virulence or bovine rotaviruses. J Hyg (Camb.) 96: 257–264
5. Broome RL, Vo PT, Ward RL, Clark HF, Greenberg HB (1993) Murine rotavirus genes encoding outer capsid proteins VP4 and VP7 are not major determinants of host range restriction and virulence. J Virol 67: 2448–2455
6. Burke B, Bridger JC, Desselberger U (1994a) Temporal correlation between a single amino acid change in the VP4 of a porcine rotavirus and a marked change in pathogenicity. Virology 202: 754–759
7. Burke B, Desselberger U (1996) Rotavirus pathogenicity. Virology 218: 299–305
8. Burke B, McCrae MA, Desselberger U (1994b) Sequence analysis of two procine rotaviruses differing in growth in vitro and in pathogenicity. Distinct VP4 sequences and conservation of NS53, VP6 and VP7 genes. J Gen Virol 75: 2205–2212
9. Das M, Dunn SJ, Woode GN, Greenberg HB, Rao CD (1993) Both surface proteins (VP4 and VP7) of an asymptomatic neonatal rotavirus strain (I321) have high levels of sequence identity with the homologous proteins of a serotype 10 bovine rotavirus. Virology 194: 374–379
10. Desselberger U, McCrae MA (1994) The rotavirus genome. In: Ramig RF (ed) Rotaviruses. Springer, Berlin Heidelberg New York Tokyo, pp 31–66
11. Dunn SJ, Cross TL, Greenberg HB (1994) Comparison of the rotavirus nonstructural protein NSP1 (NS53) from different species by sequence analysis and Northern blot hybridization. Virology 203: 178–183
12. Estes MK (1996) Rotaviruses and their replication. In: Fields BN, Knipe DM, Howley PM, Chanock RM, Melnick J, Monath TP, Roizman B, Straus SE, (eds) Fields Virology, 3rd ed. Lippincott, Raven, Philadelphia, pp 1625–1655
13. Estes MK, Cohen J (1989) Rotavirus gene structure and function. Microbiol Rev 53: 410–499

14. Estes MK, Graham DY, Mason BB (1981) Proteolytic enhancement of rotavirus infectivity. Molecular mechanisms. J Virol 39: 879–888
15. Gentsch JR, Das BK, Jiang B, Bhan MK and Glass RI (1993) Similarity of the VP4 protein of human rotavirus strain 116E to that of the bovine B223 strain. Virology 194: 424–430
16. Gombold JL, Ramig RF (1986) Analysis of reassortment of genome segments in mice mixedly infected with rotaviruses SA11 and RRV. J Virol 57: 110–116
17. Gorziglia M, Green K, Nishikawa K, Taniguchi K, Jones R, Kapikian AZ, Chanock RM (1988) Sequence of the fourth gene of human rotaviruses recovered from asymptomatic or symptomatic infections. J Virol 64: 2978–2984
18. Gorziglia M, Nishikawa K, Hoshino Y, Taniguchi K (1990) Similarity of the outer capsid protein VP4 of the Gottfried strain of porcine rotavirus to that of asymptomatic human rotavirus strains. J Virol 64: 414–418
19. Graham A, Kudesia G, Allen AM, Desselberger U (1987) Reassortment of human rotavirus possessing genome rearrangements with bovine rotavirus: evidence for host cell selection. J Gen Virol 68: 115–122
20. Greenberg, HB, Clark HF, Offit PA (1994) Rotavirus pathology and pathophysiology. Curr Top Microbiol Immunol 185: 255–283
21. Hall GA, Bridger JC, Parsons KR, Cook R (1993) Variation in rotavirus virulence: a comparison of pathogenesis in calves between two rotaviruses of different virulence. Vet Pathol 30: 223–233
22. Hoshino Y, Kapikian AZ (1994) Rotavirus antigens. In: Ramig RF (ed) Rotaviruses. Springer, Berlin Heidelberg New York Tokyo, pp 179–227
23. Hoshino Y, Saif LJ, Kang SY, Sereno MM, Chen WK, Kapikian AZ (1995) Identification of group A rotavirus genes associated with virulence of a porcine rotavirus and host range restriction of a human rotavirus in the gnotobiotic piglet model. Virology 209: 274–280
24. Hua J, Mansell EA, Patton JT (1993) Comparative analysis of the rotavirus NS53 gene: Conservation of basic and cysteine-rich regions in the protein and possible stem-loop structures in the RNA. Virology 196: 372–378
25. Kalica AR, Flores J, Greenberg HB (1983) Identification of the rotavirus gene which codes for the haemagglutinin and protease-enhanced plaque formation. Virology 125: 194–205
26. Kantharidis P, Dyall-Smith ML, Tregear GW, Holmes IH (1988) Nucleotide sequence of UK bovine rotavirus segment 4: Possible host restriction of VP3 genes. Virology 166: 308–315
27. Kapikian AZ, Chanock RM (1996) Rotaviruses. In: Fields BN, Knipe DM, Howley PM, Chanock RM, Melnick JL, Monath TP, Roizman B, Straus SE (eds) Fields Virology, 3rd ed. Lippincott Raven, Philadelphia pp 1657–1708
28. Lopez S, Lopez I, Romero P, Mendez E, Sobero X, Arias CF (1991) Rotavirus YM gene 4: Analysis of its deduced amino acid sequence and prediction of the secondary structure of the VP4 protein. J Virol 65: 3738–3745
29. Matson DO, O'Ryan ML, Velasquez FR, Calva JJ, Ruiz-Palacios GM, Pickering LK (1996) Antibody correlates of protection against rotavirus infection. Abstracts, PW03–1, p. 100. International Congress of Virology, Jerusalem, Israel
30. Mattion NM, Cohen J, Estes MK (1994) The rotavirus proteins. In: Kapikian AZ (ed) Viral Infections of the gastrointestinal tract, 2nd ed. Dekker, New York Basel Hong Kong, pp 169–249
31. Mims CA, White DO (1984) Viral Pathogenesis and Immunology. Blackwell Oxford, pp 158–198

32. Nakagomi O, Ohshima A, Aboudy Y, Shif I, Mochizuki M, Nagakomi T, Gotlieb-Stematsky T (1990) Molecular identification by RNR-RNA hybridization of a human rotavirus that is closely related to rotaviruses of feline and canine origin. J Clin Microbiol 28: 1198–1203
33. Nakagomi O, Nakagomi T (1991) Genetic diversity and similarity among mammalian rotaviruses in relation to interspecies transmission of rotavirus. Arch Virol 120: 43–55
34. Nakagomi O, Isegawa Y, Hoshino Y, Aboudy Y, Shif I, Silberstein I, Nakagomi T, Ueda S, Sears J, Flores J (1993) A new serotype of the outer capsid protein VP4 shared by an unusual human rotavirus strain Ro1845 and canine rotaviruses. J Gen Virol 74: 2771–2774
35. Offit PA, Blavat G, Greenberg HB, Clark HF (1986) Molecular basis for rotavirus virulence: role of rotavirus gene segment 4. J Virol 57: 46–49
36. Palombo EA, Bishop RF (1994) Genetic analysis of NSP1 genes of human rotaviruses isolated from neonates with asymptomatic infection. J Gen Virol 75: 3635–3639
37. Palombo EA, Brigg HC, Masendycz PJ, Coulson BJ, Barnes GL, Bishop RF (1996) Genetic and antigenic analysis of multiple gene in vivo reassortants, Abstracts, WO3–3. International Congress of Virology, Jerusalem, Israel
38. Phillips AD (1989) Mechanisms of mucosal injury: human studies. In: Farthing MJG (ed) Viruses and the Gut. Swan Press, London, pp 30–40
39. Prasad BVV, Chiu W (1994) Structure of rotaviruses. In Ramig R (ed) Rotaviruses. Springer, Berlin Heidelberg New York Tokyo, pp 9–29
40. Rott R, Orlich M, Scholtissek C (1979) Correlation of pathogenicity and gene constellation of influenza viruses. III. Non-pathogenic recombinants derived from highly pathogenic parent strains. J Gen Virol 44: 471–477
41. Ruggeri FM, Greenberg HB (1991) Antibodies to the trypsin-cleavage peptide VP8* neutralize rotavirus by inhibiting binding of virions to target cells in cell culture. J Virol 65: 2211–2219
42. Santos N, Gouvea V, Timenetsky MC, Clark HF, Riepenhoff-Talty M, Garbarg-Chenon A (1994) Comparative analysis of VP8* sequences from rotaviruses possessing M37-like VP4 recovered from children with and without diarrhoea. J Gen Virol 75: 1775–1780
43. Scholtissek C, Vallbracht A, Flehmig B, Rott R (1979) Correlation of pathogenicity and gene constellation of influenza A viruses. Virology 95: 492–500
44. Schulman JL (1983) Virus-determined differences in the pathogenicity of influenza virus infection. In: Palese P, Kingsbury D (eds) Genetics of influenza viruses. Springer, Berlin Heidelberg New York Tokyo, pp 305–320
45. Steele AD, Garcia D, Sears J, Nakagomi O, Flores J (1993) Distribution of VP4 gene alleles in human rotavirus by using probes to the hyperdivergent region of the VP4 gene. J Clin Microbiol 31: 1735–1740
46. Sweet C, Smith H (1990) The pathogenicity of viruses. In: Collier LH, Timbury MC (eds) Topley and Wilson's principles of bacteriology, virology, and immunity, 8th ed Vol. 4. Arnold, London Melbourne Auckland, pp 106–130
47. Taniguchi K Urasawa T, Urasawa S (1994) Species specificity and interspecies relatedness in VP4 genotypes demonstrated by VP4 sequence analysis of equine, feline, and canine rotavirus strains. Virology 200: 390–400
48. Tauscher GI, Bridger J, Desselberger U (1996) The molecular basis of rotavirus pathogenicity in pigs: new evidence for the involvement of VP4. Abstracts. PW04-2. p. 107. International Congress of Virology, Jerusalem, Israel.
49. Tauscher GI, Desselberger U (1997) Viral determinants of rotavirus pathogenicity in pigs: Production of reassortants by asynchronous coinfection. J Virol 71: 853–857

50. Tian P, Estes MK, Hu Y, Ball JM, Zeng CQ–Y, Schilling WP (1995) The rotavirus nonstructural glycoprotein NSP4 mobilizes Ca^{2+} from the endoplasmic reticulum. J Virol 69: 5763–5772
51. Tian P, Hu Y, Schilling WP, Lindsay DA, Eiden J, Estes MK (1994) The nonstructural glycoprotein of rotavirus affects intracellular calcium levels. J Virol 68: 251–257
52. Tyler KN, Fields BN (1996) Pathogenesis of viral infections. In: Fields BN, Knipe DM, Howley PM, Chanock RM, Melnick JL, Monath TP, Roizman B, Straus SE (eds) Fields Virology, 3rd ed. Lippincott-Raven, Philadelphia, pp 173–218
53. Tyler KN, Fields BN (1996) Reoviruses. In: Fields BN, Knipe DM. Chanock RM, Melnick JL, Monath TP, Roizman B, Straus SE (eds) Fields Virology, 3rd ed. Lippincott-Raven, Philadelphia, pp 1597–1623
54. Xu L, Tian Y, Tarlow O, Harbour D, McCrae MA (1994) Molecular biology of rotaviruses. IX. Conservation and divergence in genome segment 5. J Gen Virol 75: 3413–3421
55. Xu Z, Woode GN (1994) Studies on the influence of the VP5 gene on rotavirus replication. Virology 198: 394–398
56. Yeager M, Berriman JA, Baker TS, Bellamy AR (1994) Three-dimensional structure of the rotavirus haemagglutinin VP4 by cryo-electron microscopy and difference map analysis. EMBO J 13: 1011–1018

Authors' address: Dr. U. Desselberger, Clinical Microbiology and Public Health Laboratory, Addenbrooke's Hospital, Cambridge, CB2 QW, U.K.

Viral zoonoses and food of animal origin: caliciviruses and human disease

I. N. Clarke and **P. R. Lambden**

Molecular Microbiology Group, University Medical School,
Southampton General Hospital, Southampton, U.K.

Summary. Caliciviruses are important veterinary and human pathogens. The viruses gain their name from characteristic cup-shaped structures seen on the virion surface by negative stain electron microscopy. In humans caliciviruses are a major cause of diarrhoeal disease. There are two fundamentally different genome structures amongst human caliciviruses. The Norwalk-like or small round structured viruses (SRSVs) are viruses that have an amorphous structure when viewed by EM, they have a genome composed of 3 major open reading frames (ORFs). These viruses cause epidemic gastroenteritis amongst all age groups. In contrast, the 'classic' human caliciviruses (HuCVs) display the typical calicivirus surface structure and have their capsid ORF fused to and contiguous with the non structural proteins forming one giant polyprotein. HuCVs are predominantly associated with paediatric infections and are only a minor cause of disease in humans. Spread of disease for both SRSVs and HuCVs is usually by faecal oral transmission. SRSVs are a major cause of foodborne gastroenteritis especially linked to the consumption of sewage-contaminated shellfish. However, there is no evidence that these viruses replicate in shellfish or that they originate from an animal source.

Introduction

Caliciviruses are important veterinary and human pathogens and have been isolated from a wide range of species. The viruses gain their name from characteristic cup-shaped depressions (Calyx = L. cup) observed on the virion by negative stain electron microscopy. The viruses exhibiting typical calicivirus morphology display 2-, 3- and 5-fold axes of symmetry [12]. However, many calicivirus isolates do not exhibit the classic morphology but have a very similar underlying structure to typical caliciviruses as revealed by cryoelectron microscopy and 3D image reconstruction [50, 51]. The application of molecular technology has demonstrated that EM appearance should no longer be considered as the main criterion for the classification of these viruses.

Caliciviruses are small non-enveloped viruses (27–35 nm) with a buoyant density in the range 1.33 to 1.40 gm/cm^3 in CsCl and contain a single-stranded

Table 1. Properties of caliciviruses

Virus	Cell culture	Genome organisation	Species specificity	Disease
SRSVs	-	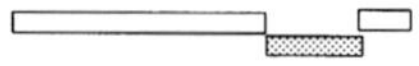	Human	Gastroenteritis
HuCV	-		Human	Gastroenteritis
SMSV	+	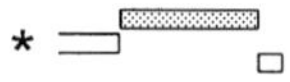	Broad	Vesicular
FCV	+		Cats (dog)	Pneumonia
RHDV	-	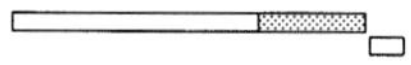	European rabbit	Haemorrhagic
EBHSV	-	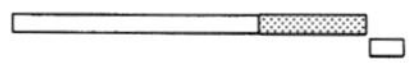	European brown hare	Haemorrhagic

The genome organisations were determined from sequence data deposited in the EMBL/Genbank database. Accession numbers are as follows: SRSVs, M87661, L07418, X86557; HuCV, X86560; SMSV, M87481, M87482; FCV, M86379; RHDV, M67473; EBHSV, Z69620. The asterisk shows the limit of known sequence for SMSV

positive sense RNA genome encapsidated by a single major polypeptide of 59–70 kDa [13]. Viruses that fulfil the morphological and biophysical characteristics of the family and for which genome sequence is available are summarised in Table 1. These genome structures clearly distinguish caliciviruses from the *Picornaviridae*. In contrast to picornaviruses, caliciviruses have a single capsid gene which is located to the 3′ terminus of the genome with the non-structural genes located to the 5′ terminus [25]. In addition, caliciviruses synthesise a 3′ co-terminal subgenomic RNA of 2.2 to 2.4 kb which is thought to enhance expression of the capsid protein during viral replication [22, 43, 47]. Since the last meeting of the ICTV in 1993 further new sequence data have become available showing that the European brown hare syndrome virus (EBHS) virus has a similar genome structure to RHDV and that it should also be recognised as a calicivirus [37]. Recently the human classic calicivirus genome sequence was determined and confirmed that taxonomic status of this virus as a member of the *Caliciviridae* [39]. Human hepatitis E virus (HEV) remains a "candidate" calicivirus because of major differences in the organisation of identifiable motifs in the viral genome [32]. A number of other viruses have some of the morphological and biophysical properties of the *Caliciviridae* but must remain as 'candidate' caliciviruses until further molecular data become available to support their inclusion within the family. The molecular and clinical aspects of caliciviruses have been the subject of several recent reviews [7, 13, 15, 17, 20, 30, 31]. The purpose of this report is to provide an overview and background on the zoonotic potential of caliciviruses.

Human caliciviruses

History

A non-bacterial cause for epidemics of gastroenteritis had long been suspected. A viral aetiology was established between the 1940's and early 1970's through the use of human volunteers and infection with bacteria-free faecal filtrates [1, 11, 16, 18, 28, 52]. The first description of a virus associated with an outbreak of gastroenteritis was of the Norwalk virus from an elementary school in Norwalk, Ohio [29]. In the following decade a large number of morphologically different enteric viruses were identified by electron microscopy. Some of these viruses were easily identified by the large number of particles shed and their characteristic morphology e.g. rotaviruses and enteric adenoviruses. However, a number of the small round faecal viruses were difficult to classify. In an attempt to bring order to this confused state of enteric virus taxonomy an interim classification scheme was devised [10]. This scheme was based on virus morphology and allowed differentiation of small round faecal viruses (27–35 nm) displaying clear surface structures into three distinct groups; (i) the small round structured viruses (SRSVs), (ii) classical caliciviruses (HuCVs) and (iii) astroviruses. Astroviruses belong to a separate family (*Astroviridae*) and therefore do not come within the scope of this review.

Small round structured viruses (SRSVs)

SRSVs are viruses that have a ragged edge and an amorphous structure (Fig. 1). The prototype virus is the Norwalk virus but many different isolates from around the world have now been described. A proper classification scheme for members of the *Caliciviridae* has not been established. Thus SRSVs have traditionally been given geographic names indicating the place of their initial isolation. Research into SRSVs has been difficult because of the lack of a cell culture system and the very low numbers of virus particles shed during infection. Initially, virus from volunteers was used in the molecular cloning of the Norwalk virus genome [24, 42]. Based on these preliminary sequence data rapid progress has been made on the molecular characterisation of SRSVs obtained directly from clinical samples. Sequence comparisons of different isolates have revealed that SRSVs can be divided into two distinct genetic groups [35]. Complete genome sequences [7.5–7.7kb) are now available for Norwalk and Southampton viruses (group I) [26, 33, 34] and Lordsdale virus (group II) [14] and show that the basic overall genome organisation is preserved between the two groups.

Epidemiology

SRSVs are now established as a major cause of diarrhoeal disease in humans [30]. SRSVs have been administered to a wide range of mammalian species including mice, guinea pigs, rabbits, kittens, calves, baboons, chimpanzees and monkeys [19, 30]. Although some chimpanzees showed a serological response attempts to establish infection with SRSVs in animals were not successful [63]. Spread of disease in humans is usually by faecal-oral transmission and large outbreaks have been reported in semi-closed communities such as schools, day-care centres, hospitals, cruise ships and hotels. Secondary transmission of the infection is common because of the highly infectious nature of the viruses with as few as 10–100 virus particles required to initiate infection. The extreme stability of SRSVs often results in secondary spread associated with environmental contamination.

SRSVs and food

SRSVs are a major cause of water and foodborne gastroenteritis especially linked to the consumption of molluscan shellfish [21]. It is important to note that SRSVs do not replicate in foods but may contaminate this medium by two main routes [3]. The bivalve molluscs (oysters, cockles and mussels) concentrate SRSVs by filtration from sewage-contaminated waters. Although there is no evidence that SRSVs replicate in these shellfish or that these viruses originate from an animal source, contaminated shellfish are major sources of epidemic gastroenteritis with several major outbreaks of shellfish-associated gastroenteritis linked to SRSVs [2]. Food may also become contaminated by infected food handlers shedding viruses. Foods involved in these kinds of outbreaks are those that usually require a large amount of handling, e.g. cold foods, sandwiches and cooked meats.

New technologies

Availability of SRSV genome sequences and expression of recombinant viral capsids in heterologous systems has greatly facilitated the development of new molecular tools to study these important viruses [14, 25, 27, 38]. The development of new diagnostic reagents (antisera to baculovirus-expressed capsids) and detection methods (PCR amplification of SRSV genomes from food matrices) will in the future allow the detection of virus contamination of food. Isolation of infected individuals and strict disinfection of the immediate environment are currently the only way for controlling outbreaks [9].

Classic caliciviruses

HuCVs were first described in 1976 by Madeley and Cosgrove [41]. These viruses were found in the stools of infants and have the classic cup-shaped morphology of caliciviruses when observed by EM (Fig. 1). In contrast to

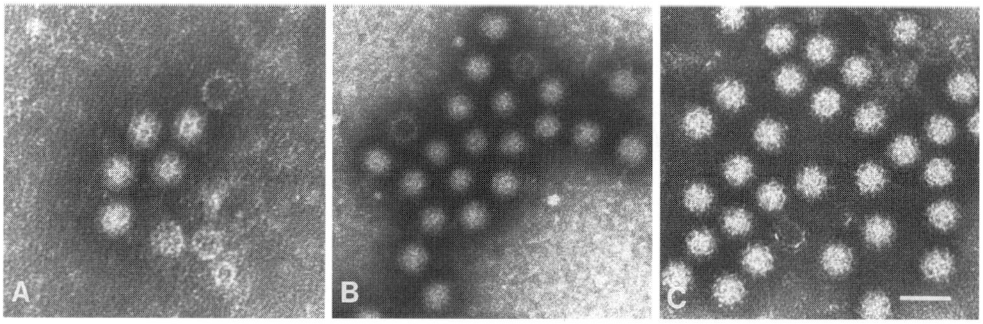

Fig. 1. Electron micrographs of negatively stained caliciviruses. **A** Classic HuCV in stool: **B** SRSV in stool; **C** FCV purified from a CsCl gradient. Bar: 50 nm

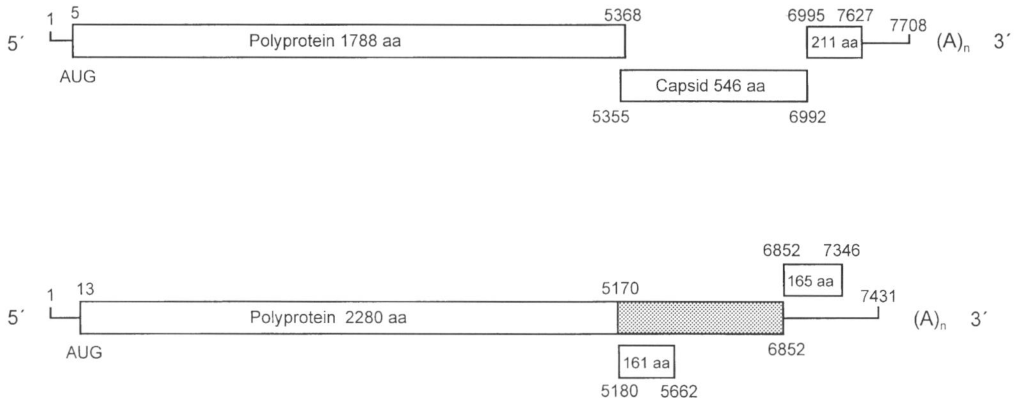

Fig. 2. Diagrammatic representation of the two different genome organisations of SRSVs and classic human caliciviruses. The SRSV (Southampton virus) genome (top) has three potential open reading frames whereas in classic HuCV (Manchester virus) the non-structural protein open reading frame is fused to and contiguous with the capsid encoding region generating a single large polypeptide

SRSVs, the classic HuCVs (although occasionally infecting older adults) are predominantly pathogens of infants and young children. In addition to the morphological and epidemiological differences with SRSVs classic HuCVs are also phylogenetically distinct and have a fundamentally different genome organisation [39]. The 'classic' human caliciviruses have their capsid ORF fused to and contiguous with the non-structural protein ORF forming one giant polyprotein (Fig. 2). The Manchester human calicivirus genome is 7431nt in length [39a] with a similar genome arrangement to the hepatotropic rabbit virus RHDV [44] (Table 1).

As with the SRSVs progress on the molecular characterisation of the classic human caliciviruses has been greatly impeded by the lack of a cell culture system. There is only limited evidence for the involvement of classic HuCV's in food borne disease. A single report describes a potential zoonotic infection from a dog that was violently sick in an old people's home in Exeter, UK. An outbreak of gastroenteritis and vomiting occurred 24 h later involving staff, their children and residents of the home. Suggestions of a zoonotic infection were inferred from cross-reactivity between sera from the dog and affected people. However, stool samples were not available from the dog to prove the association [23].

Animal caliciviruses

Caliciviruses have been isolated from mammals, birds reptiles, amphibians, fish and invertebrates [13]. They are a major cause of disease in cats (feline calicivirus – FCV) and rabbits (rabbit haemorrhagic disease virus – RHDV) and they have also been extensively reported in marine mammals (San Miguel Sealion virus – SMSV) found on the channel islands off the Californian coast. Immunological and phylogenetic analyses have shown these three groups of animal viruses are genetically and antigenically distinct [6].

Marine caliciviruses

Several fascinating and detailed reviews describe the 'marine calicivirus story' and the scientific evidence that has linked the causal agent of vesicular disease in sealions (SMSV) with a vesicular disease of pigs [4–6, 57]. Vesicular exanthema of swine virus (VESV) is the prototypic calicivirus. This virus causes vesicular lesions identical to foot-and-mouth disease in swine and was first described infecting domestic swine herds in Southern California in the spring of 1932. It is thought that the virus was spread by feeding uncooked garbage to pigs. Unlike FMDV this virus could not be transmitted to guinea pigs, horses or cattle and was thus originally classified as an atypical FMDV.

Outbreaks of VESV occurred sporadically in California over the following two decades and of became known as the California disease. Initially the virus was spread from contaminated feed but in 1936 a direct pig-to-pig transmission cycle was established. In the most notorious episode, VESV was spread from California over large areas of the United States. This nationwide outbreak in June 1952 occurred because of an illegal disposal of uncooked pork trimmings in Cheyenne, Wyoming from a train originating in San Francisco. National control measures required cooking of all garbage fed to pigs and led to eradication of the disease in 1955. VESV is a disease peculiar to the USA except for one report of an outbreak amongst pigs fed raw pork scraps at a US military base in Iceland. Following eradication, VESV was designated a 'foreign' exotic animal disease by the Secretary of Agriculture.

VESV was recently shown at the genetic level to be very closely related to SMSV strongly supporting the view that the VES virus was originally introduced to pigs from a marine mammal source [48]. However, it has been argued that the sealion population is not large enough to generate the diversity of SMSV serotypes and that a larger marine reservoir may be involved [5].

SMSV has been isolated from the opal eye perch [56]. This fish shares the same marine environment as Californian sealions and it is possible that the perch offer a greater reservoir for these viruses. In addition, SMSVs have been isolated from the sealion lungworm which is a nematode parasite using the opal eye perch as an intermediate host [5]. Experimental transmission of SMSV between these three species has established the possibility of a cycle of infection in the marine food chain.

SMSV has a broad host range in vivo infecting many different domestic animal species including pigs, cattle, and dogs as well as primates, fish, nematodes, seabirds, reptiles, shellfish, cetaceans and pinnipeds (A.W. Smith, pers. comm.). In cell culture the viruses can also infect a range of cell lines including some of human origin [54].

Specific antibodies to SMSV have been detected in three laboratory workers [55] and one case of ocular vesicular disease occurred in a biologist who had been handling seals with flipper lesions [54]. Although this virus clearly has the potential for zoonotic infection the only other documented human infection occurred in a laboratory worker (Berry ES, Skilling DE, Degner M, Smith AV (1987) Abstracts of the Annual Meeting, ASM, T3, p 315).

FCV

Feline calicivirus is an important pathogen of cats. The virus was first isolated in 1957 but in contrast to SMSV there appears to be only one serotype. Four FCV strains are well described [8, 46, 47, 49, 59, 61]. The main clinical presentations of infection are respiratory illness, tongue ulcers, chronic stomatitis and lameness. Asymptomatic carriage in cats is common. A live attenuated FCV vaccine is available based on the F9 strain. The vaccine protects against disease but not infection or the carrier state. FCV has been isolated from cheetahs [53] and also dogs but appears to have a much narrower host range than SMSV.

RHDV

RHDV causes a disease with rapid onset and high mortality in *Oryctolagus cuniculus* (European rabbit). The first description came from China in 1984 [40] and since then RHDV has spread rapidly across the globe. This virus has gained some notoriety for caliciviruses following its accidental escape from biological control experiments conducted on Wardang island off the coast of South Australia [62]. The virus in Australia originates from the Czech strain of RHDV V-351 and has an experimental lethality greater than 99% in Australian rabbits [45]. Contact transmission of the virus does not appear to account for the rapid spread of the virus on the Australian continent. RHDV is specific to the European rabbit as it does not cause disease in other lagomorphs including the Eastern cotton tail, volcano rabbit, black tailed jack rabbit and the European brown hare (EBH). The EBH is also susceptible to a related calicivirus infection (European Brown Hare syndrome Virus – EBHSV) but EBHSV does not cause disease in rabbits. Prior to the test experiments on Wardang Island a number of species were tested for susceptibility to RHDV including farm animals, immigrant rodents, marsupials, birds, lizards and bats. Under the conditions used none of these species could be infected with RHDV. In addition there is no evidence that RHDV can infect man.

Like human enteric caliciviruses RHDV does not grow in cell culture therefore virus purified from infected rabbit livers has been used to develop an effective killed vaccine. Using baculovirus vectors the RHDV capsid protein has also been expressed as virus-like particles in insect cells. In an experimental context these VLPs provide effective protection against RHDV [36, 45, 58]. The very narrow host range, high mortality and availability of a vaccine make this virus an appealing biological control agent for the European rabbit.

Conclusions

The inability of human caliciviruses to infect animals and the lack of a cell culture system strongly suggests that these viruses are highly species-specific and have a fastidious tissue tropism. There is no direct evidence for the natural transmission of either FCV or RHDV to man. FCV has a slightly broader host range (including dogs and big cats) than RHDV and can be grown in cell culture. RHDV is specific for the European rabbit, is hepatotropic and cannot be grown in cell culture. Experimental infection with high doses of virus has not even been achieved with closely related lagomorphs. It thus seems unlikely that FCV and RHDV will transmit across the species barrier to man. In contrast, SMSV is a virus that is not species-specific, it has wide host range and will replicate in a variety of cultured cells. Human vesicular disease has not been extensively investigated but the isolation of SMSV from both the pygmy chimpanzee and lowland gorilla has confirmed that these viruses can infect some of the closest relatives to humans. Infection of humans with caliciviruses of animal origin offers a promising avenue for future investigation.

Acknowledgements

We thank our colleagues in the molecular virology group at the University of Southampton and Dr. E. O. Caul (Bristol Regional Virus Laboratory) for helpful discussions. Genome analysis studies were supported by The Wellcome Trust grant no. 038076. We are also grateful to Ms. Sue Cox for assistance with photography.

References

1. Adler I, Zickl R (1969) Winter vomiting disease. J Infect Dis 119: 668–673
2. Ando T, Jin Q, Gentsch JR, Monroe SS, Noel JS, Dowell SF, Cicirello HG, Kohn MA, Glass RI (1995) Epidemiologic applications of novel molecular methods to detect and differentiate small round structured viruses (Norwalk-like viruses). J Med Virol 47: 145–152

3. Appleton H, Caul EO, Clewley JP, Cruickshank JG, Cubitt, WD, Kurtz JB, McSwiggan DA, Palmer SR, Riordan T (1988) Foodborne viral gastroenteritis. PHLS Microbiol Digest 5: 69–75
4. Bankowski RA (1981) Vesicular exanthema of swine: San Miguel sea lion viral infection. In: Beran GW (ed) Viral zoonoses. CRC Press, Boca Raton, pp 176–182
5. Barlough JE, Berry ES, Skilling DE, Smith AW (1986a) The marine calicivirus story – Part II. Conpend Contin Educ Pract Vet 8: F75–F82
6. Barlough JE, Berry ES, Skilling DE, Smith AW (1986b) The marine calicivirus story – Part I. Compend Contin Educ Pract Vet 8: F5–F14
7. Carter MJ, Milton ID, Madeley CR (1991) Caliciviruses. Rev Med Virol 1: 177–186
8. Carter MJ, Milton ID, Meanger J, Bennett M, Gaskell RM, Turner PC (1992) The complete nucleotide sequence of a feline calicivirus. Virology 190: 443–448
9. Caul EO (1994) Small round structured viruses-airborne transmission and hospital control. Lancet 343: 1 240–1 242
10. Caul EO, Appleton H (1982) The electron microscopical and physical characteristics of small round human fecal viruses: and interim scheme for classification. J Med Virol 9: 257–265
11. Clarke SKR, Cook GT, Egglestone SLI, Hall TS, Miller DL, Reed SE, Rubenstein D, Smith AJ, Tyrrell DAJ (1972) A virus from epidemic vomiting disease. Br Med J 3: 86–89
12. Cubitt D, Bradley DW, Carter MJ, Chiba S, Estes MK, Saif LJ, Schaffer FL, Smith AW, Studdert MJ, Thiel HJ (1995) Caliciviridae. In: Murphy FA, Fauquet CM, Bishop DHL, Ghabrial SA, Jarris AW, Martelli GP, Mayo MA, Summers MD (eds) Virus Taxonomy. Classification and Nomenclature of Viruses. Sixth Report of the International Committee on Taxonomy of Viruses. Springer, Wien New York, pp 359–363 (Arch Virol [Suppl] 10)
13. Cubitt WD (1994) Caliciviruses. In: Kapikian AZ (ed) Viral infections of the gastrointestinal tract. Marcel Dekker, New York, pp 549–568
14. Dingle KE, Lambden PR, Caul EO, Clarke IN (1995) Human enteric *Caliciviridae*: the complete genome sequence and expression of virus-like particles from a genetic group II small round structured virus. J Gen Virol 76: 2 349–2 355
15. Dinulos MB, Matson DO (1994) Recent developments with human caliciviruses. Pediatr Infect Dis J 13: 998–1 003
16. Dolin R, Blacklow NR, DuPont H, Formal S, Buscho RF, Kasel JA, Chames RP, Hornick R, Chanock RM (1971) Transmission of acute infectious nonbacterial gastroenteritis to volunteers by oral administration of stool filtrates. J Infect Dis 123: 307–312
17. Estes MK, Hardy ME (1995) Norwalk virus and other enteric caliciviruses. In: Blaser MJ, Smith PD, Ravdin JI, Greenberg HB, Guerrant RL (eds) Infections of the gastrointestinal tract. Raven Press, New York, pp 1 009–1 034
18. Gordan I, Ingraham HS, Korns RF (1947) Transmission of epidemic gastroenteritis to human volunteers by oral administration of fecal filtrates. J Exp Med 86: 409–422
19. Greenberg HB, Skaar M, Monroe SS (1990) The 22 to 30 nm gastroenteritis agents of man. In: Saif LJ, Theil KW (eds) Viral diarrheas of man and animals. CRC Press, Boca Raton, pp 137–159
20. Greenberg HB, Matsui SM (1992) Astroviruses and caliciviruses: emerging enteric pathogens. Infect Agents Dis 1: 71–91
21. Hedberg CW, Osterholm MT (1993) Outbreaks of food-borne and waterborne viral gastroenteritis. Clin Microbiol Rev 6: 199–210

22. Herbert TP, Brierly I, Brown TDK (1996) Detection of the ORF3 polypeptide of feline calicivirus in infected cells and evidence for its expression from a single, functionally bicistronic, subgenomic mRNA. J Gen Virol 77: 123–127
23. Humphrey TJ, Cruickshank JG, Cubitt WD (1984) An outbreak of calicivirus associated gastroenteritis in an elderly persons home. A possible zoonosis? J Hyg 92: 293–299
24. Jiang X, Graham DY, Wang K, Estes MK (1990) Norwalk virus genome cloning and characterization. Science 250: 1 580–1 583
25. Jiang X, Wang M, Graham DY, Estes MK (1992) Expression, self-assembly, and antigenicity of the Norwalk virus capsid protein. J Virol 66: 6 527–6 532
26. Jiang X, Wang M, Wang K, Estes MK (1993) Sequence and genomic organization of Norwalk virus. Virology 195: 51–61
27. Jiang X, Matson DO, Ruiz-Palacios GM, Hu J, Treanor J, Pickering LK (1995) Expression, self-assembly, and antigenicity of a Snow Mountain Agent-like calicivirus capsid protein. J Clin Microbiol 33: 1 452–1 455
28. Jordan WS, Gordan I, Dorrance WR (1953) A study of illness in a group of Cleveland families. VII. Transmission of acute nonbacterial gastroenteritis to volunteers: evidence for two different etiologic agents. J Exp Med 98: 461–475
29. Kapikian AZ, Wyatt RG, Dolin R, Thornhill TS, Kalica AR, Chanock RM (1972) Visualisation by immune electron microscopy of a 27-nm particle associated with infectious nonbacterial gastroenteritis. J Virol 10: 1 075–1 081
30. Kapikian AZ (1994) Norwalk and Norwalk-like viruses. In: Kapikian AZ (ed) Viral infections of the gastrointestinal tract. Marcel Dekker, New York, pp 471–518
31. Kapikian AZ, Estes MK, Chanock RM (1996) Norwalk Group of Viruses. In: Fields BN, Knipe DM, Howley PM, Chanock RM, Melnick JL, Monath TP, Roizman B, Strauss SE (eds) Fields Virology. Lippincott-Raven, Philadelphia, pp 783–810
32. Koonin EV, Gorbalenya AE, Purdy MA, Rozanov MN, Reyes GR, Bradley DW (1992) Computer-assisted assignment of functional domains in the non-structural polyprotein of hepatitis E virus: delineation of an additional group of positive-strand RNA plant and animal viruses. Proc Natl Acad Sci USA 89: 8 259–8 263
33. Lambden PR, Caul EO, Ashley CR, Clarke IN (1993) Sequence and genome organization of a human small round-structured (Norwarlk-like) virus. Science 259: 516–519
34. Lambden PR, Liu BL, Clarke IN (1995) A conserved sequence motif at the 5′ terminus of the Southampton virus genome is characteristic of the *Caliciviridae*. Virus Genes 10: 149–152
35. Lambden PR, Clarke IN (1995) Genomic organization in the Caliciviridae. Trends Microbiol 3: 261–265
36. Laurent S, Vautherot J, Madelaine M, Le Gall G, Rasschaert D (1994) Recombinant rabbit hemorrhagic disease virus capsid protein expressed in baculovirus self-assembles into viruslike particles and induces protection. J Virol 68: 6 794–6 798
37. Le Gall G, Huguet S, Vende P, Vautherot J, Rasschaert D (1996) European brown hare syndrome virus: molecular cloning and sequencing of the genome. J Gen Virol 77: 1 693–1 697
38. Lew JF, Kapikian AZ, Jiang X, Estes MK, Green KY (1994) Molecular characterization and expression of the capsid protein of a Norwalk-like virus recovered from a Desert Shield troop with gastroenteritis. Virology 200: 319–325
39. Liu BL, Clarke IN, Caul EO, Lambden PR (1995) Human enteric caliciviruses have a unique genome structure and are distinct from the Norwalk-like viruses. Arch Virol 140: 1 345–1 356

39a. Liu BL, Clarke IN, Caul EO, Lambden PR (1997) The genomic 5′ terminus of Manchester calicivirus. Virus Genes (in press)
40. Liu SJ, Xue HP, Pu BQ, Qian NH (1984) A new viral disease in rabbits. Anim Husb Vet Med 16: 253–255
41. Madeley CR, Cosgrove BP (1976) Caliciviruses in man. Lancet i: 199–200
42. Matsui SM, Kim JP, Greenberg HB, Su W, Sun Q, Johnson PC, Dupont HL, Oshiro LS, Reyes GR (1991) The isolation and characterization of a Norwalk virus-specific cDNA. J Clin Invest 87: 1456–1461
43. Meyers G, Wirblich C, Thiel H (1991a) Genomic and subgenomic RNAs of rabbit hemorrhagic disease virus are both protein-linked and packaged into particles. Virology 184: 677–686
44. Meyers G, Wirblich C, Thiel H (1991b) Rabbit hemorrhagic disease virus-molecular cloning and nucleotide sequencing of a calicivirus genome. Virology 184: 664–676
45. Nagesha HS, Wang LF, Hyatt AD, Morrissy CJ, Lenghaus C, Westbury HA (1995) Self-assembly, antigenicity, and immunogenicity of the rabbit haemorrhagic disease virus (Czechoslovakian strain V-351) capsid protein expressed in baculovirus. Arch Virol 140: 1095–1108
46. Neill JD (1990) Nucleotide sequence of a region of the feline calicivirus genome which encodes picornavirus-like RNA-dependent RNA-polymerase, cysteine protease and 2C polypeptides. Virus Res 17: 145–160
47. Neill JD, Reardon IM, Heinrikson RL (1991) Nucleotide sequence and expression of the capsid protein gene of feline calicivirus. J Virol 65: 5440–5447
48. Neill JD, Meyer RF, Seal BS (1995) Genetic relatedness of the Caliciviruses: San Miguel sea lion and Vesicular exanthema of swine viruses constitute a single genotype within the *Caliciviridae*. J Virol 69: 4481–4488
49. Oshikamo R, Tohya Y, Kawaguchi Y, Tomonaga K, Maeda K, Takeda N, Utagawa E, Kai C, Mikami T (1994) The molecular cloning and sequence of an open reading frame encoding for non-structural proteins of feline calicivirus F4 strain isolated in Japan. J Vet Med Sci 56: 1093–1099
50. Prasad BVV, Matson DO, Smith AW (1994a) Three-dimensional structure of calicivirus. J Mol Biol 240: 256–264
51. Prasad BVV, Rothnagel R, Jiang X, Estes MK (1994b) Three-dimensional structure of baculovirus-expressed Norwalk virus capsids. J Virol 68: 5117–5125
52. Reimann HA, Price AH, Hodges JA (1945) The cause of epidemic diarrhea, nausea and vomiting (viral dysentery?). Proc Soc Exp Biol Med 59: 8–9
53. Sabine M, Hyne RHJ (1970) Isolation of a feline picornavirus from cheetahs with conjunctivitis and glossitis. Vet Rec 87: 794–796
54. Smith AW, Madin SH, Vedros NA, Bankowski RA (1977) Host range comparisons of five serotypes of caliciviruses. Am J Vet Res 38: 101–105
55. Smith AW, Prato C, Skilling DE (1978) Caliciviruses infecting monkeys and possibly man. Am J Vet Res 39: 287–289
56. Smith AW, Skilling DE, Dardiri AH, Latham AB (1980) Calicivirus pathogenic for swine: A new serotype isolated from opaleye Girella nigricans, an ocean fish. Science 209: 940–941
57. Smith AW (1981) Marine reservoirs for caliciviruses. In: Beran Gw (ed) Viral zoonoses. CRC Press, Boca Raton, pp 182–190
58. Soledad Marin M, Martin Alonso JM, Perez Ordoyo Garcia LI, Antonio Boga J, Arguello-Villares JL, Casais R, Venugopal K, Jiang W, Gould EA, Parra F (1995) Immunogenic properties of rabbit haemorrhagic disease virus structural protein VP60 expressed by a recombinant baculovirus: an effective vaccine. Virus Res 39: 119–128

59. Sosnovtsev S, Green KY (1995) RNA transcripts derived from a cloned full-length copy of the feline calicivirus genome do not require VpG for infectivity. Virology 210: 383–390
60. Taniguchi K, Urasawa S, Urasawa T (1979) Virus-like particles, 35 to 40 nm, associated with an institutional outbreak of acute gastroenteritis in adults. J Clin Microbiol 10: 730–736
61. Tohya Y, Taniguchi Y, Takahashi E, Utagawa E, Takeda N, Miyamura K, Yamazaki S, Mikami T (1991) Sequence analysis of the 3′-end of feline calicivirus genome. Virology 183: 810–814
62. Tribe DE (1996) Run rabbit run, from calicivirus disease. Aust Biotechnol 5: 335–336
63. Wyatt RG, Greenberg HB, Dalgard DW, Allen WP, Sly DL, Thornhill TS, Chanock RM, Kapikian AZ (1978) Experimental infection of chimpanzees with the Norwalk agent of epidemic viral gastroenteritis. J Med Virol 2: 89–96

Authors' address: Dr. I. N. Clarke, Molecular Microbiology Group, University Medical School, Southampton General Hospital, Southampton SO16 6YD, U.K.

The role of human caliciviruses in epidemic gastroenteritis

K. Y. Green

Laboratory of Infectious Diseases, National Institute of Allergy and Infectious Diseases, National Institutes of Health, Bethesda, Maryland, U.S.A

Summary. Members of the *Caliciviridae* family of small, positive-sense RNA viruses exhibit a broad host range. The Norwalk and Norwalk-like caliciviruses in this family are major etiologic agents of epidemic gastroenteritis in humans. This illness characteristically lasts 24–48 h and often occurs in group settings such as families, schools, institutions, or communities. The spread of the human caliciviruses is considered to be predominantly by person-to-person contact via the fecal-oral-route. However, the ingestion of calicivirus-contaminated food or water can result in large-scale common-source outbreaks. Many basic features concerning the biology and replication of the human caliciviruses are not known because they have not yet been grown in cell culture and the virus does not appear to replicate in animal models other than the chimpanzee. Sequence analysis of RT-PCR-generated DNA fragments derived from serotypically distinct reference strains (such as the Norwalk, Hawaii, and Snow Mountain viruses) and other circulating strains associated with gastroenteritis has provided evidence for marked genetic diversity among these viruses. Moreover, analysis of the antigenic relationships among these viruses using paired sera from individuals infected with well-characterized reference strains or from animals immunized with recombinant "virus-like particles" (VLPs) suggests that several serotypes of these viruses are circulating worldwide. The availability of molecular techniques for the detection of these fastidious viruses has enabled epidemiologic studies that have strengthened the association of human caliciviruses with acute gastroenteritis and has demonstrated a potential role for antigenic diversity in the natural history of these pathogens.

Introduction

Norwalk virus is the prototype strain in the *Caliciviridae* family of positive-sense single-strand RNA viruses that represents a genetically and antigenically diverse group of viruses associated with epidemic gastroenteritis in humans. The onset of this disease is characteristically acute with symptoms including nausea, diarrhea, vomiting, abdominal cramping, fever and malaise [25, 26].

Table 1. Food and water sources commonly implicated in outbreaks of the Norwalk-like human caliciviruses

Raw oysters and shellfish	Cold meats
Lettuce	Bakery products (frosting)
Fruit salad	Cooked ham
Potato salad	Sandwiches
Cole slaw	Water (drinking, swimming pools)
Melon	Commercial ice
Celery	

Compiled from reviewed literature in [22]

The illness is usually self-limited and lasts approximately 24–48 h. In certain settings, this disease may impose a special burden such as in the elderly who are at increased risk for hospitalization and mortality from diarrheal illness [6, 32]. The spread of these viruses can occur by person-to-person contact through the fecal-oral route [2, 25, 26] and several studies have reported evidence for airborne transmission [3, 7]. The highly infectious nature of these agents can result in sharp outbreaks under conditions where individuals are in close contact, such as family, community, school, hospital, cruise ship, recreational camp, military, and institutional settings [25]. However, an important aspect of the epidemiology of these viruses is their efficient transmission by common-source exposure via contaminated food or water that can result in large-scale outbreaks. Some common food and water sources associated with Norwalk virus gastroenteritis are shown in Table 1. Of special note is the strong association of acute gastroenteritis with the ingestion of raw or under-cooked shellfish [2, 53]. Fifty per cent of Norwalk virus-confirmed gastroenteritis outbreaks reported to the Centers for Disease Control (U.S.) from 1976–1980 were linked to shellfish [26]. Taken together, the economic loss due to the morbidity from epidemic gastroenteritis is undoubtedly high, but estimates of the yearly costs for medical treatment, reduced work productivity, and outbreak management are not available. There is presently no evidence that zoonotic transmission plays a role in the epidemiology of the human caliciviruses.

Research obstacles in the study of human caliciviruses

Extensive efforts have been made to grow the human caliciviruses in cell culture since the discovery of Norwalk virus in 1972 [23]. However, all attempts to cultivate these viruses have failed, and many basic features of the human caliciviruses remain poorly understood. Viruses in the *Caliciviridae* exhibit a broad host range and much of what is known about calicivirus replication has been learned from the study of animal caliciviruses such as feline calicivirus (FCV), San Miguel sea lion virus (SMSV), and rabbit hemorrhagic disease virus (RHDV). In our laboratory, we have used FCV as a model for the study of mechanisms responsible for the strong growth restriction of FCV in nonpermissive cells. Feline calicivirus grows efficiently in permissive feline kidney cells and exhibits a characteristic cytopathic effect (CPE) that often appears within 24 h following infection [51]. Early studies demonstrated that FCV, in general, showed a strong restriction for growth in animal cell lines that were not of feline origin [31, 51]. We recently analyzed cell cultures of human origin such as foreskin, embryonic kidney, intestinal, Caco-2, and 293 in order to determine whether such cells could support efficient FCV growth and replication. None of these cells exhibited characteristic FCV CPE when inoculated with virus that was infectious in feline kidney cells. Furthermore, examination of the human cells by immunofluorescence using a hyperimmune serum prepared against FCV showed no evidence for virus growth (data not shown). Consistent with the early studies by others, cells appeared to be either permissive or nonpermissive for the growth of FCV. Feline calicivirus studies by Kreutz et al. [27] suggested that the cellular receptor may play an important role in such restriction for growth. We recently developed a recombinant DNA-based infectious RNA system for FCV that allows direct manipulation of the viral RNA genome [50] and we are using this system in efforts to map genetic determinants in FCV that may be involved in cell tropism. These studies may give insight into the development of experimental approaches for the growth of human caliciviruses in cultured cells.

Several attempts have been made to identify a practical animal model for the study of human calicivirus infection and disease. Animals that have been challenged with an infectious bacteria-free stool filtrate containing Norwalk virus include: rabbits, mice, monkeys, cats, calves, guinea pigs, marmosets, baboons and chimpanzees [25]. We recently failed to find evidence for Norwalk virus infection following oral challenge of cotton rats (unpubl. studies). Of the animals tested, only chimpanzees have shown evidence for infection with Norwalk virus as determined by the demonstration of a serologic response to the virus following oral challenge [56]. However, the chimpanzees did not develop symptoms of gastroenteritis. Virus particles were not observed in the stool material of infected chimpanzees in these early studies, but a 33K soluble antigen was detected in the stool by immunoprecipitation of radioiodinated viral protein [14]. We recently observed virus-like particles morphologically similar to Norwalk virus in the stool material from one of these animals in a screening test with immune serum globulins using the technique of immune electron microscopy (IEM) (Fig. 1). This observation suggested that chimpanzees may support full viral replication. The failure to readily infect many different animal species with Norwalk virus suggests a strong species tropism for the human caliciviruses. However, extensive seroepidemiologic studies in wild and domestic animals for evidence of infection with the human caliciviruses have not been conducted. Such studies will be important in the identification of a possible role, if any, for zoonotic transmission of the human caliciviruses.

A major impediment in the elucidation of the natural history of the human caliciviruses has been the inability to establish serotypic differences by viral

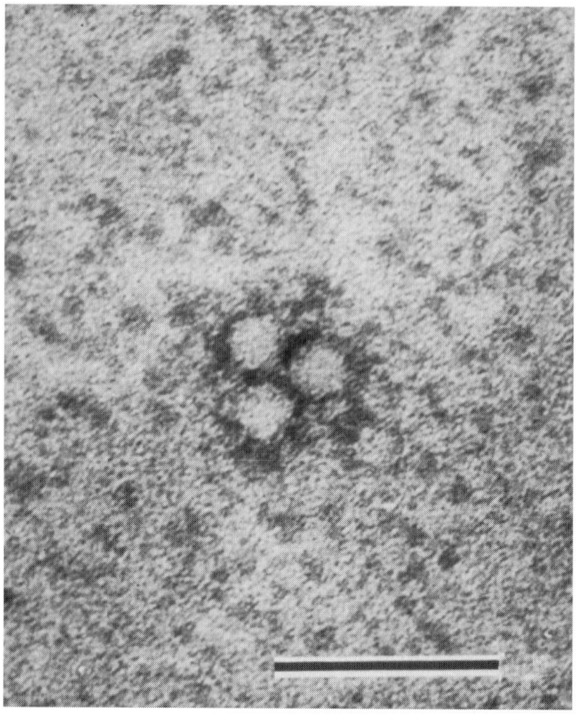

Fig. 1. Virus particles (27–32 nm) in the stool material of a chimpanzee that underwent challenge with a stool filtrate containing Norwalk virus by the alimentary route. The chimpanzee is the only known animal model for human calicivirus infection. Bar: 100 nm

neutralization assays. Alternative approaches that do not require cell culture propagation have been used for the identification of antigenic differences among strains. Initially, adult human volunteer cross-challenge studies were conducted that examined antigenic relationships. Studies such as these identified the Norwalk and Hawaii viruses as antigenically distinct because infection with one virus did not induce protective immunity against the other [55]. In addition, adult volunteer challenge studies were the major source of virus-positive stool specimens that were used in the development of the first-generation diagnostic assays and in the subsequent cloning of the Norwalk virus genome [14, 16, 55]. The volunteer studies also provided pre- and post-challenge infection sera. These sera were used to establish antigenic relationships among selected strains by IEM, which has been the "gold-standard" technique [24]. In this technique, specific antibodies, if present in a serum sample, react with surface viral antigens, resulting in virus particles coated with immunoglobulin molecules. The reactivity of antibodies with virus is scored by EM and given a visual rating on a scale from 0 to 4. The IEM technique identified Norwalk, Hawaii, Snow Mountain and Taunton viruses as "serotypically" distinct. A numbering system for these four putative serotypes of human caliciviruses has been proposed by Lambden and colleagues in which these viruses represent serotypes 1, 2, 3, and 4, respectively, based on their historical precedence [26, 29]. Classification systems have been proposed also by researchers in the U.K. and Japan that are based on IEM or solid phase IEM (SPIEM) [37, 45]. Because these techniques required the use of an electron microscope along with limited reagents from outbreaks or volunteer studies, they were not practical for widespread use in most diagnostic laboratories.

Recent molecular biologic approaches for study of the human caliciviruses

The recent molecular characterization of Norwalk virus and other human calicivirus genomes has allowed the development of molecular biologic approaches for the study of these viruses [5, 16, 20, 28, 39, 42]. Sequence analysis of reverse transcriptase (RT)-PCR generated DNA fragments derived from the RNA-dependent RNA polymerase region of the RNA viral genome is now a widely used method for diagnosis and examination of diversity among circulating human calicivirus strains [1, 8, 42, 54]. Strains associated with epidemic gastroenteritis presently segregate into two major genetic groups, or "genogroups", represented by Norwalk virus (Genogroup I) and Snow Mountain virus (Genogroup II) based on sequence analysis of the polymerase region (reviewed in [30]). These viruses have been described as the "small round structured viruses" (SRSVs) because of their morphological appearance by EM or IEM in which they characteristically lack the well-defined "classical" cup-like surface structure common to many animal caliciviruses, the first caliciviruses discovered and from which the name (calyx=cup) originated. An additional genetically distinct group of human caliciviruses were originally designated as "classical" human enteric caliciviruses because of their striking morphological similarity to the animal caliciviruses by EM. The role of the "classical" human caliciviruses in gastrointestinal disease has not been fully elucidated, but they have characteristically been associated with mild gastroenteritis in infants and young children [21, 22, 40, 41]. Sequence analysis of the polymerase region of the Sapporo virus and the entire genome of the Manchester virus (representative strains with "classical" calicivirus morphology) indicated that the actually share a closer genetic relationship to the animal caliciviruses and show major differences in genome organization when compared to the human "SRSVs" [29, 39, 41]. The sequence analysis of these "classical" human caliciviruses suggests that interspecies transmission among the caliciviruses may have occurred at a distant time point. However, further studies will be needed to fully delineate the genetic and evolutionary relationships among the human and animal caliciviruses.

The antigenic structure of a calicivirus is constructed from multiple copies of a major structural protein that forms the viral capsid. The diagnostic technique of RT-PCR presently focuses primarily on analysis of the conserved RNA polymerase region of the human calicivirus genome and does not give conclusive information on the antigenic specificity of the virus. Thus, considerable effort has been directed toward the molecular characterization of the gene encoding the capsid protein from various human calicivirus strains. Capsid sequence comparisons of "SRSVs" representing Genogroups I and II have shown that these human caliciviruses segregate into two genetic groups that are analogous to those determined by analysis of the RNA polymerase region [55, 12, 35, 54] (Fig. 2). The overall amino acid identities of capsid proteins from strains belonging to different genogroups is generally

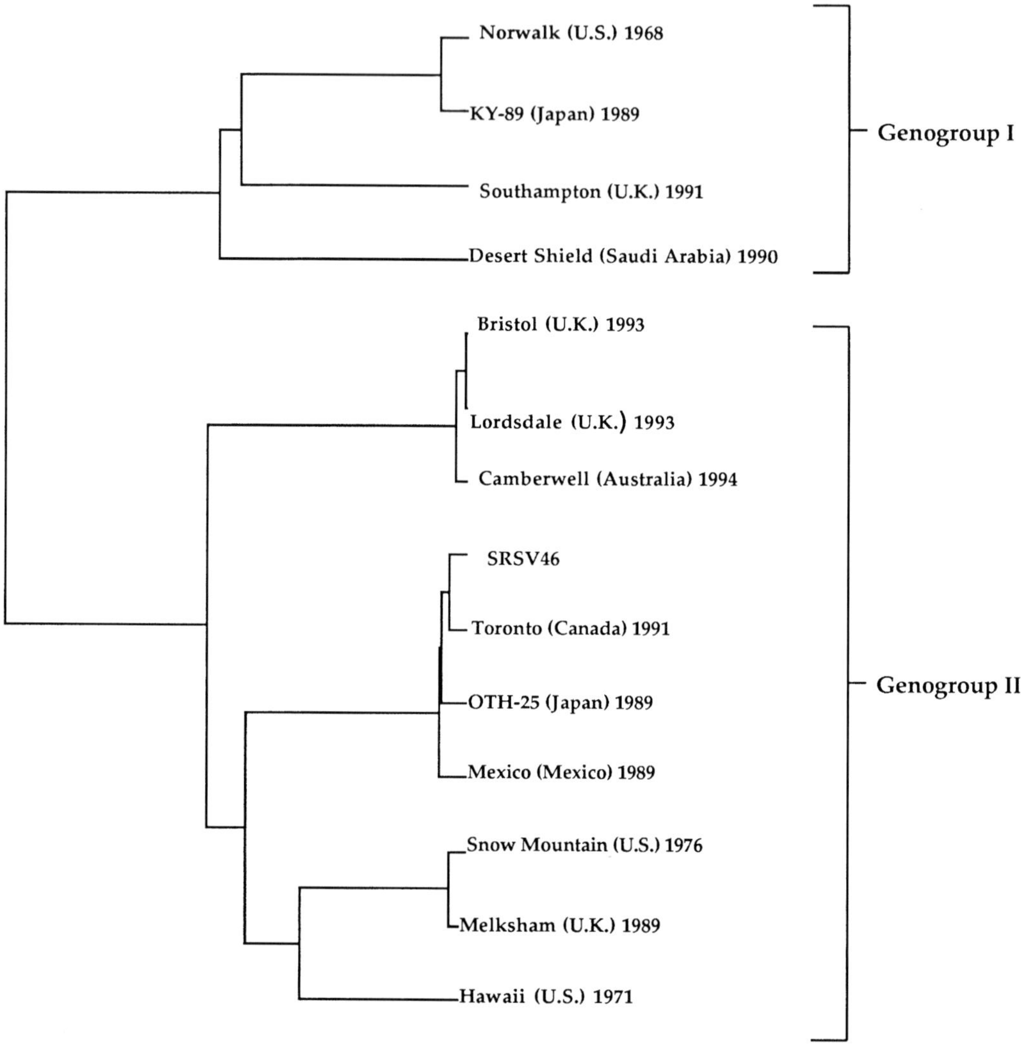

Fig. 2. Dendogram illustrating clustering relationships of capsid proteins from Norwalk and Norwalk-like "small round structured viruses" in the *Caliciviridae* associated with epidemic gastroenteritis. The distance along the horizontal axis is proportional to the difference between sequences; the vertical axis has no significance. The capsid proteins of these viruses (often named according to the location of the original outbreak) segregate into two major genetic groups designated Genogroup I and Genogroup II. Fourteen sequences were available at the time of this analysis and the Genbank accession numbers were as follows: Norwalk (M87661), KY-86 (L23828), Southampton (L07418), Desert Shield (U04469), Bristol (X76716), Lordsdale (X86557), Camberwell (U46500), SRSV46 (U46039), Toronto 24 (U02030), OTH25 (L23830), Mexico (U22498), Snow Mountain (U70059), Melksham (X81879), and Hawaii (U07611). The PILEUP program of GCG was used to generate the dendogram [4]

less than 50% in pairwise comparisons. Among the capsid sequences determined thus far, amino acid identities between strains within a genogroup range from approximately 62–99.8%. The genetic diversity within the capsid protein sequences among the human caliciviruses is greater than that observed within the polymerase region, undoubtedly reflecting antigenic diversity. A major question in the natural history of these viruses is whether this genetic diversity in the capsid protein reflects epidemiologically important antigenic diversity.

An important advance in the development of diagnostic reagents for the noncultivatable human caliciviruses was the finding by Jiang et al. [19] that expression of a recombinant Norwalk virus (rNV) capsid protein in the baculovirus system resulted in the production of self-assembled virus-like particles (VLPs). These rNV VLPs are antigenically indistinguishable from native Norwalk virions present in human stool specimens when used as the antigen in techniques such as enzyme-linked immunosorbent assay (ELISA) or IEM [9, 19, 46, 48]. We have developed rVLPs in our laboratory for the Toronto (TV), Hawaii (HV), and Desert Shield (DSV) viruses derived from gastroenteritis episodes that occurred in a hospital, family, and military setting, respectively [10, 33, 34, 38, 52] Lew, in prep.). Hyperimmune sera specific for each of the VLPs were raised in guinea pigs in order to examine the antigenic relationships among these viruses by IEM. The IEM technique was performed as described previously by Kapikian et al. [24]. Briefly, 0.2 ml of phosphate buffered saline (PBS) containing a 1:10 dilution of pre or post-guinea pig hyperimmune serum was mixed with 0.8 ml of culture fluid removed directly from recombinant baculovirus-infected insect cells five days post-infection. The antibody and VLP-containing culture fluid mixture was incubated at room temperature for 1 h, after which the mixture was pelleted by centrifugation at 35 000 × **g** for 90 min. The pellet was resuspended in distilled water and stained with 3% phosphotungstic acid. Experiments were performed under code. A difference between the pre- and post-antibody rating of a 1 or more was considered a significant serologic response that reflected serotypic relatedness with the immunizing VLP. Our first goal was to determine whether we could find evidence for serotypic diversity among viruses belonging to the same genogroup using the IEM technique and the rVLPs. The capsid protein of the Toronto virus (TV24) shares 76% amino acid identity with the Hawaii virus capsid and segregates into the same genogroup (II) as the Hawaii virus by sequence analysis of the polymerase region in which they share 90% identity [34, 35]. Paired sera from a guinea pig immunized with rHV VLPs or rTV VLPs were examined for a serologic response to the rHV VLPs and rTV VLPs by IEM (Table 2). A guinea pig immunized with rHV VLPs showed a serologic response to the rHV VLPs, but not to the rTV VLPs. The guinea pig immunized with rTV VLPs showed a serologic response to the rTV VLPs, but not show a response to the rHV VLPs. These results are consistent with the classification of these viruses as "serotypically" distinct by IEM. A similar result was seen between Norwalk virus and Desert Shield virus, both belonging to Genogroup I

Table 2. Relationship of Hawaii and Toronto recombinant virus-like particles by immune electron microscopy (IEM)

Guinea pig immunized with VLPs from	Guinea pig serum sample	IEM rating of guinea pig serum with: rHawaii VLPs	rToronto VLPs
Hawaii virus	Pre	0	0
	Post	**2–3+**	0
Toronto virus	Pre	0	0
	Post	0	**4+**

Numbers in bold represent a significant serologic response by IEM as defined in the text

(data not shown). Studies such as these will be essential in establishing an interim "serotyping" system until antigenic relationships can be established using the classical technique of neutralization.

The rVLPs for the Hawaii, Toronto, and Desert Shield viruses were used to develop ELISAs for the detection of serologic responses to each of these viruses. The assays were examined for specificity using paired sera from a chimpanzee undergoing infection with Norwalk virus or from individuals infected with either the Desert Shield virus or Hawaii virus. The specificity of each rVLP ELISA with these reference sera is shown in Table 3. The chimpanzee infected with Norwalk virus showed a serologic response to both rNV and rDSV VLPs in the ELISA and the adult infected with Desert Shield virus showed a serologic response to both of these VLPs as well. The Norwalk and Desert Shield viruses both belong to Genogroup I. However, the chimpanzee infected with Norwalk virus and the individual infected with Desert Shield virus did not show a serologic response to the Genogroup II rHV and rTV VLPs. This observation was consistent with major antigenic differences between viruses belonging to Genogroups I and II. Similarly, adult

Table 3. Serologic responses of individuals and a chimpanzee infected with representative human calicivirus strains to various recombinant virus-like particles as determinined by ELISA

Source of paired sera	Infecting virus		Antibody titer to following recombinant antigens as measured by ELISA			
			rNV (Genogroup I)	rDSV (Genogroup I)	rHV (Genogroup II)	rTV (Genogroup II)
Chimpanzee 14G	Norwalk virus	Pre	800	400	50	800
	(Genogroup I)	Post	**≥102 400**	**3200**	<200	1 600
U.S. soldier	Desert Shield 395	Pre	400	<200	3 200	800
	(Genogroup I)	Post	**6 400**	**12 800**	1 600	1 600
Adult volunteer	Hawaii virus	Pre	≥200	200	400	800
	(Genogroup II)	Post	200	200	**25 600**	<1600

Numbers in bold represent an ≥4-fold increase in antibody titer between pre and post sera. Paired sera from the individual infected with TV was not available

volunteers challenged with Hawaii virus developed a serologic response with the rHV VLPs, but not with the rNV, rDSV, or rTV VLPs. It thus appears that the VLP serologic assays are, in general, highly specific for the genogroup of the virus used to generate the VLPs [9, 10, 17, 19, 38, 48]. However, the inability of the rTV VLPs to detect evidence of infection with Hawaii virus indicates that a rVLP ELISA may not detect serologic evidence of infection with all viruses belonging to the same genogroup. The repertoire of antigens required in serologic assays for the efficient detection of infection with the multiple antigenic types of circulating human caliciviruses will require further study.

New epidemiologic studies using recombinant protein-based assays

The important role of Norwalk and Norwalk-like human caliciviruses in acute gastroenteritis was established soon after the discovery of Norwalk virus using first-generation diagnostic assays such as the radioimmunoassay [13, 14, 26]. However, these initial assays relied on the use of virus-positive stool specimens that were often limited in quantity and availability, making the diagnosis of infection with these viruses difficult for many years. Recombinant VLPs have now been developed for several antigenically distinct human calicivirus strains and diagnostic assays using these VLPs are increasing the ability to identify the etiology of human calicivirus-associated gastroenteritis outbreaks [5, 10, 15, 17, 19, 38]. For example, we recently demonstrated that Genogroup II human caliciviruses were the major viral agents associated with epidemic gastroenteritis in the elderly in a Maryland (U.S.) nursing home study using rTV and rHV ELISAs [11]. Large-scale seroprevalence studies in varying age groups have confirmed that these viruses are ubiquitous and endemic in the human population [9, 18, 36, 44, 47, 49]. However, the role of human caliciviruses as agents of pediatric diarrhea will require further study.

The human caliciviruses are the major viruses associated with food or water-borne gastroenteritis, but the "reservoir" for these viruses is not known. The availability of broadly-reactive recombinant DNA-based diagnostic assays should facilitate studies to examine the distribution and stability of these viruses in the environment. In addition, these reagents should allow the development of rapid diagnostic assays for wide-spread use by health care workers that can be applied to improved control of calicivirus-associated gastroenteritis.

Acknowledgements

I thank Dr. Albert Kapikian, Dr. Judy Lew, Dr. Stanislav Sosnovtsev, Dr. Adriene King, and Mr. Jose Valdesuso of the National Institutes of Health in Bethesda, Maryland for their active role in the work described here from our laboratory. The human foreskin, embryonic kidney, and intestinal cell cultures were generously provided by Monroe Vincent of the Uniformed Services University of the Health Sciences, Bethesda, Maryland and the rNV VLPs were a kind gift from Dr. Mary Estes, Baylor College of Medicine, Houston, Texas.

References

1. Ando T, Monroe S, Gentsch JR, Jin Q, Lewis DC, Glass, RI (1995) Detection and differentiation of antigenically distinct small-round structured viruses (Norwalk-like viruses) by reverse transcription-PCR and southern hybridization. J Clin Microbiol 33: 67–71
2. Centers for Disease Control (1990) Viral agents of gastroenteritis: public health importance and outbreak management. Morbid Mortal Weekly Rep 39 (RR-5): 1–24
3. Chadwick PR, Walker M, Rees AE (1994) Airborne transmission of a small round structured virus. Lancet 343: 171
4. Devereux J, Haeberli P, Smithies O (1984) A comprehensive set of sequence analysis programs for the VAX. Nucleic Acids Res 12: 387–395
5. Dingle KE, Lambden PR, Caul EO, Clarke IN (1995) Human enteric Caliciviridae: the complete genome sequence and expression of virus-like particles from a genetic II small round structured virus. J Gen Virol 76: 2 349–2 355
6. Gangarosa RE, Glass RI, Lew JF, Boring JR (1992) Hospitalizations involving gastroenteritis in the United States, 1985: the special burden of the disease among the elderly. Am J Epidemiol 135: 280–290
7. Gellert GA, Waterman SH, Ewert D, Oshiro L, Giles MP, Monroe SS, Gorelkin L, Glass RI (1990) An outbreak of acute gastroenteritis caused by a small round structured virus in a geriatric convalescent facility. Infect Control Hosp Epidemiol 11: 459
8. Green J, Norcott JP, Lewis D, Arnold C, Brown DW (1993) Norwalk-like viruses: demonstration of genomic diversity by polymerase chain reaction. J Clin Microbiol 31: 3 007–3 012
9. Green KY, Lew JF, Jiang X, Kapikian AZ, Estes MK (1993) Comparison of the reactivities of baculovirus-expressed recombinant Norwalk virus capsid antigen with those of the native Norwalk virus antigen in serologic assays and some epidemiologic observations. J Clin Microbiol 31: 2 185–2 191
10. Green KY, Kapikian AZ, Valdesuso J, Sosnovtsev S, Lew JF (1997) Expression and self-assembly of recombinant capsid protein from the antigenically distinct Hawaii human calicivirus. J Clin Microbiol 35: 1909–1914
11. Green KY, Taylor JL, Valdesuso J, Lew JF, Kapikian AZ, Lin F-L (1997) Identification of Norwalk-like caliciviruses as an important cause of epidemic gastroentertitis in nursing homes for the elderly (submitted)
12. Green SM, Lambden PR, Caul EO, Ashley CR, Clarke IN (1995) Capsid diversity in small round-structured viruses: molecular characterization of an antigenically distinct human enteric calicivirus. Virus Res 37: 271–283
13. Greenberg HB, Valdesuso J, Yolken RH, Gangarosa E, Gary W, Wyatt RG, Konno T, Suzuki H, Chanock RM, Kapikian AZ (1979) Role of Norwalk virus in outbreaks of nonbacterial gastroenteritis. J Infect Dis 139: 564–568
14. Greenberg HB, Wyatt RG, Kalica AR, Yolken RH, Black R, Kapikian AZ, Chanock RM (1981) New insights in viral gastroenteritis. Persp Virol 11: 163–187
15. Hyams KC, Malone JD, Kapikian AZ, Estes MK, Jiang X, Bourgeios AL, Paparello S, Hawkins RD, Green KY (1993) Norwalk virus infection among Desert Storm troops. J Infect Dis 167: 986–987
16. Jiang X, Graham DY, Wang K, Estes MK (1990) Norwalk virus genome: cloning and characterization. Science 250: 1 580–1 583.
17. Jiang X, Matson DO, Ruiz-Palacios GM, Hu J, Treanor J, Pickering LK (1995) Expression, self-assembly, and antigenicity of a Snow Mountain agent-like calicivirus capsid protein. J Clin Microbiol 33: 1 452–1 455

18. Jiang X, Matson DO, Velazquez FR, Calva JJ, Zhong WM, Hu J, Ruiz-Palacious GM, Pickering LK (1995) Study of Norwalk-related viruses in Mexican children. J Med Virol 47: 309–316
19. Jiang X, Wang M, Graham DY, Estes MK (1992) Expression, self-assembly, and antigenicity of the Norwalk virus capsid protein. J Virol 66: 6 527–6 532
20. Jiang X, Wang M, Wang K, Estes MK (1993) Sequence and genomic organization of Norwalk virus. Virology 195: 51–61
21. Kapikian AZ (1993) Viral gastroenteritis. JAMA 269: 627–630
22. Kapikian AZ (1994) Norwalk and Norwalk-like viruses. In: Kapikian AZ (ed) Viral infections of the gastrointestinal tract. Marcel Dekker, New York, pp 471–518
23. Kapikian AZ, Wyatt RG, Dolin R, Thornhill TS, Kalica AR, Chanock RM (1972) Visualization by immune electron microscopy of a 27 nm particle associated with acute infectious non-bacterial gastroenteritis. J Virol 10: 1 075–1 081
24. Kapikian AZ, Feinstone SM, Purcell RH, Wyatt RG, Thornhill TS, Kalica AR, Chanock RM (1975) Detection and identification by immune electron microscopy of fastidious agents associated with respiratory illness, acute nonbacterial gastroenteritis, and hepatitis A. Persp Virol 9: 9–47
25. Kapikian AZ, Estes MK, Chanock RM (1996) Norwalk group of viruses. In: Fields BN, Knipe DM, Howley PM (eds) Virology. Lippincott-Raven, Philadelphia, pp 783–810
26. Kaplan JE, Gary GW, Baron RC, Singh N, Schonberger LB, Feldman R, Greenberg HB (1982) Epidemiology of Norwalk gastroenteritis and the role of Norwalk virus in outbreaks of acute nonbacterial gastroenteritis. Ann Int Med 96: 756–761
27. Kreutz LC, Seal BS, Mengeling WL (1994) Early interaction of feline calicivirus with cells in culture. Arch Virol 136: 19–34
28. Lambden PR, Caul EO, Ashley CR, Clark IN (1993) Sequence and genome organization of a human small round-structured (Norwalk-like) virus. Science 259: 516–519
29. Lambden PR, Caul EO, Ashley CR, Clarke IN (1994) Human enteric caliciviruses are genetically distinct from small round structured viruses. Lancet 343: 666–667
30. Lambden PR, Clark IN (1995) Genome organization in the Caliciviridae. Trends Microbiol 3: 261–265
31. Lee KM, Kniazeff AJ, Fabricant CG, Gillespie JH (1969) Utilization of various cell culture systems for propagation of certain feline viruses and canine herpesvirus. Cornell Vet 59: 539–547
32. Lew JF, Glass RI, Gangarosa RE, Cohen IP, Bern C, Moe CL (1991) Diarrheal deaths in the United States, 1979–1987. JAMA 265: 3 280–3 284
33. Lew JF, Kapikian AZ, Jiang X, Estes M, Green KY (1994) Molecular characterization and expression of the capsid protein of a Norwalk-like virus recovered from a desert shield troop with gastroenteritis. Virology 200: 319–325
34. Lew JF, Petric M, Kapikian AZ, Jiang X, Estes M, Green KY (1994) Identification of minireovirus as a Norwalk-like virus in pediatric patients with gastroenteritis. J Virol 68: 3 391–3 396
35. Lew JF, Kapikian AZ, Valdesuso J, Green KY (1994) Molecular characterization of Hawaii virus and other Norwalk-like viruses: evidence for genetic polymorphism among human caliciviruses. J Infect Dis 170: 535–542
36. Lew JF, Valdesuso J, Vesikari T, Kapikian AZ, Jiang X, Estes MK, Green KY (1994) Detection of Norwalk virus or Norwalk-like virus infections in Finnish infants and young children. J Infect Dis 169: 1 364–1 367
37. Lewis D (1990) Three serotypes of Norwalk-like virus demonstrated by solid-phase immune electron microscopy. J Med Virol 30: 78–81

38. Liete JPG, Ando T, Noel JS, Jiang B, Humphrey C, Lew JF, Green KY, Glass RI, Monroe SS (1996) Characterization of Toronto virus capsid protein expressed in baculovirus. Arch Virol 141: 865–875
39. Liu BL, Clarke IN, Caul EO, Lambden PR (1995) Human enteric caliciviruses have a unique genome structure and are distinct from the Norwalk-like viruses. Arch Virol 140: 1 345–1 356
40. Madely CR, Cosgrove BP (1976) Caliciviruses in man. Lancet i: 199–200
41. Matson DO, Zhong W-M, Nakata S, Numata K, Jiang X, Pickering LK, Chiba S, Estes MK (1995) Molecular characterization of a human calicivirus with sequence relationships closer to animal caliciviruses than other known human caliciviruses. J Med Virol 45: 215–222
42. Matsui SM, Kim JP, Greenberg HB, Su W, Sun Q, Johnson PC, Dupont HL, Oshiro LS, Reyes GR (1991) The isolation and characterisation of a Norwalk virus-specific cDNA. J Clin Invest 87: 1 456–1 461
43. Moe CL, Gentsch J, Ando T, Grohmann G, Monroe SS, Jiang X, Wang J, Estes MK, Seto Y, Humphrey C, Stine S, Glass RI (1994) Application of PCR to detect Norwalk virus in fecal specimens from outbreaks of gastroenteritis. J Clin Microbiol 32: 642–648
44. Numata K, Nakata S, Jiang X, Estes MK, Chiba S (1994) Epidemiological study of Norwalk virus infections in Japan and Southeast Asia by enzyme-linked immunosorbent assays with Norwalk virus capsid protein produced by the baculovirus expression system. J Clin Microbiol 32: 121–126
45. Okada S, Sekine S, Ando T, Hayeshi Y, Murao M, Yabuuchi K, Miki T, Ohashi M (1990) Antigenic characterization of small, round, structured viruses by immune electron microscopy. J Clin Microbiol 28: 1 244–1 248
46. Parker S, Cubitt D, Jiang JX, Estes M (1993) Efficacy of a recombinant Norwalk virus protein enzyme immunoassay for the diagnosis of infections with Norwalk virus and other human 'candidate' caliciviruses: J Med Virol 41: 179–184
47. Parker SP, Cubitt WD, Jiang X (1995) Enzyme immunoassay using baculovirus-expressed human calicivirus (Mexico) for the measurement of IgG responses and determining its seroprevalence in London, UK. J Med Virol 46: 194–200
48. Parker SP, Cubitt WD, Jiang X, Estes MK (1993) Efficacy of a recombinant Norwalk virus protein enzyme immunoassay for the diagnosis of infection with Norwalk virus and other "candidate" caliciviruses. J Med Virol 41: 108–118
49. Parker SP, Cubitt WD, Jiang XJ, Estes MK (1994) Seroprevalence studies using a recombinant Norwalk virus protein enzyme immunoassay. J Med Virol 42: 146–150
50. Sosnovtsev S, Green KY (1995) RNA transcripts derived from a cloned full-length copy of the feline calicivirus genome do not require VPg for infectivity. Virology 210: 383–390
51. Studdert MJ, Martin MC, Peterson JE (1970) Viral diseases of the respiratory tract of cats: Isolation and properties of viruses tentatively classified as picornaviruses. Am J Vet Res 31: 1 723–1 732
52. Thornhill TS, Wyatt RG, Kalica AR, Dolin R, Chanock RM, Kapikian AZ (1977) Detection by immune electron microscopy of 26–27 mm virus-like particles associated with two family outbreaks of gastroenteritis. J Infect Dis 135: 20–27
53. Viral Gastroenteritis Sub-committee of the PHLS Virology Committee (1993) Outbreaks of gastroenteritis associated with SRSVs. PHLS Microbiol Dig 10: 2–8
54. Wang J, Jiang X, Madore P, Gray J, Desselberger U, Ando T, Seto Y, Oishi I, Lew J, Green K, Estes M (1994) Sequence diversity of small, round-structured viruses in the Norwalk virus group. J Virol 68: 5 982–5 990

55. Wyatt RG, Dolin R, Blacklow NR (1974) Comparison of three agents of acute infectious nonbacterial gastroenteritis by cross-challenge in volunteers. J Infect Dis 129: 709–714
56. Wyatt RG, Greenberg HB, Dalgard DW, Allen WP, Sly DL, Thornhill TS, Chanock RM, Kapikian AZ (1978) Experimental infection of chimpanzees with the Norwalk agent of epidemic viral gastroenteritis. J Med Virol 2: 89–96

Author's address: Dr. K. Y. Green, National Institutes of Health, Building 7, Room 129, Rockville Pike, Bethesda, MD 20892, U.S.A.

Clinical similarities and close genetic relationship of human and animal Borna disease virus

L. Bode[1] and **H. Ludwig**[2]

[1] Department of Virology, Robert Koch-Institut, Berlin, Federal Republic of Germany
[2] Institute of Virology, Free University of Berlin, Berlin, Federal Republic of Germany

Summary. Borna disease virus (BDV) is the prototype genus of a new family, *Bornaviridae*, within the order *Mononegavirales*. BDV naturally infects animals and man. The symptomatology in animals ranges from subclinical infection to rare cases of encephalitis. Asymptomatic infection seemed more frequent than expected, based on antibody data from 100 healthy horses derived from different stables with a history of diseased cases (30–40% carriers). Likewise, phasic episodes of a neurobehavioral syndrome followed by recovery were much more common than fatal neurologic disease. They were paralleled by expression of BDV antigens (N-protein p40, P-protein p24) and RNA transcripts in peripheral blood mononuclear cells, indicating viral activation. Representative longitudinal studies showed that episodes of depressive illness in humans as well as apathetic phases in infected horses were accompanied by antigen expression and followed a similar clinical course. After recovery, BDV antigen disappeared. This temporal congruence, together with the recent isolation of infectious BDV from such patients, points to a contributory role of this virus in human affective disorders. Successful amelioration of BDV-induced neurobehavioral disease in horses with antidepressants applied in psychiatry, supported a common viral pathomechanism, involving reversible disturbances of the neurotransmitter network in the limbic system. Sequences of genetic material amplified from infected animal tissue and human PBMCs revealed a close interspecies relationship and high sequence conservation of the BDV genome. In human BDV isolates, however, single unique mutations were prominent in four genes. This finding supports the hypothesis that despite of high genomic conservation, species-specific genotypes may be definable, provided the sequences are derived from RNA of infectious virus.

Introduction

Borna disease virus (BDV) is an enveloped, nonsegmented, negative- and single-stranded (NNS) RNA virus within the order *Mononegavirales* [10, 12, 31].

Analogously, its genome consists of five genes. Open reading frame (ORF) I (1 110 nucleotides [nt]) encodes the N-protein (p40), ORF II (603 nt) the P-protein (p24), ORF III (426 nt) the M-proteine (gp 16), ORF IV (1509 nt) the putative G-protein (gp 56), and ORF V (5145 nt) the L-polymerase of BDV. Based on similarities of the L-polymerase, BDV has a relatively close relationship to rabiesvirus. However, in contrast to all other animal NNS viruses, Bornavirus replicates in the nucleus of the infected cell [9, 11], a unique property, which led to its classification as the prototype of a new family (*Bornaviridae*). Extraordinary genetic features, including a complex RNA splicing machinery [29, 13], contribute to the outstanding biological properties of this infectious agent.

BDV induces a persistent infection in various cell types (brain and body) of a broad spectrum of vertebrate hosts [20]. Infection causes more or less transient functional disturbances at the emotional and behavioral level but with generally reversible impairments. In humans, meanwhile, a considerable body of evidence points to a contributory role of BDV in recurrent affective disorders [2, 7, 8]. In this paper, we present a unifying pathogenetic concept for BDV infection in animals and humans, as based on previous and recent new data. Our hypothesis elucidates the evolutionary extremely successful strategy of this agent. It addresses both worldwide prevalence and unusual genetic stability in different hosts and why BDV is likely to survive in the limbic system, in one of the most vulnerable areas of the vertebrate brain.

Materials and methods

Diagnosis of BDV infection in vivo

Anticoagulated blood samples from humans (1ml 0.11 mol/l sodiumcitrate and 9 ml blood) were separated into plasma and peripheral blood mononuclear cells (PBMCs) by density-gradient centrifugation in Ficoll-Paque (1.077) [5, 7]. For equally pretreated samples from horses, we used a Ficoll medium with higher density (1.09).

BDV antibodies were monitored by a double-stain immunofluorescence test (IF) [1]. BDV antigens in PBMCs were determined by either flow cytometry [5, 7] or (for routine use) by a double-sandwich-enzyme immuno assay (EIA) [8], using monoclonal antibodies (mabs) against N-protein p40 and P-protein p24 (W1 and Kfu2, respectively [21]). BDV nucleic acid in PBMCs (aliquot samples) was analyzed using RNA extraction and nested reverse transcriptase (RT) polymerase chain reaction (PCR) methods, as described previously [7, 8, 14]. Appropriate primers are summarized in Table 1. Routinely, primers were applied which amplify a 441-bp fragment of BDV ORF p40 (first description [32]).

Subjects

During a one-year-period, a cohort of 190 patients, hospitalized (mean duration 4 weeks) in the Psychiatric Department (Crisis Intervention Center) of the Benjamin Franklin Hospital of the Free University in Berlin, were investigated. They presented with acute mental disorders of all types, including recurrent major mood disorders (unipolar or bipolar type). Of a representative subgroup, a detailed description of clinical diagnoses and follow-up screening for BDV infection parameters had been given previously [8]. In this study, a longterm follow-up of BDV monitoring is presented in selected patients with a representative course of acute major depression.

Animals

During a three-year-period, approximately 500 horses, presenting with clinical signs of a neurobehavioral disorder and originating from all over Germany, were investigated by at least 2 follow-up blood samples, in parallel to acute symptoms. In this study, data of a case with a BDV induced, typical course of a disease episode are given. This case was selected because longitudinal monitoring over a 6-month-period was possible, and detailed clinical data had been provided. In addition, a hundred healthy horses from 7 different stables all over Germany, where previous cases of Borna disease had been reported, were included in our studies.

Results and discussion

Prevalence of natural Borna disease versus prevalence of BDV infection

Prior to any information on the nature of the etiologic agent, a large body of data on Borna disease (BD), a rare progressive type of non-purulent encephalomyelitis in horses and sheep, had been collected [33]. Due to the clustering of this horse disease in certain geographic areas in Saxony (East Germany) around the town Borna, its severity, and its economic importance at the end of the last century, the causative agent was called Borna disease virus (BDV). The transmissibility of BD to other mammals, especially rabbits, by intracerebral (i.c.) inoculation of brain material from diseased horses had been demonstrated early [25]. Around the turn of the century, the incidence of fatal BD of horses in Saxony was fairly high by 1% based on a total number of about 150 000 horses. Although severe cases of "classical" (full-blown) BD are still occurring up to date, their incidence has remarkedly dropped, with 0.1% in Saxony in the Sixties to $< 0.01\%$ nowadays, based on a total number of 100 000 and 20 000 horses, respectively (Dürrwald and Ludwig, unpubl.).

Verification of BD cases were solely based on clinical record and *post mortem* histology of the brain, the latter mainly founding on Joest-Degen inclusion bodies [17] and perivascular infiltrates. However, their absence does not exclude BDV etiology, as recently proven by molecular biological techniques [32]. Likewise, absence of BDV serum antibodies, the only available *intra vitam* diagnostic parameter till 1994, does not exclude BDV infection. In contrast to experimentally infected animals (rabbits, rats) [20, 21] which usually develop high persisting serum antibody titers (IF) of $> 1:1\,000$ to $1:10\,000$, natural BDV infections, in general, elicit only low mean IF antibody levels of 1 : 20 to 1 : 80 (range 1 : 10 to 1 : 320) which do not persist. Even in horses with fatal acute BD, as later confirmed by nested RT-PCR amplification of BDV RNA in brain material, only one third presented with serum antibodies, one week prior to death [32]. Moreover, in all naturally infected animals (horses, sheep, cattle, cats), development of BDV antibodies is independent of whether or not clinical symptoms had been reported [4, 21–23].

In humans, a low percentage of antibody carriers (1–2%) is known in the normal population ([1, 27], review: [6]). In German horses, previous seroepidemiological investigations revealed a prevalence of at least 10% in healthy animals [19]. Our recent studies in 7 stables (total number: 100 horses) with a history of typical clinical cases of BD, came up with 30–40% of asymptomatic carriers.

These antibody studies clarify that an unexpectedly high background of subclinical BDV infections must exist in nature, and in turn, the textbook view of an always deleterious outcome of BDV infection in horses is wrong.

A significant step forward towards re-evaluation of natural BDV infections was possible by the recent discovery of direct *intra vitam* infection markers, namely BDV antigens in PBMCs. Antigen positive cells were first identified by flow cytometry ([15], review: [6]) and later routinely detected by an improved EIA technique [8, 23], both methods using mabs against N- and P-proteins of BDV [21]. Meanwhile, nucleic acid was detected in PBMCs, particularly from psychiatric patients, by our group [7] and others [18, 28], and human isolates were achieved [8].

The significance of the peripheral infection markers, especially of antigen, became evident by follow-up studies in psychiatric patients [7, 8] as well as diseased horses [3]. Expression of BDV proteins in PBMCs frequently appeared in parallel to depressive/apathetic disease episodes, and disappeared during recovery. These observations strongly suggested that persistent BDV infection alternates between "latent" and activated phases, the latter recognizable by the presence of antigens/transcripts in PBMCs. This, together with the observation of antibodies following antigen appearance, means that reliable diagnostic statements depend upon the time point of sampling, and the investigation of both antigens/and or transcripts and antibodies. As demonstrated below by a few representative case reports, longitudinal monitoring is essential to evaluate the relationship between BDV infection and disease.

Clinical significance of BDV infection in humans and animals

Since the first report on BDV antibodies in psychiatric patients [26], the question arose whether BDV plays a role in certain mental disorders which resemble virus-induced neurobehavioral syndromes in animals [15, 24]. Although seroepidemiological studies indicated higher antibody prevalence in neuropsychiatric than in other patients or healthy controls [6], these point prevalence analyses (one sample/patient) did not allow any conclusions on the potential clinical significance of human BDV infections to be made. Our concept of follow-up studies in acute psychiatric patients revealed seroconversion and significantly higher antibody rates in patients with major depressive illness than with other disorders [2], and highest frequencies of BDV antigen expression in PBMCs of such patients [5, 7, 8].

Figure 1 illustrates two representative human patients. In *case 1* (Fig. 1, upper graph), a 73-year old female patient was hospitalized 8 weeks after onset of severe major depression (DSM-III-R: 296.33) [7], patient III). Between the

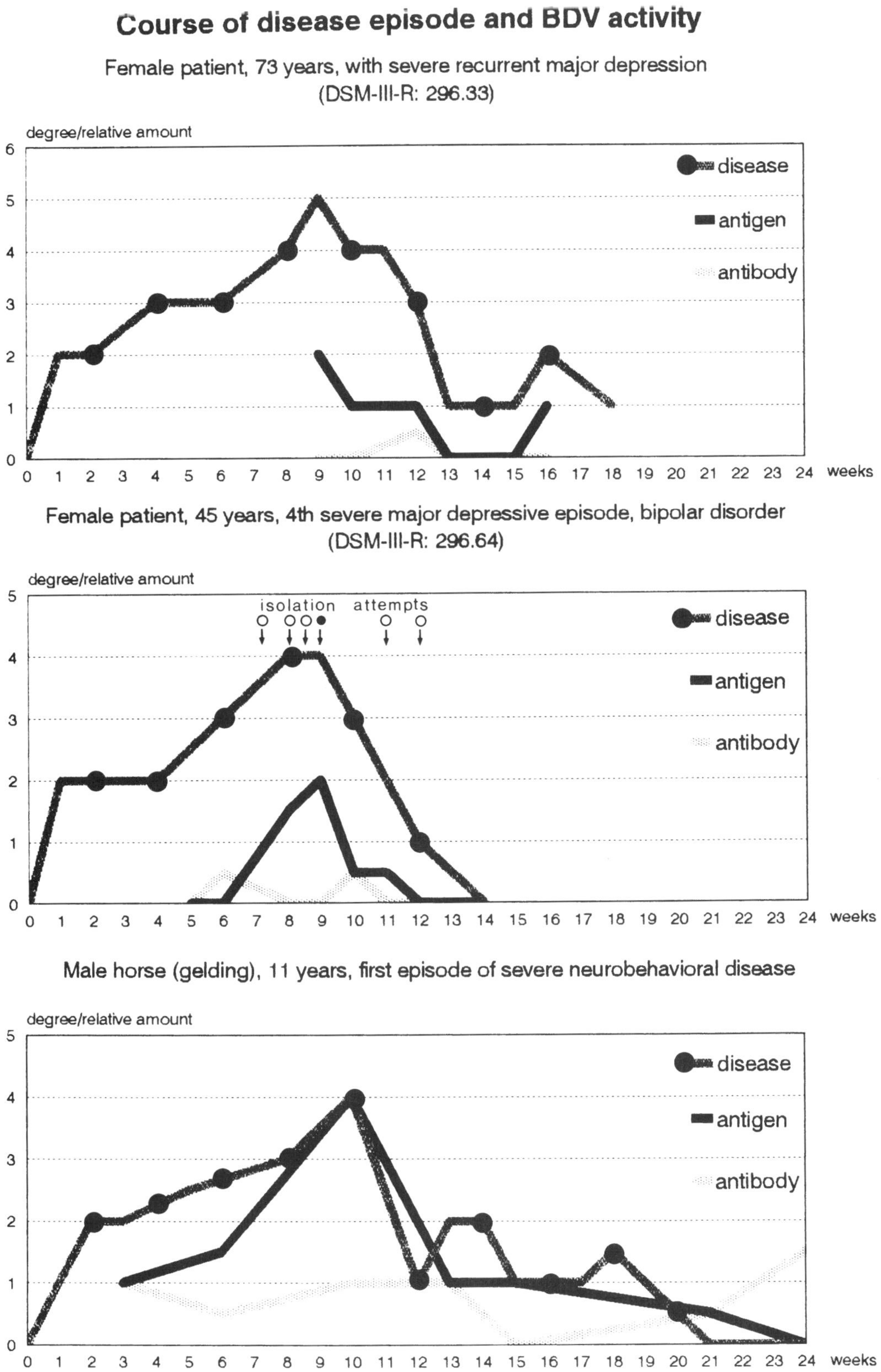

Fig. 1 (*caption on p. 172*)

8th and 9th week, at the peak of her illness, the first blood analysis yielded strong BDV antigen expression, but no antibodies. Activated BDV infection lasted until the 12th week of illness. At that time, a simultaneous analysis for genomic transcripts was also positive, and low IF antibodies appeared. Parallel to a significant decline in her depressive symptoms between the 12th and 13th week, any sign of BDV activity (antigen/transcripts) disappeared. A clinically observed short relapse of depressive symptoms in the 16th week, was again accompained by antigen positivity in PBMCs.

Case 2 (Fig. 1, central graph), a 45-year-old female patient, was hospitalized 4 weeks after onset of her 4th depressive episode ([8], patient H1). Her blood samples taken in the 5th, 6th and 7th week of illness were antigen-negative, but a low antibody titer (6th week) indicated the infection. Between the 8th and 9th week of illness, at the peak of depressive symptoms, high BDV activity was measured. Moreover, infectious virus could be isolated from the PBMC sample taken during the 9th week, after co-cultivation and longterm passaging in a human oligodendroglia cell line ([8], isolate BDV-Hu-H1), whereas isolation attempts with samples taken prior or later were unsuccessful. At the 10th week of illness, antigen expression already declined, but BDV nucleic acid remained and antibodies could be shown. As in case 1, her depression considerably improved around the 12th week, when no BDV activity was present. Since this patient experienced her first manic episode six months after recovery from depression, she was reclassified as having a bipolar illness (DSM-III-R: 296.64).

These representative cases did not only point to a contribution of activated BDV infection to affective disorder, but also documented the importance of an adequate investigation period. From our experience, multiple samples of a single patient are a prerequisite for any meaningful evaluation of the relationship between disease and virus infection.

Fig. 1 (p. 171). Temporal sequence of clinical symptoms and BDV infection. Follow-up studies of BDV antigen (in PBMCs) and antibodies (in plasma) were done parallel to acute major depressive illness in two representative infected patients (upper graph: patient III in [7]; central graph: patient H1 in [8]), in comparison to an infected horse with phasic neurobehavioral disease (lower graph). Blood samples were taken weekly between the 9th and 16th week (upper graph) and between the 4th and 14th week (central graph). The horse (lower graph) was monitored between the 3rd and 25th week in blood samples taken every third to fourth week. Clinical signs of patients according to DSM-III-R-criteria: (nearly every day) depressed mood, psychomotor retardation, apathy, loss of interest, hypersomnia, loss of energy, indecisiveness, diminished ability to concentrate; clinical symptomatology in the horse (scoring by the owner): degree 1 and 2: body weight loss, mild hindlimb ataxia, hyperesthesia, headshaking, wobbling; degree 3 and 4: abnormal eating behavior (intake of stones or cat-food; switching between increased/decreased amounts of food and water), strong ataxia with straddle-legged standing, wobbling, frequent yawning, severe depression, inappetence, unresponsiveness, "stupid look"; sudden improvement in the 12th week within a few days. Note: the owner's intensive daily care most probably helped to prevent recumbence and further progression, especially of the pronounced "depressive" symptoms

In contrast to human infections, there is no doubt about the etiological role of BDV infection in behavioral and neurological disorders of animals. However, in natural infections, this causative relationship was only established in terms of acute BD with fatal outcome. Our recent investigations in horses revealed that phasic neurobehavioral disorder (resembling human affective disorder) followed by complete recovery, is more common than the rare fatal outcome. As in humans, longitudinal observation on virus activity, parallel to disease, is useful and of prognostic value [3].

This is documented in detail by a representative *case 3* (Fig. 1, lower graph). An 11-year old Hanoverian horse (gelding) presented with initial symptoms of body weight loss, mild ataxia, wobbling, and headshaking. Three weeks after onset of disease, antigen expression and BDV antibodies were present (first available blood sample). Between the 6th and 11th week, the symptomatology worsened. The horse showed severe ataxia and straddle-legged standing (typical for BD, Fig. 2b), pronounced wobbling, abnormal eating behavior, deep apathy (depression) (Fig. 2a), unresponsiveness, loss of interest, and sensitivity to light and sound. Recumbence could probably be prevented by daily intensive care and emotional support provided by the owner. In parallel to disease progression, BDV antigen in PBMCs increased with a peak in the 10th week. In the 12th week, the owner reported a sudden considerable improvement in clinical symptomatology. This was accompanied by a significant decline of BDV antigen in PBMCs (13th week). Mild clinical relapses occurred in the 14th and 18th week, but from the 21th week on, the horse had recovered completely (Fig. 2c). This process was paralleled by the disappearance of BDV antigen. In the last blood samples, 6 and 12 months after disease onset, no antigen but antibodies were present, as a remaining sign of "latent" BDV infection. In other less severe cases, a shorter period of illness and BDV activity (2–4 weeks) was observed.

A comparison between the course of affective disorder in human patients with BDV induced neurobehavioral disease in horses clearly shows a high temporal congruence (Fig. 3), with peaks of illness between the 8th and 11th week. The coincidence of illness and BDV activity in both human patients and naturally infected horses emphasizes the likelihood of a similar pathogenic role of the virus. BDV is not cytopathogenic [20]. A primary pathomechanism could be that BDV infection, once activated by unknown conditions, produces complete virus, (more likely) RNPs, or only proteins which cause functional reversible disturbances in the limbic system by interaction with neurotransmitter receptors [8]. From animal data, the glutamate system seems to be involved [16]. Any disturbances of this key system will influence the whole neurotransmitter network in the limbic system (e.g. serotonin, dopamine). This hypothesis of a reversible viral action fits with the fact that antidepressants which act by increasing the presence of (mainly) serotonin in the synaptic cleft, contribute to earlier amelioration of depressive symptoms in patients. Restoration of the neurotransmitter balance may provide a condition for reduced viral activity. The similarity between human affective disorders and

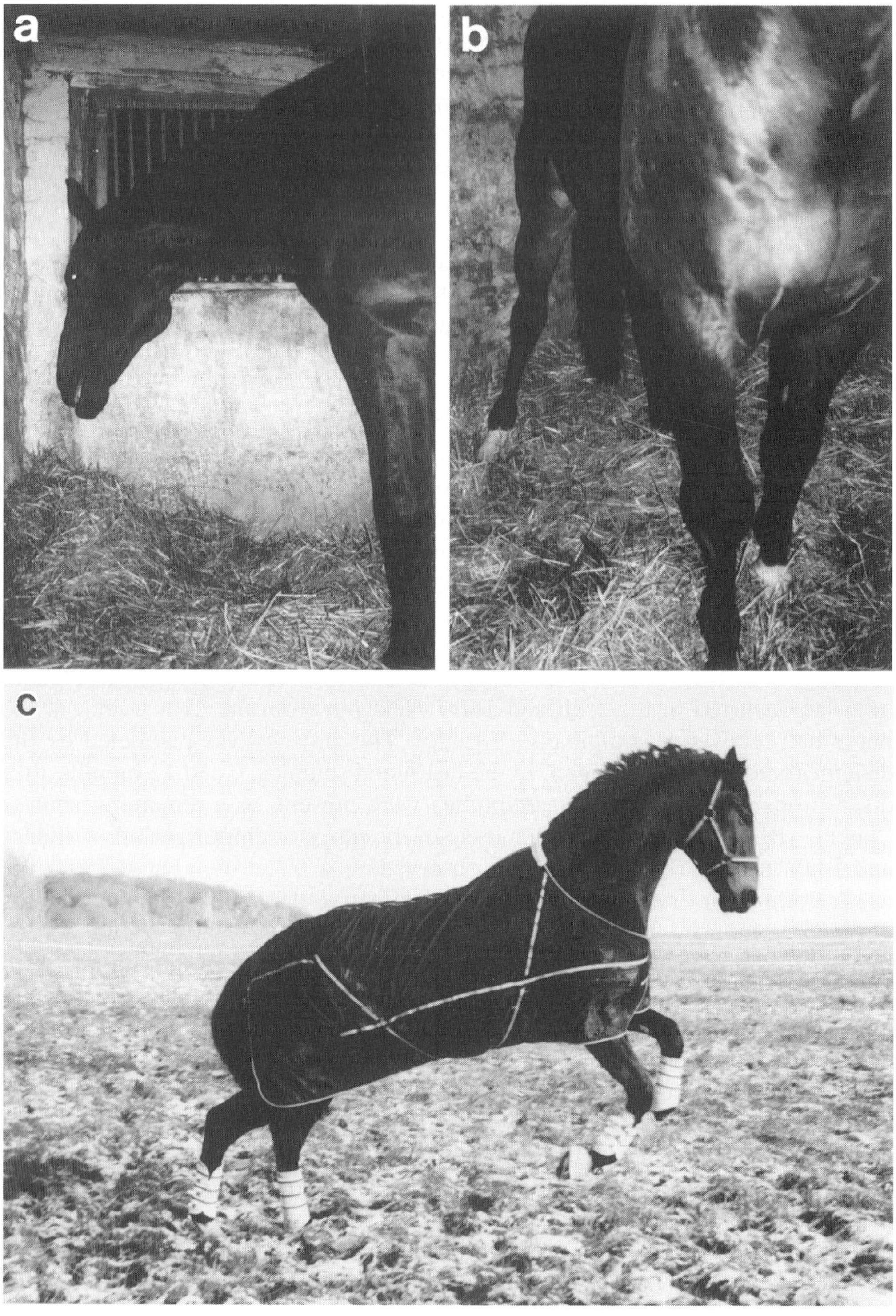

Fig. 2. Neurobehavioral Borna disease and recovery of an 11-year-old Hanoverian horse (gelding). **a** shows severe apathy (unresponsiveness, "stupid look"), and **b** ataxia (typical straddle-legged standing) in the 11th weak of illness (peak of clinical symptoms); **c** shows the same horse in the 21th week after onset of illness, completely recovered

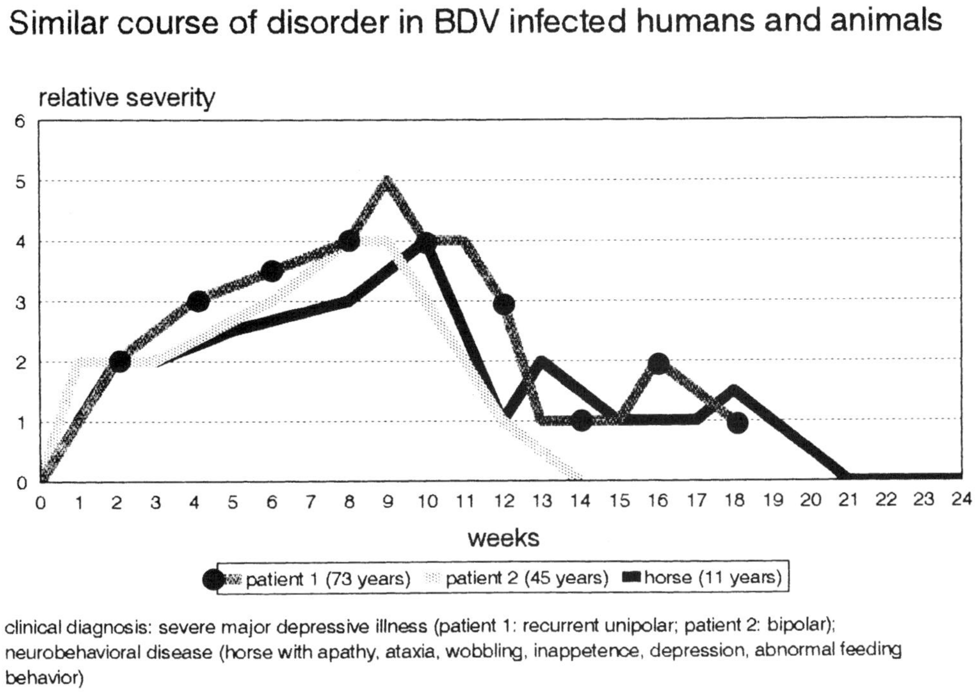

Fig. 3. Comparison of severe depressive episodes in BDV infected human patients with a typical phase of severe apathetic neurobehavioral disease in a BDV infected horse

neurobehavioral BD in horses, as shown by parallels in the clinical course and symptomatology, is further supported by successful treatment of diseased BDV infected horses with a combination of a human antidepressant (nortriptyline) and a mood stabilizer (carbamazepine) (data not shown). Although the data are still preliminary (low number of cases), the duration and severity of acute apathy accompanied by hindlimb ataxia could be ameliorated, using a dosage recommended for human patients. Even long-term maintenance medication (in a single case: 18 months) did not produce any side effects and prevented relapses in infected (stress-sensitive) animals.

In horses, development of disease or subclinical infection are likely to depend upon the individual animal's resistance rather than the virus strain, because BDV displayed a very low genetic divergence ($< 5\%$ on the nucleotide level) in different animals strains [30]. Likewise, the assumed morbidity risks in humans may be considerably different, namely low or zero in healthy people (who show almost no BDV activity), but relatively high in individuals with inherited vulnerability and stress or other factors influencing mood disorders.

Genetic relationship between human and animal strains

In 1994, two complete sequences of the BDV genome have been published [10, 12]. One sequence is originating from released infectious virions of animal reference strain V [10], whereas the other is based on ribonucleoproteins (RNPs) of another animal strain, C6BV [12]. Both strains were originally obtained from diseased horses and underwent several adaptations in other animals (rabbits, rats) and in cells. In 1996, BDV strain V [21] was officially acknowledged by the International Committee on Taxonomy of Viruses (ICTV), as prototype strain of the genus *Bornavirus* in the family *Bornaviridae*. The genomes of strain V and C6BV differ less than 5% at the nucleotide level [10, 12]. Likewise, sequence comparison (ORF I and II) of wildtype equine BDV, amplified from brain by nested RT-PCR, with these reference strains, revealed an unusually high level of conservation at the nucleotide and amino acid level (> 95% and > 98%, respectively) [30].

The first sequences (fragment of ORF I) of human BDV, amplified from PBMCs of psychiatric patients, showed a close genetic relationship to animal strain V, with a maximum divergence of 3.6% at the nucleotide level [7]. A similarly low range of divergence of human sequences with high inter- and intraspecific conservation has meanwhile been reported by others who also found a close relationship to animal BDV sequences [28]. Considerably higher sequence variations including deletions and insertions, reported for Japanese patients [18], are likely to be due (at least in part) to methodological conditions (high-error-rate polymerase) [28].

Point mutations found in a 439-bp fragment of BDV ORF I (p40) of human [7] and animal [30] derived material are compared in Table 2. They are randomly distributed, none of these changes involves an amino acid change, and in several nucleotides, the same substitutions are present between individual human patients, as well as between patients' and horse virus RNAs.

Table 1. BDV-specific primers used for nested RT-PCR amplification

Primer	Sequence	Polarity	Nucleotide positions BDV RNA genome	ORF (Gene)	Ref.
277F	5′-GCCTTGTGTTTCTATGTTTG-3′	anti-genomic	277–296	I(p40)	7, 32
717R	5′-ATTCTTTACCTGGGGACTCA-3′	genomic	717–698		
259F	5′-TTCATACAGTAACGCCCAGC-3′	anti-genomic	259–278	I(p40)	8
829R	5′-GCAACTACAGGGATTGTAAGGG-3′	genomic	829–808		
2.1	5′-CAGGAGGCTCAATGGCAACG-3′	anti-genomic	1 261–1 280	II(p24)	14
2.2	5′-TTTATGGTATGATGTCCCAC-3′	genomic	1 878–1 859		
3.1	5′-ATCGAATCACCATGAATTCAAAGC-3′	anti-genomic	1 882–1 905	III(p16)	14
3.2	5′-GTCAGTATTGCAACTAAGGC-3′	genomic	2 334–2 315		
4.4	5′-GCACGCAATTAATGCAGC-3′	anti-genomic	2 225–2 242	IV(p56)	14
4.3	5′-CGGTACGGTTTATTCCTGC-3′	genomic	3 756–3 738		
2962F	5′-AAGTTGAGAAGGCGGCGTAG-3′	anti-genomic	2 962–2 981	IV(p56)	14
3030R	5′-CAGTGTAGGCCTAAGCTTGTG-3′	genomic	3 030–3 010		

Primer sequences are based on the BDV strain V genome [10] in [7, 32], and on the C6BV genome [12] in [14]

Table 2. Mutations in a fragment of ORF I (p40) (nucleotides 226–664; amino acids 76–221) of BDV from different species

Nucleotide position		297	301	313	315	324	349	369	372	393	408	435
	Ref.	•	•	•	•	•	•	•	•	•	•	•
Animal strain V (virions)	10	TCC	CTG	GTC	GTC	GGA	AAG	ACG	ACG	CTG	ATC	ATA
Patient I (human PBMC)	7											
Patient II (human PBMC)	7							A			A	T
Patient III (human PBMC)	7	T	T		T	G			A	A		T
Hu-H1 (human isolate)	8											
Hu-H2 (human isolate)	8			A			G					
Hu-H3 (human isolate)	8			A								
WT-1 (horse brain)	30	T	T		T	G			A	A		T
Animal strain C6BV (RNPs)	12	T			T	G		A	A		A	T
Amino acid codon		99	101	105	105	108	117	123	124	131	136	145
Amino acid		Ser	Leu	Val	Val	Gly	Lys	Thr	Thr	Leu	Ile	Ile
Change		–	–	Ile	–	–	Glu	–	–	–	–	–

Nucleotide position		453	456	465	473	480	517	528	555	558	600	625
	Ref.	•	•	•	•	•	•	•	•	•	•	•
Animal strain V (virions)	10	TCG	TCG	ATC	GGA	GAG	TTA	CCA	CTC	CAG	CAA	TCT
Patient I (human PBMC)	7											
Patient II (human PBMC)	7			T							G	
Patient III (human PBMC)	7	A	A	T		A		G	T	A	G	
Hu-H1 (human isolate)	8											C
Hu-H2 (human isolate)	8				A		G					
Hu-H3 (human isolate)	8											
Wt-1 (horse brain)	30	A	A	T		A		G	T	A		
Animal strain C6BV (RNPs)	12		A	T					T		G	
Amino acid codon		151	152	155	158	160	173	176	185	186	200	209
Amino acid		Ser	Ser	Ile	Gly	Glu	Leu	Pro	Leu	Gln	Gln	Ser
Change		–	–	–	Glu	–	Val	–	–	–	–	Pro

In contrast to the mutations found in BDV-RNAs directly amplified from infected tissue, almost all mutations found in the first three isolates of infectious human BDV, are in different nucleotides and comprise non-conservative amino acid changes in each of the ORFs I-IV [8, 14]. Each isolate displays an individual pattern of changes (Table 3). These mutations are unique, as shown in Table 2 (p40-fragment), except one single change of BDV-HU-H2 in amino acid codon 117 (Lys → Glu) of BDV ORF I (Table 3) that has also been detected in the PBMCs of another psychiatric patient [28]. Neither deletions nor insertions were recorded in the human isolates [8, 14]. With the exception of ORF I (50% sequenced), the point mutations were clustered, especially in ORF III and IV (Fig. 4). It is noteworthy that the human viruses remained genetically stable during long-term passaging in OL cells. This was demonstrated by comparing the RNA from original patients' PBMCs with RNA of the isolates [14].

Table 3. Nucleotide mutations and amino acid changes in three human BDV isolates [8, 14] with respect to the reference sequence of animal strain V [10]

Isolate	ORF	Nucleotide position in ORF	Nucleotide position in genome	Nucleotide change	Amino acid codon	Amino acid change
BDV-Hu-H1	I(p40)	625	678	T>C	209	Ser>Pro
	II(p24)	10	1 281	C>G	4	Arg>Glu
		11	1 282	G>*A*	4	
		76	1 347	C>T	26	Pro>Ser
		379	1 650	C>T	127	His>Tyr
		588	1 859	A>G	196	
	III(p16)	12	1 904	A>G	4	
	IV(p56)	976	3 211	G>A	326	Ala>Thr
		1 234	3 469	G>A	412	Ala>Thr
		1 502	3 737	G>T	501	Trp>Leu
BDV-Hu-H2	I(p40)	313	366	G>A	105	Val>Ile
		349	402	A>G	117	Lys>Glu
		473	526	G>A	158	Gly>Glu
		517	570	T>G	173	Leu>Val
	II(p24)	10	1 281	C>G	4	Arg>Gly
		76	1 347	C>T	26	Pro>Ser
		588	1 859	A>G	196	
	III(p16)	12	1 904	A>G	4	
		35	1 927	A>G	12	Asp>Gly
	IV(p56)	976	3 211	G>A	326	Ala>Thr
BDV-Hu-H3	I(p40)	313	366	G>A	105	Val>Ile
	II(p24)	10	1 281	C>G	4	Arg>Glu
		11	1 282	G>A	4	
		76	1 347	C>T	26	Pro>Ser
		378	1 649	T>C	126	
		588	1 859	A>G	196	
	III(p16)	12	1 904	A>G	4	
		35	1 927	A>G	12	Asp>Gly
		80	1 972	A>G	27	Asp>Gly
	IV(p56)	976	3 211	G>A	326	Ala>Thr

The unique individual mutations are likely to define distinct genotypes of human BDV strains, because they were derived from infectious virions, present only at a particular time point during each patient's mood disorder [8]. This condition has also been met with animal strain V so far, the sequences of which are derived from infectious viral particles, as mentioned above [10]. In contrast, all other sequences, either from animal [30] or from human origins [7, 18, 28], were read from viral RNA directly amplified from infected tissue (brain in case of animals, PBMCs in case of human patients). This material, however, is likely

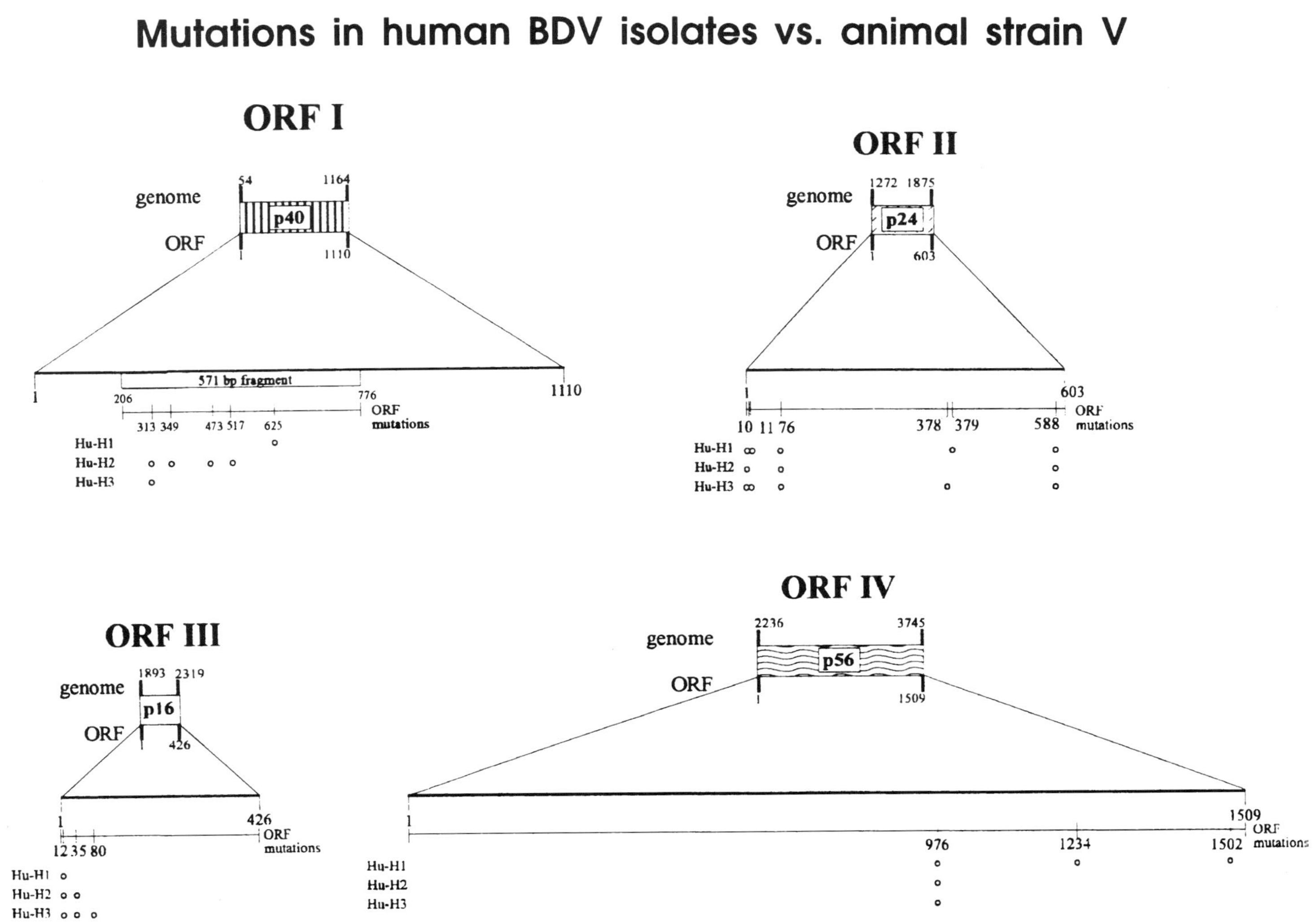

Fig. 4. Localization of nucleotide changes in the BDV genome recorded in three human isolates [8, 14] with respect to the reference sequence of animal strain V [10]

to contain a mixture of RNAs from infectious and defective viruses, with undetermined proportions of both. Only in case of successful virus isolation [8] can one be sure that the RNAs from infectious particles prevail and provide the basis for differing sequences (as demonstrated [14]). In the other cases, only a random selection of viral sequences can be expected. In this respect, the distinct individual sequence differences found in our human isolates, are most probably the result of stable mutations within the otherwise highly conserved genome, and thus may represent "true" human BDV strains. Sequences obtained directly from PBMCs or tissues are valuable in terms of diagnostic evaluations in an individual subject, but may be of less significance in terms of identifying genotypes with potential species-specificity. The biological properties of these few defined human strains support their individual character. Their pathogenicity in animals was similar but not identical to that of strain V, and their neutralizability clearly differed [8]. Further isolates from PBMCs of other species are necessary to investigate the possibility of different natural animal genotypes. Such defined isolates will, furthermore, be essential tools to address epidemiological questions.

Acknowledgements

We are grateful to P. Reckwald for excellent technical assistance, A. Muluneh for IF antibody tests, R. Ferszt and E. Severus for patients' clinical data (Fig. 1), M. Schoss for kindly providing pictures (Fig. 2) and a detailed description of the clinical course/ symptomatology of her horse, and R. Stoyloff for computer-assisted drawing of Fig. 4. This work was in part supported by a grant of the Deutsche Forschungsgemeinschaft (Lu 142/5-2 and - 3) and a contract of the European Community (BMH1-CT94-1791) to H. Ludwig. Financial support of the Federal Ministry of Health (BMG) 325–4 471-02/22) for the improvement of diagnostic tools are gratefully acknowledged.

References

1. Bode L, Riegel S, Lange W, Ludwig H (1992) Human infections with Borna disease virus: seroprevalence in patients with chronic diseases and healthy individuals. J Med Virol 36: 309–315
2. Bode L, Ferszt R, Czech G (1993) Borna disease virus infection and affective disorders in man. Arch Virol [Suppl] 7: 159–167
3. Bode L, Dürrwald R, Koeppel P, Ludwig H (1994) Neue Aspekte der equinen Borna-Virus-Infektion mit und ohne Krankheit. Diagnoseschema und Einsendeprotokoll. Prakt Tierarzt 75: 1 065–1 068
4. Bode L, Dürrwald R, Ludwig H (1994) Borna virus infections in cattle associated with fatal neurologic disease. Vet Rec 135: 283–284
5. Bode L, Steinbach F, Ludwig H (1994) A novel marker for Borna disease virus infection. Lancet 343: 297–298
6. Bode L (1995) Human infections with Borna disease virus and potential pathogenic implications. In: Koprowski H, Lipkin WI (eds) Borna disease. Curr Top Microbiol Immunol 190: 103–130
7. Bode L, Zimmermann W, Ferszt R, Steinbach F, Ludwig H (1995) Borna disease virus genome transcribed and expressed in psychiatric patients. Nature Med 1: 232–236

8. Bode L, Dürrwald R, Rantam FA, Ferszt R, Ludwig H (1996) First isolates of infectious human Borna disease virus from patients with mood disorders. Mol Psych 1: 200–212
9. Briese T, De la Torre JC, Lewis A, Ludwig H, Lipkin WI (1992) Borna disease virus, a negative-strand RNA virus, transcribes in the nucleus of infected cells. Proc Natl Acad Sci USA 89: 11486–11489
10. Briese T, Schneemann A, Lewis AJ, Park Y-S, Kim S, Ludwig H, Lipkin WI (1994) Genomic organization of Borna disease virus. Proc Natl Acad Sci USA 91: 4362–4366
11. Cubitt B, De la Torre JC (1994) Borna disease virus (BDV), a nonsegmented RNA virus, replicates in the nuclei of infected cells where infectious BDV ribonucleoproteins are present. J Virol 68: 1371–1381
12. Cubitt B, Oldstone C, De la Torre JC (1994) Sequence and genome organization of Borna disease virus. J Virol 68: 1382–1396
13. De la Torre JC (1994) Molecular biology of Borna disease virus: prototype of a new group of animal viruses. J Virol 68: 7669–7675
14. De la Torre JC, Bode L, Dürrwald R, Cubitt B, Ludwig H (1996) Sequence characterization of human Borna disease virus. Virus Res 44: 33–44
15. Dittrich W, Bode L, Ludwig H, Kao M, Schneider K (1989) Learning deficiencies in Borna disease virus-infected but clinically healthy rats: Biol Psych 26: 818–828
16. Gosztonyi G, Ludwig H (1995) Borna disease – neuropathology and pathogenesis. In: Koprowski H, Lipkin WI (eds) Borna disease. Curr Top Microbiol Immunol 190: 39–73
17. Joest E, Degen K (1909) Über eigentümliche Kerneinschlüsse der Ganglienzellen bei der enzootischen Gehirn-Rückenmarksentzündung der Pferde. Z Infkrankh Haustiere 9: 1–98
18. Kishi M, Arimura Y, Ikuta K, Shoya Y, Lai PK, Kakinuma M (1996) Sequence variability of Borna disease virus open reading frame II found in human peripheral blood mononuclear cells. J Virol 70: 635–640
19. Lange H, Herzog S, Herbst W, Schliesser T (1987) Seroepidemiologische Untersuchungen zur Bornaschen Krankheit (ansteckende Gehirn-Rückenmarkentzündung) der Pferde. Tierärztl Umschau 42: 938–946
20. Ludwig H, Bode L, Gosztonyi G (1988) Borna disease: a persistent virus infection of the central nervous system. Prog Med Virol 35: 107–151
21. Ludwig H, Furuya K, Bode L, Klein N, Dürrwald R, Lee DS (1993) Biology and neurobiology of Borna disease viruses (BDV), defined by antibodes, neutralizability and their pathogenic potential. Arch Virol [Suppl] 7: 111–133
22. Lundgren A-L, Czech G, Bode L, Ludwig H (1993) Natural Borna disease in domestic animals other than horses and sheep. J Vet Med B 40: 298–303
23. Lundgren A-L, Zimmermann W, Bode L, Czech G, Gosztonyi G, Lindberg R, Ludwig H (1995) Staggering disease in cats: isolation and characterization of the feline Borna disease virus. J Gen Virol 76: 2215–2222
24. Narayan O, Herzog S, Frese K, Scheefers H, Rott R (1983) Behavioral disease in rats caused by immunpathological responses to persistent Borna virus in the brain. Science 220: 1401–1403
25. Nicolau S, Galloway IA (1928) Borna disease and enzootic encephalomyelitis of sheep and cattle. MRC Spec Rep Series 121: 7–90
26. Rott R, Herzog S, Fleischer B, Winokur A, Amsterdam JD, Dyson W, Koprowski H (1985) Detection of serum antibodies to Borna disease virus in patients with psychiatric disorders. Science 228: 755–756

27. Rott R, Herzog S, Bechter K, Frese K (1991) Borna disease, a possible hazard for man? Arch Virol 118: 143–149
28. Sauder C, Müller A, Cubitt B, Mayer J, Steinmetz J, Trabert W, Ziegler B, Wanke K, Mueller-Lantzsch N, De la Torre JC, Grässer FA (1996) Detection of Borna disease virus (BDV) antibodies and BDV RNA in psychiatric patients: evidence for high sequence conservation of human blood-derived BDV RNA. J Virol 70: 7713–7724
29. Schneemann A, Schneider PA, Lamb RA, Lipkin WI (1995) The remarkable coding strategy of Borna disease virus: a member of the nonsegmented negative strand RNA viruses. Virology 210: 1–8
30. Schneider PA, Briese T, Zimmermann W, Ludwig H, Lipkin WI (1994) Sequence conservation in field and experimental isolates of Borna disease virus. J Virol 68: 63–68
31. Zimmermann W, Breter H, Rudolph M, Ludwig H (1994) Borna disease virus: immunoelectron microscopic characterization of cell-free virus and further information about the genome. J Virol 68: 6755–6758
32. Zimmermann W, Dürrwald R, Ludwig H (1994) Detection of Borna disease virus RNA in naturally infected animals by a nested polymerase chain reaction. J Virol Methods 46: 133–143
33. Zwick W (1939) Bornasche Krankheit und Enzephalomyelitis der Tiere. In: Gildenmeister E, Haagen E, Waldmann O (eds) Handbuch der Viruskrankheiten, vol 2. Fischer, Jena, pp 254–354

Authors' address: Dr. L. Bode, Department of Virology, Robert Koch-Institut, Nordufer 20, D-13353 Berlin, Federal Republic of Germany.

Molecular characterization of Borna disease virus from naturally infected animals and possible links to human disorders

S. Herzog[1], I. Pfeuffer[1], K. Haberzettl[1], H. Feldmann[2], K. Frese[3], K. Bechter[4], and J. A. Richt[1]

[1] Institut für Virologie, Giessen, Günzburg Germany
[2] Institut für Virologie, Marburg, Günzburg Germany
[3] Institut für Veterinär-Pathologie, Giessen, Günzburg Germany
[4] Bezirkskrankenhaus Günzburg, Universität Ulm, Günzburg, Günzburg Germany

Summary. In this review data are presented which indicate a high degree of genetic stability of BDV in his natural host, the horse. Despite this high degree of sequence conservation, variation in antigenicity was found, which did not influence the pathogenic properties of the virus. In addition, the correlation between BDV-seropositivity and a variety of psychiatric and neurological disorders in humans is discussed. In diagnostically unselected psychiatric patients we found a similar distribution of psychiatric disorders in BDV seropositives compared to seronegatives. Investigations of cerebrospinal fluid revealed cases of BDV encephalitis in BDV seropositive psychiatric and neurological patients. In contrast to others, we have found no evidence for the presence of BDV-RNA or BDV in human peripheral blood leucocytes.

Introduction

Borna disease (BD) is an infectious, immunopathological disease of the central nervous system (CNS), characterized by a disseminated meningoencephalomyelitis [26]. BD occurs as a natural infection mainly in horses and sheep. Natural infections in other *Equidae*, ruminants, rabbits, cats and ostriches have also been described [27].

The aetiological agent, Borna disease virus (BDV), has been characterized recently as a non-segmented, negative-stranded RNA virus with the property of a nuclear site for replication and transcription [15, 30]. Correspondingly, RNA splicing contributes to the generation of mature mRNAs [14, 32]. Six major open reading frames (ORFs I, II, III, IV, V, x1) are predicted in the genome sequence [15, 30, 34]. Only 3 ORFs correspond to previously identified proteins with molecular weights of 18 (ORF III: gp18), 24 (ORF II: p24) and 38/39 kd (ORF I: p38). The genomic organization is similar to that displayed by members of the *Mononegavirales* order, and therefore a new family *Bornaviridae* within the order *Mononegavirales* was established.

Extensive epizootiological studies have shown that infection with BDV among horses is more widespread than previously thought. According to recent surveys, Borna disease is rare, but occurs in all states of the western part of Germany; this indicates that the occurrence of BD extends beyond the classical endemic regions [19]. Furthermore, BDV-specific antibodies were found in horses in several European countries as well as in Israel and USA. BDV infections in the majority of horses, however, run an inapparent course, since BDV-specific antibodies are found frequently in clinically healthy horses [19]. Such inapparent infected horses can excrete BDV and may therefore represent a virus reservoir and a potential source of infection [24].

Molecular characterization of BDV isolates from naturally infected horses

In order to study the genetic variability of BDV, virus isolates were derived from three horses with clinical Borna disease (BD) from different locations in Germany (Table 1). BDV-specific antibodies were detected in the sera, infectious virus, viral proteins and viral RNA in the CNS of all three horses. Five different ORFs were cloned and sequenced. The coding regions of ORFs I (p38), II (p24), III (gp18), IV (p57) and x1 (p10; 34) were amplified by the RT-PCR technique using total RNA isolated from the CNS. Per sample two to four clones were sequenced on both strands. The sequences were aligned and compared to viral sequences derived from cell culture adapted virus from strain V. No insertions or deletions were observed in any of the analyzed clones. However, similarly as already described for cell adapted viruses [14, 32], introns within ORF III (intron#1) and ORF IV (intron#2) were identified in all three horse isolates [22]. The degree of homology between the analyzed BDV isolates and reference strain V [13] is summarized for the p38 BDV gene in Table 2. There is a maximum divergence of 1.5% at the nucleotide level and 0.8% at the amino acid level. These data are in accordance with recently published observations, which also found a high degree of homology within the p38 BDV gene of wildtype and cell adapted viruses [31]. When the other 4 ORFs were analyzed in similar ways, a homology $> 95\%$ at the nucleotide level and $> 97\%$ at the amino acid level were found, indicating that besides the

Table 1. Origin of Borna disease virus isolates

Isolate	Year of isolation	Area of origin	Tissue source of virus
Strain V	1929	Lower Saxony	Oligo/TL cells[a]
Horse#215	1989	Rhineland-Palatine	CNS
Horse#640	1993	Lower Saxony	CNS
Horse#2300	1995	Bavaria	CNS

[a]Molecular characterization in [13]

Table 2. Percentage of the p38 BDV gene sequence homology

	Strain V	Horse #215	Horse #640	Horse #2300
Strain V		99.4	99.0	99.3
Horse #215	99.7		99.3	99.0
Horse #640	99.4	99.2		98.5
Horse #2300	99.7	99.2	99.2	

Numbers above the diagonal represent the % homology among nucleotides of the ORF of the p38 gene. Numbers below the diagonal are % homology at the amino acid level

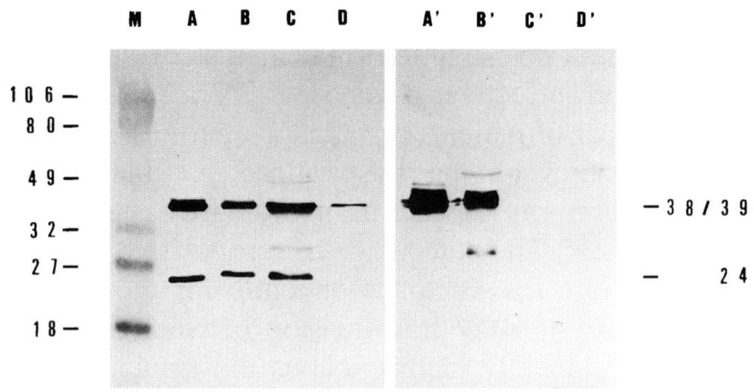

Fig. 1. Western blot analysis with brain and retinal homogenates from horses with clinical BD. *A–D* Stained with polyclonal BDV-specific rat serum; *A′–D′* stained with the mouse mAb Bo 18. *A* Brain homogenate from wildtype horse BDV strain. *B* Retinal homogenate from wildtype horse BDV strain. *C* Brain homogenate from the p38 mutated horse BDV strain. *D* Retinal homogenate from the p38 mutated horse BDV strain. *M* Molecular weight marker

previously analyzed p24 and p38 BDV genes [31, 7], these other genes (ORF III, IV, x1) exhibited also a relatively low mutation rate [22].

Nevertheless, antigenic variations of BDV isolates from horses were found when brain material from ca. 50 horses with BD were screened by immunocytochemistry and immunoblot analysis using various monoclonal antibodies. We were able to identify two horse isolates which show a remarkable difference in antigenicity. These two isolates were not recognized by the p38 specific monoclonal antibody (mAb) Bo18 [18], which is routinely used for diagnostic investigations. Polyclonal reconvalescent serum, however, reacted with this BDV-protein of both isolates in immunoblots (Fig. 1), indicating that the p38 BDV-protein was synthesized, the recognition site for the mAb Bo18, however, was altered. In order to get detailed information of the location of these changes, we cloned and analyzed the p38 BDV gene of these two horse isolates. Sequence analysis revealed several nucleotide substitutions within both p38 BDV genes, only one, however, leading to an amino acid replacement. These amino acid replacements were located adjacent to each other within the N-terminal part of the p38 protein. When the mutated p38 BDV-proteins were expressed as fusion proteins with glutathione-S-transferase

(GST), they were easily recognized by polyclonal BDV-specific sera. In contrast, mAb Bo18 did not react with the mutated p38 BDV proteins, indicating that the amino acid replacements of the mutants are located within the epitope of mAb Bo18.

Borna disease virus – a possible cause of human neuropsychiatric disorders?

A number of studies have demonstrated BDV-specific antibodies in sera and cerebrospinal fluids (CSF) of psychiatric and neurological patients in Germany, Japan and the USA. Initial investigations indicated a correlation between BDV-specific antibodies and affective psychoses [29]. Recent studies with a comparably large number of patients revealed a significantly higher prevalence of BDV serum antibodies among hospitalized psychiatric patients and a moderately increased seroprevalence in neurological patients compared to surgical controls [2, 4, 25]. The percentage of seropositive normal controls was higher in areas which are known to be endemic for the disease in animals, indicating the possibility of BDV transmission between animals and man. In addition, the fact that clinically inconspicuous individuals have BDV-specific antibodies in their sera resembles the situation in horses where the majority – as mentioned above – was found to be inapparently infected [19].

The pathogenic role of BDV for humans was addressed by epidemiological and clinical studies. Epidemiological dynamics of BDV seroprevalence in surgical and neurological groups showed a continuous increase in older age groups. Interestingly, the increase of seropervalence in neurological patients was paralleled by an increase of cases suffering from acute or chronic meningoencephalitis and an increase in psychiatric comorbidity among the BDV seropositives [4]. Epidemiological dynamics were quite different in psychiatric patients. Here, the youngest quartile (17–30 years of age) showed a 6-fold increase of seroprevalence compared to surgical controls [6], indicating a role of BDV in the pathogenesis of psychiatric disorders especially of young patients. In diagnostically unselected psychiatric patients we found a similar distribution of psychiatric disorders in BDV seropositives compared to BDV seronegatives, i.e. an unequivocal classification to a definite psychiatric disorder was not possible.

Only psychiatric comorbidity and brain atrophy was increased in seropositives with a tendency to worse clinical courses [1, 6]. We, therefore, cannot confirm the conclusions from studies in diagnostically selected psychiatric or neurological patients, where a prominent role of BDV in the pathogenesis of either depressive [9] or schizophrenic [35] disorders or multiple sclerosis [10] has been proposed [4, 21], although a slight increase of affective and schizophrenic psychoses in later stages of BDV infections would be compatible with our data [5, 6]. The situation might resemble findings in HIV encephalitis where non-specific psychiatric disorders occur in early stages of infection, whereas specific psychoses become more prevalent in later stages [33].

Factors like genetic susceptibility, age, immune status and other endogenous or exogenous factors may considerably influence the type of the resulting psychiatric or neurological disorder in the individual BDV-infected patient.

In addition to serological studies, cerebrospinal fluid (CSF) from several patients was analyzed for the presence of infectious BDV and BDV-specific antibodies. From the CSF of 3 seropositive patients, BDV could be isolated (in acute episodes of disease) as found by virus-specific immunoreactivity of inoculated rabbits or cell cultures; the "human BDV isolates", however, could not be passaged in rabbit embryo brain cells or laboratory rodents. They apparently had properties different from BDV isolates of horses [28]. In addition, BDV-specific antibodies were detected in the CSF of seropositive patients [3]. The question whether these antibodies were transported through the blood liquor barrier or synthesized within the CNS was evaluated. Using a method described by Felgenhauer [17], 25–30% of seropositive patients were found to synthesize BDV-specific antibodies intrathecally, indicating an active inflammatory process in the CNS of these patients [3]. Most suffered from affective and schizophrenic psychoses, some from personality disorders [3]. The CSF of a schizophrenic patient which obviously contained infectious BDV [28], had unusually increased BDV-specific immunoglobulin G; when reexamined after five years, the patient was in good mental health and the antibody titer in the CSF was negative [5]. It might be assumed that in this patient the psychotic phase correlated with BDV encephalitis, documented by the presence of BDV and intrathecally produced BDV-specific antibodies. Recently, BDV-specific protein and RNA were detected in human autopsy brain samples from patients with hippocampal sclerosis in 4 out of 5 patients [16]; this should be easily confirmed by other laboratories. Since extensive *intra vitam* studies based on brain biopsy samples are not feasible, more focus should be addressed to the analysis of CSF samples from seropositive neurological and psychiatric patients.

Analysis of peripheral blood mononuclear cells for the presence of BDV or BDV-specific RNA

Recently, BDV-specific proteins [11], BDV-specific RNA [12, 20] and infectious BDV [8] have been found in peripheral blood mononuclear cells (PBMCs) of psychiatric patients. Since these studies have important implications for the safety of blood and blood products, we have conducted a double blind investigation of psychiatric patients. Samples were aliquoted prior to being distributed to two different laboratories for RNA-isolation and RT-PCR analysis. Each sample was tested independently in two different laboratories.

In this study 42 seropositive psychiatric patients and 4 seronegative controls from a BDV endemic area in Germany were analyzed. Consecutive samples obtained over a nine month period were also studied. The nested RT-PCR method was used with primers specific for the p24 BDV-gene, the internal primers identical to the one used by Kishi and coworkers [20]. Positive results for the presence of BDV-specific RNA were obtained only in the laboratory where BDV is routinely studied. Simultaneous analysis of the samples in another laboratory, where there had been no prior work on BDV, yielded negative results [23]. This observation forced us to conclude that contamination had a serious influence on the outcome of studies using nested RT-PCR methodology to probe for BDV-specific sequences in PBMCs of psychiatric patients. Therefore, such investigations should be conducted in a double blind manner including laboratories with no prior exposure to BDV.

In addition, cocultivation of PBMCs from 47 samples of seropositive patients over a period of 2 passages and from 8 samples over a period of 8–10 passages with rabbit embryonic brain (REB) cells failed to provide any evidence for the presence of BDV or BDV-specific antigens [23]. Immunofluorescence analysis was used to examine the possibility of BDV-replication and immunoblot techniques employing monoclonal and polyclonal BDV-specific antisera for detection of BDV-antigens.

In conclusion, an association between BDV-infection and psychiatric and neurological disorders was proposed on the basis of finding BDV-reactive antibodies in the sera of such patients. The more recent detection of BDV-proteins and RNA in human brain tissue [16] strongly indicates that humans can be infected by BDV. Whether BDV or BDV-RNA is present in human PBMCs remains still questionable. Conclusions implicating BDV infection as an aetiological factor in human mental disorders should not be drawn too quickly. Future molecular, seroepidemiological and clinical studies will determine the possible contribution of BDV to the pathophysiology of neuropsychiatric disorders.

Acknowledgements

The work done by the authors I.P., K.H. and J.A.R. was supported by the German Research Foundation (SFB 272), by the Ewald and Hilde Berge Stiftung (J.A.R.) and the Bundesministerium für Gesundheit (S.H. and J.A.R.). We would like to thank Prof. R. Rott for his continuous support and for critically reading the manuscript.

References

1. Bechter K, Bauer M, Estler HC, Herzog S, Schüttler R, Rott R (1994) Erweiterte kernspintomographische Untersuchungen bei Borna Disease Virus-seropositiven psychiatrischen Patienten und Kontrollen. Nervenarzt 65: 169–174
2. Bechter K, Herzog S (1990) Über Beziehungen der Borna'schen Krankheit zu endogenen Psychosen. In: Kaschka WP, Aschauer HN (eds) Psychoimmunologie. Thieme, Stuttgart, pp 133–141
3. Bechter K, Herzog S, Behr W, Schüttler R (1995) Investigations of cerebrospinal fluid in Borna disease virus seropositive psychiatric patients. Eur Psych 10: 250–258
4. Bechter K, Herzog S, Schüttler R (1992) Possible significance of Borna disease for humans. Neurol Psych Brain Res 1: 23–29
5. Bechter K, Herzog S, Schüttler R (1996a) Borna disease-possible cause of human neuropsychiatric disorders. Neurol Psych Brain Res 4: 45–52

6. Bechter K, Herzog S, Estler HC, Schüttler R (1996b) Increased psychiatric comorbidity in Borna disease virus seropositive patients. Eur Psych 11: 311
7. Binz T, Lebelt J, Niemann H, Hagenau K (1994) Sequence analysis of the p24 gene of Borna disease virus in naturally infected horse, donkey and sheep. Virus Res 34: 281–289
8. Bode L, Dürrwald R, Rantam FA, Ferszt R, Ludwig H (1996) First isolates of infectious human Borna disease virus from patients with mood disorders. Mol Psych 1: 200–212
9. Bode L, Ferszt R, Czech G (1993) Borna disease virus infection and affective disorders in man. Arch Virol [Suppl] 7: 159–167
10. Bode L, Riegel S, Lange W, Ludwig H (1992) Human infections with Borna disease virus: seroprevalence in patients with chronic diseases and healthy individuals. J Med Virol 36: 309–315
11. Bode L, Steinbach F, Ludwig H (1994) A novel marker for Borna disease virus infection. Lancet 343: 297–298
12. Bode L, Zimmermann W, Ferszt R, Steinbach F, Ludwig H (1995) Borna disease virus genome transcribed and expressed in psychiatric patients. Nat Med 1: 232–236
13. Briese T, Schneemann A, Lewis AJ, Park YS, Kim S, Ludwig H, Lipkin WI (1994) Genomic organization of Borna disease virus. Proc Natl Acad Sci USA 91: 4 362–4 366
14. Cubitt B, Oldstone C, Valcarcel J, de la Torre JC (1994) RNA splicing contributes to the generation of mature mRNAs of Borna disease virus, a non-segmented negative strand RNA virus. Virus Res 34: 69–79
15. de la Torre JC (1994) Molecular biology of Borna disease virus: prototype of a new group of animal viruses. J Virol 68: 7 669–7 675
16. de la Torre JC, Gonzalez-Dunia D, Cubitt B, Mallory M, Mueller-Lantzsch N, Grässer FA, Hansen LA, Masliah E (1996) Detection of Borna disease virus antigen and RNA in human autopsy brain samples from neuropsychiatric patients. Virology 223: 272–282
17. Felgenhauer K (1991) Das Konzept der Blut-Hirn-Schranke und die Bestimmung lokal synthetisierter Antikörper. In: Huffmann G, Braune HJ (eds) Infektionskrankheiten des Nervensystems. Einhorn, Reinbek, pp 20–24
18. Haas B, Becht H, Rott R (1986) Purification and properties of an intranuclear virus-specific antigen from tissues infected with Borna disease virus. J Gen Virol 67: 235–241
19. Herzog S, Frese K, Richt JA, Rott R (1994) Ein Beitrag zur Epizootiologie der Bornaschen Krankheit beim Pferd. Wien Tierärztl Monatsschr 81: 374–379
20. Kishi M, Nakaya T, Nakamura Y, Zhong Q, Ikeda K, Senjo M, Kakinuma M, Kato S, Ikuta K (1995) Demonstration of human Borna disease virus RNA in human peripheral blood mononuclear cells. FEBS Lett 3645: 293–297
21. Kitze B, Herzog S, Rieckmann P, Poser S, Richt JA (1996) No evidence of Borna disease virus-specific antibodies in multiple sclerosis patients in Germany (letter). J Neurol 234: 660–661
22. Pfeuffer I, Herzog S, Haberzettl K, Richt JA (1996) Molecular characterization of Borna disease virus isolates from naturally infected animals (submitted)
23. Richt JA, Alexander RC, Herzog S, Hooper DC, Kean R, Spitsin S, Bechter K, Schüttler R, Feldmann H, Heiske A, Fu ZF, Dietzschold B, Rott R, Koprowski H (1996) Failure to associate human psychiatric disorders with Borna disease virus infection. J Neurovirol 3: 174–178
24. Richt JA, Herzog S, Haberzettl K, Rott R (1993a) Demonstration of Borna disease virus-specific RNA in secretions of naturally infected horses by the polymerase chain reaction. Med Microbiol Immunol 182: 293–304

25. Richt JA, Herzog S, Pyper J, Clements JE, Narayan O, Bechter K, Rott R (1993b) Borna disease virus: nature of the etiologic agent and significance of infection in man. Arch Virol [Suppl] 7: 101–109
26. Richt JA, VandeWoude S, Zink MC, Clements JE, Herzog S, Stitz L, Rott R, Narayan O (1992) Infection with Borna disease virus: molecular and immunobiological characterization of the agent. Clin Infect Dis 14: 1240–1250
27. Rott R, Becht H (1995) Natural and experimental Borna disease in animals. In: Koprowski H, Lipkin WI (eds) Borna disease. Springer, Berlin Heidelberg New York Tokyo, pp 17–30
28. Rott R, Herzog S, Bechter K, Frese K (1991) Borna disease, a possible hazard for man? Arch Virol 118: 143–149
29. Rott R, Herzog S, Fleischer B, Winokur A, Amsterdam J, Dyson W (1985) Detection of serum antibodies to Borna disease virus in patients with psychiatric disorders. Science 228: 755–756
30. Schneemann A, Schneider PA, Lamb RA, Lipkin WI (1995) The remarkable coding strategy of Borna disease virus: A new member of the nonsegmented negative strand RNA viruses. Virology 210: 1–8
31. Schneider PA, Briese T, Zimmermann W, Ludwig H, Lipkin WI (1994a) Sequence conservation in field and experimental isolates of Borna disease virus. J Virol 68: 63–68
32. Schneider PA, Schneemann A, Lipkin WI (1994b) RNA splicing in Borna disease virus, a nonsegmented, negative-strand RNA virus. J Virol 68: 5007–5012
33. Sewell DD, Jeste DV, Atkinson JH, Heaton RK, Hesselink JR, Wiley C, Thal L, Chandler JL, Grant I, and the San Diego HIV Neurobehavioral Research Center Group (1994) HIV-associated psychosis: a study of 20 cases. Am J Psychiatry 151: 237–242
34. Wehner T, Ruppert A, Frese K, Herden C, Becht B, Richt JA (1996) Detection of a novel Borna disease virus encoded protein of 10 kilodalton in infected cells and tissues. J Gen Virol (in press)
35. Waltrip II RW, Buchanan RW, Summerfelt A, Breier A, Carpenter WT, Bryant NL, Rubin SA, Carbone KM (1995) Borna disease virus and schizophrenia. Psych Res 56: 33–44

Authors' address: Dr. J. A. Richt, Institut für Virologie, Frankfurterstrasse 107, D-35392 Giessen, Federal Republic of Germany.

Haemorrhagic fevers and ecological perturbations

B. Le Guenno

WHO Collaborating Center for Arboviruses and Hemorrhagic Fever Viruses, Institut Pasteur, Paris, France

Summary. Hemorrhagic fever is a clinical and imprecise definition for several different diseases. Their main common point is to be zoonoses. These diseases are due to several viruses which belong to different families. The *Flaviviridae* have been known for the longest time. They include the Amaril virus that causes yellow fever and is transported by mosquitoes. Viruses that have come to light more recently belong to three other families: *Arenaviridae*, *Bunyaviridae*, and *Filoviridae*. They are transmitted by rodents (hantaviruses and arenaviruses) or from unknown reservoirs (Ebola Marburg). The primary cause of most outbreaks of hemorrhagic fever viruses is ecological disruption resulting from human activities. The expansion of the world population perturbs ecosystems that were stable a few decades ago and facilitates contacts with animals carrying viruses pathogenic to humans. Another dangerous human activity is the development of hospitals with poor medical hygiene. Lassa, Crimean-Congo or Ebola outbreaks are mainly nosocomial. There are also natural environmental changes: the emergence of Sin Nombre in the U.S. resulted from heavier than usual rain and snow during spring 1993 in the Four Corners.

Biological industries also present risks. In 1967, collection of organs from monkeys allowed the discovery in Marburg of a new family of viruses, the *Filoviridae*. Hemorrhagic fever viruses are cause for worry, and the avenues to reduce their toll are still limited.

Definition of the hemorrhagic fevers

Hemorrhagic fever is a clinical and imprecise definition for several different diseases. The common syndrome is characterized by an incubation period of three days to three weeks followed by a gradual or sudden onset of systemic signs. The main one is high fever generally accompanied by headache, myalgia, arthralgia and nausea. There is no difference to influenza or malaria during the two or three first days and mild forms exist, without specific symptoms. In the severe cases, this period is followed by a general deterioration in health during which bleeding may occur. But it is important to note that even for Ebola, hemorrhagic signs are absent in 40% of the cases. Superficial bleeding reveals itself through skin signs, such as petecchiae, purpura. The heavier cases present with epistaxis, melena, hematemesis. Other cardiovascular, digestive, renal and neurological complications can follow. In the most serious cases, the patient dies of shock due to plasma leakage at the capillary level or more rarely of internal hemorrhages or multiple organ failure. All these diseases are zoonoses.

The viruses and their natural hosts

Hemorrhagic fever viruses are divided into several families (Table 1). The *Flaviviridae* family has been known for the longest time. It includes the Amaril virus that causes yellow fever and is transported by mosquitoes, also other viruses responsible for mosquito- and tick-borne diseases, such as dengue or Kyasanur forest disease. Viruses that became known more recently belong to three other families: *Arenaviridae*, *Bunyaviridae*, and *Filoviridae*. They have names taken from the places where they first caused recognized outbreaks of disease, towns like Junin, Marburg or Guanarito, a lake like Puumala or rivers like Hantaan, Machupo or Ebola.

All the *Arenaviridae* and the bunyaviridae responsible for hemorrhagic fevers circulate naturally in various populations of arthropods and/or vertebrates. It is uncommon for them to spread directly from person

Table 1. Virological and ecological classification of the main hemorrhagic fevers

Family	Genus	Virus	Disease	Transmission
				Arthropods
Flaviviridae	Flavivirus	Amaril	Yellow fever	Aedes sp.
		Dengue 1 2 3 4	Dengue	Aedes aegypti
Bunyaviridae	Phlebovirus	Rift Valley	Rift Valley fever	Mosquitoes sp.
	Nairovirus	Crimean-Congo	Crimean-Congo HF	Ticks
				Rodents
	Hantavirus	Hantaan	H.F.R.S.	A agrarius
		Seoul	H.F.R.S.	R norvegicus
		Puumala	Nephropathia Epidemica	C glareolus
		Sin nombre	HPS	P maniculatus
Arenaviridae	Arenavirus	Junin	Argentine HF	C laucha
		Machupo	Bolivian H.F.	C callosus
		Guanarito	Venezuelan H.F.	S hispidus
		Sabia	Brazilian H.F.	?
		Lassa	Lassa fever	M natalensis
				Unknown
Filoviridae	Filovirus	Marburg	Marburg disease	?
		Ebola Zaïre	Ebola disease	?
		Ebola Sudan	Ebola disease	?
		Ebola Reston	Ebola disease (monkey)	?
		Ebola Côte d'Ivoire	Ebola disease	?

to person in natural conditions. Epidemics are linked to the presence of animals that serve as reservoirs for the viruses and sometimes as vectors that help to transfer it to people. Various species of rodents are both excellent reservoirs and vectors for these viruses, as they do not suffer from this chronic infection and they shed viral particles throughout their lives in feces and in urine. The filoviruses are still a mystery: we do not know how they are transmitted, but our last results allow to establish new hypotheses that we present farther.

Hemorrhagic fever viruses are among the most threatening examples of what are sometimes called emerging pathogens. They are not really new. Mutations or genetic recombination between existing viruses can increase virulence, but what appear novel viruses are generally viruses that have existed for millions of years and merely come to light when environmental conditions change.

Changes in environmental conditions

These changes may lead to the emergence of these diseases by:

- Permitting contact between human and reservoirs in their natural area like war or deforestation.
- Increasing the reservoir population, like building of dams change in the agricultural patterns or climate perturbations.
- Allowing spread of the viruses from a sporadic case to the general population like bad hygiene conditions in hospitals.

Ecological disruption resulting from human activities

The expansion of the world population perturbs ecosystems that were stable a few decades ago and eases contacts with animals carrying viruses pathogenic to humans. For example. in South America, four hemorrhagic fevers have been recognized during the last forty years. They are due to closely related viruses and their natural hosts are rodents. However, the epidemiological patterns of the diseases vary.

Changes in agricultural patterns

Junin causes Argentine hemorrhagic fever, which was recognized in the pampas west of Buenos Aires in 1953 [2] and causes epidemics every year since 1958. The cultivation of large areas of maize supported huge populations of Calomys laucha and C. musculinus and multiplied contacts between these rodents and agricultural workers. They are only field rodents so the population at risk is mainly adult males and a rodent control program is not feasible. Today mechanization has put the operators of agricultural machinery on the front line: combine harvesters not only suspend clouds of infectious dust, they also create an aerosol of infectious blood when they accidentally crush the animals.

Deforestation

Other arenaviruses responsible for hemorrhagic fevers have been discovered later, e.g. Machupo, which appeared in Bolivia in 1959 [12] and Guanarito [15] in Venezuela in 1989.

Machupo resides in the same genus of rodents as Junin, but the behaviour of the species is different. The Bolivian species, Calomys callosus enters human dwellings. People of all age and sex are at risk but a rodent control has been efficient until recently. It had prevented any human infections with Machupo since 1974. After a lull of 20 years, however, this virus has reappeared, in the same place: seven people, all from one family, were infected during the summer of 1994. The cause of the emergence of this disease is not clear, but it is limited to a small part of the Beni region. The infected villages were at the border between the pampas and the Amazon forest where subsistence crops were raised in areas called "chacos" cut out from the forest by the slash and burn method.

Deforestation was clearly the cause in 1989 of an epidemic in Venezuela due to an arenavirus now called Guanarito. The first 15 cases belonged to a rural community that had started to clear a forested region in the center of the Portugesa state. Two species of rats have been found infected, Sigmodon hispidus and Zygodontomis brevicauda. Subsequently, more than 100 additional cases were diagnosed in the same area.

The war

Several diseases have been recognized during wars. Soldiers living in the field and more particularly in trenches have contacts with animals rarely met, as nocturnal rodents. Two main serotypes of hantaviruses exist in Eurasia. The Asian type is called Hantaan [10] when the north western European type is called Puumala. [5] Hantaan typically causes an illness known as hemorrhagic fever with renal syndrome; it was described in a Chinese medical text 1 000 years ago. The reservoir is Apodemus agrarius, a field mouse and the agricultural workers are the main victims. The West first became interested in this illness during the Korean War, when more than two thousand United Nations troops suffered from it between 1951 and 1953 [3]. A non fatal form exists in Europe. It was described in Sweden in 1934 as the "nephropathia epidemica" [4] but medical records of French and English military physicians report similar disease called "néphrite de guerre" or "trench nephritis" during the 1914–18 war. The largest epidemic with about four thousand cases occurred among Russian and German troops fighting during 1942 in Lapland. The main reservoir of the hantavirus Puumala, the bank vole, is a woodland animal. In western Europe, during peace time, the most frequent route of contamination is inhalation of contaminated dust while handling wood gathered in the forest or while working in sheds and barns.

In the Balkans, both viruses exist, a Hantaan like serotype called Dobrava and the Puumala type. About seven hundred cases have been reported among Bosnian and Republica serbska soldiers around Sarajevo, Tuzla, and Bihac during the 1994–96 period. Several soldiers from the Italian, French, English and Canadian troops were also infected.

Another disease that was recognized after a large outbreak among soldiers living in the field was the Crimean hemorrhagic fever. It is due to a tick borne virus, now called Congo-Crimean hemorrhagic fever virus. Sporadic cases are frequent in the Eurasian steppes but epidemics need many people exposed simultaneously to these arthropods. This happened in 1945 among Soviet soldiers. The main outbreaks are related to blood transmission in hospitals or in slaughter houses [8].

Water management and economic development

Some *Bunyaviridae* are carried by mosquitoes rather than by rodents. Consequently, ecological perturbations such as the building of dams and the expansion of irrigation can favour these viruses. Dams raise the water table, which allows the multiplication of the insects and attracts humans and their herds together in new population centers. These two factors probably explain two epidemics of Rift Valley fever in Africa: one in 1977 in Egypt and the other in 1987 in Mauritania [1, 9].

The responsible virus was recognized as long ago as 1931 as the cause of several epizootics among sheep in Western and South Africa. Some breeders in contact with sick or dead animals became infected, but at that time the infection was not serious in humans. The situation became more grim in 1970. After the construction of the Aswan Dam, there were major losses of cattle; of the 200 000 people infected, 600 died. In 1987 another epidemic followed the damming of the Senegal River in Mauritania and reached about 200 cases.

The Rift Valley fever virus is found in several species of mosquitoes, notably those of the genus Aedes. The females transmit the virus to their eggs that may survive several dry seasons. Under dry conditions the mosquitoes' numbers are limited, but abundant rain or irrigation allows them to multiply rapidly. While feeding on blood, they then transmit the virus to cattle and humans. There is also direct infection from infected cattle to humans during abortion or slaughtering.

An other disease rapidly spreading in all the tropical countries is Dengue. Thousands of cases appear every year. If hemorrhagic forms represent less than 5% of the reported cases, their absolute number is high due to the enormous number of infections. The natural cycle of this zoonosis in Asia involved mosquitoes and monkeys and this cycle still exists in Western Africa. However, human beings are now the reservoir in Asia, Central and South America. The main vector is *Aedes aegypti*. Poor hygiene conditions allow the insects to breed in collections of water like tires or cans in the growing cities of the developing countries. Once controlled in Central and South America during the fight against yellow fever, this mosquito swarms again. These conditions that have favored the emergence of Dengue are a risk for the reemergence of urban yellow fever in the Americas.

Climate perturbations

Humans are not always the cause of dangerous environmental changes. The emergence of Sin Nombre in the U.S. resulted from heavier than usual rain and snow during spring 1993 in the mountains and deserts of New Mexico, Nevada and Colorado. The principal animal host of Sin Nombre is the deer mouse *Peromyscus maniculatus*, which lives on pine kernels: the exceptional humidity favored a particularly abundant crop, and so the mice proliferated. The density of the animals multiplied 10-fold between 1992 and 1993 [13].

Risks associated with health management

Although important, ecological disturbances are not the only cause of the emergence of novel viruses. Biological industries and uncontrolled hospitals also represent risks.

Biopharmaceutical products

Vaccines are produced in animal cells. If the cells are contaminated, there is a danger that an unidentified virus may infect workers or be transmitted to those vaccinated. It was in this way that in 1967 the collection of monkeys kidneys for vaccine production allowed the discovery of a new hemorrhagic fever and a new family of viruses, the *Filoviridae* [17].

The place was Marburg, Germany, where 25 people fell ill. Seven died. Other cases were reported simultaneously in Frankfurt and in Yugoslavia, all in laboratories that had received monkeys from Uganda. The monkeys themselves also died, suggesting that they are not the natural reservoir of the Marburg virus. Four cases of natural infection with Marburg virus have been reported in Africa, but neither the reservoir nor the natural modes of transmission have been discovered. What is clear is that Marburg can propagate in hospitals: secondary cases have occurred among health workers.

Nosocomial transmission

Poor medical hygiene can foster epidemics. In January 1969 in Lassa, Nigeria, an American volunteer who worked as a nurse fell ill at work [6]. She infected, before she died, two other nurses, one of whom died. A year later an epidemic broke out in the same hospital. An epidemiological survey found that 17 of the 25 persons infected had probably been in the room where the first victim had been hospitalized. Lassa is classed as an arenavirus and its reservoir is *Mastomys natalensis*, a rodent that enters the huts of exposed African population of Nigeria, Sierra Leone, Liberia and Guinea.

In 1976 two epidemics of hemorrhagic fever occurred two months apart in the south of Sudan and in northern Zaire [14]. In Sudan, the epidemic developed by contact between patients and their family during four months in Nzara. Two patients from Nzara were treated in the hospital of Maridi and lead to the infection of 229 persons, mainly nurses. In Zaire, around Yambuku Hospital by the Ebola River, 318 cases were counted and 280 persons died. Eighty-five of them had received an injection in this hospital. The epidemic led to the identification of a new virus, Ebola.

After these two epidemics a surveillance system was set up. It permitted the identification of an isolated case in Tandala, Zaire. An other small outbreak occurred in Nzara in 79. In 1989 specialists at the USAMRIID were put in a panic when they learned that macaques from the Philippines housed in an animal quarantine facility in Reston, VA, were dying from an infection caused by an Ebola-type filovirus. The virus was also isolated from other animal facilities that had received monkeys from the Philippines. No human illnesses were recorded despite four confirmed infections.

In January 1995 I isolated a previously unknown type of Ebola from a patient who had been infected during the autopsy of a wild chimpanzee belonging to a troop that was decimated by a strange epidemic [11]. I have also diagnosed two other Ebola outbreaks in Gabon, one from November 1994 to February 1995 and the second in February 1996. The Gabonese viruses are very close to the Zaire 76 (Yambuku) and the Zaire 95 (Kikwit) strains. In both epidemics like in Côte d'Ivoire, the humans were infected through contacts with dead or sick chimpanzees. Data obtained mainly through the survey of the chimpanzee in Côte d'Ivoire and compared with the three other outbreaks allow more precise hypotheses for the search of a reservoir.

Hypotheses about the natural cycle of the Ebola virus

Natural history of Ebola Cote d'Ivoire

In the Taï National Park, Côte d'Ivoire, the behavior of a community of free living chimpanzees has been studied since 1979 by ethologists. The data have permitted an analysis of the mortality on a long term period and precise epidemiological investigation of the 1994 outbreak.

The corpse of only one individual was in good enough conditions to allow collection of tissues, but the Ebola specific immuno-histochemical staining of the liver proved that this animal was infected by this virus.

Temporal cycles

A demographic analysis revealed that the community, after a period of stability with 80 individuals has recently suffered from a constant decrease to a present number of 32 individuals. During November 1992 eight cases were recorded (attack rate 17%) and twelve cases in October–November 1994 (attack rate 28%). The geographical distribution of the corpses showed a clustered distribution within a radius of 1.5 km inside the home range. Consuming meat during September-October was the highest risk factor and the risk was increasing with the quantity of meat eaten. Chimpanzees hunt regularly monkeys throughout the year and Western Red Colobus (Colobus badius) is the major prey. A hunting party occurred on October 19 when a young red colobus was eaten, six days before the first wave and the main consumers were among victims of this wave.

The canopy hypothesis

Colobus could have been the source of Ebola infection for apes. This does mean that colobus are Ebola carriers. If so, we would face with epidemics among chimpanzees all the year. Colobus might be an intermediate host, itself contaminated by the true reservoir during October/November. Ecological behavior studies of Colobus would identify what close species could be the candidate reservoir. Colobus are strictly vegetarian monkeys (10% fruits, 90% leaves), their home range is about one square km and they live in multi-male groups of 60 to 100 individuals, spending most of their time in the canopy and emergent trees. They could be contaminated from food, from a specific arthropod of the upper strata or through contact with excretions from rodents, bats or other vertebrates.

The W.H.O. Tai Project: perspectives for the reservoir search

Our results have permitted to define a small area where Ebola circulates. Other key features are the annual and seasonal cycles of infection. This suggests that the reservoir abundance vary quickly or that the conditions for contact between monkeys and the reservoir exist only during a short period. This cycle appears also among the last outbreaks. The Gabon 94 outbreak began in November in Mekouka, the first case in Kikwit was infected by the end of December 1994, the Liberian case in November 1995 and the chimpanzee responsible for the Gabon 1996 outbreak in Mayibout become infected by the end of December 1995. Surveys of the mortality among chimpanzees and colobus are going on. Platforms have been built at different levels in the forest to study the relations and population dynamics of the species close to these primates.

A long-term ecological study of the canopy is the only way to reduce the number of candidate reservoir species to trap and to test. The identification of the Ebola reservoir is necessary to estimate the risks of emergence of this deadly infection. We cannot say for the moment if the increasing numbers of Ebola outbreaks are only due to better surveillance or to ecological perturbation in the shrinking African rain forests.

References

1. Abdel-Wahab K, El Baz L, El Tayeb E, Omar H, Ossman M Yasin W (1978) Rift Valley fever virus infections in Egypt: pathological and virological findings in man Trans Roy Soc Trop Med Hyg 72: 392–396
2. Arribalzaga R (1955) Una nueva enfermedad epidémica a german desconocido Hipertermia nefrotoxica, leucopenica y enantematica. Dia Med 27: 1 204–1 210
3. Bernstein C (1952) Haemorrhagic fever in Korea. Lancet ii: 1273
4. Myhrman G (1934) En njursjukdom med egenarrtad symptombild. Nord Med Tidskr 7: 793–794

5. Brummer-Korvenkontio M, Vaheri A, Hovi T, Von Bonsdorff CH, Vuorimies J, Manni T, Penttinen K, Oker-Blom N, Lähdevirta J (1980) Nephropathia epidemica: detection of antigen in bank vole and serologic diagnosis in human infection. J Infect Dis 141: 131–134
6. Buckley S, Casals J, Downs W (1970) Isolation and antigenic characterization of Lassa virus. Nature 227: 174
7. Hennessen W, Bonin O, Mauler R (1968) Zur Epidemiologie der Erkrankung von Menschen durch Affen. Dtsch Med Wochenschr 93: 582–589
8. Hoogstraal H (1979) The epidemiology of tick-borne Crimean-Congo hemorrhagic fever in Asia, Europe and Africa. J Med Entomol 15: 304–417
9. Jouan A, Le Guenno B, Digoutte JP, Philippe B, Riou O, Adam F (1988) A RVF epidemic in Southern Mauritania. Ann Virol Inst Pasteur 139: 3 067–308
10. Lee HW, Lee PW, Johnson K (1978) Isolation of the etiologic agent of Korean hemorrhagic fever. J Infect Dis 137: 289–308
11. Le Guenno B, Formenty P, Wyers M, Gounon P, Walker F, Boesch C (1995) Isolation and partial characterization of a new strain of Ebola virus. Lancet i: 664–666
12. Mackenzie R, Beye H, Valverde L, Garron H (1964) Epidemic hemorrhagic fever in Bolivia. Am J Trop Med Hyg 13: 620–625
13. Nichol S, Spiropoulou C, Morzunov S, Rollin P, Ksiazek T, Feldman H, Sanchez A, Childs J, Zaki S, Peters CJ (1993) Genetic identification of a Hantavirus associated with an outbreak of acute respiratory illness. Science 262: 914–917
14. Report of WHO/International study team. Ebola haemorrhagic fever in Zaire, 1976. Ebola haemorrhagic fever in Sudan, 1976. (1978) Bull World Health Organ 56: 247–293
15. Sales R, De Manzione N, Tech R, Ricco-Hessse R, Shope R, Betancourt A, Godoy O, Bruzual R, Pacheco M, Ramos B, Taibo M, Tamayo J, Jaimes E, Vasquez C, Araoz F, Querales J (1991) Venezuelan haemorrhagic fever. Lancet 338: 1 033–1 036

Authors' address: Dr. B. Le Guenno, WHO Collaborating Center for Arboviruses and Hemorrhagic Fever Viruses, Institut Pasteur, 25 Rue du Dr. Roux 75724 Paris Cedex 15, France.

Transmission, species specificity, and pathogenicity of Aujeszky's disease virus

D. Sawitzky

Institut für Klinische und Experimentelle Virologie, Universitätsklinikum Benjamin Franklin, Freie Universität Berlin, Berlin, Federal Republic of Germany

Summary. Aujeszky's disease virus (ADV), also known as pseudorabies virus (PrV), is an alphaherpesvirus that causes fatal infections in a wide range of animal species. The virus shares a variety of biological properties with human pathogenic herpesviruses like herpes simplex virus or varicella-zoster virus. Although only limited data are available, it seems unlikely that PrV causes disease in immunocompetent humans, but may pose a risk for immunocompromised patients.

Aujeszky's disease virus

Aujeszky's disease virus (ADV), also known as Pseudorabies Virus (PrV), is the causative agent of Aujeszky's disease (AD). The virus belongs to the family *Herpesviridae*, subfamily of *Alphaherpesvirinae*. Its formal taxonomic name is suid herpesvirus 1 (SHV-1; [10]).

Aujeszky's disease

Pseudorabies virus causes a mild and usually unrecognized disease in adult swine, but causes a fatal infection in piglets and many non-porcine animals, like cattle, sheep, goats, dogs, cats, and other. Horses, reptiles and higher primates are thought to be resistant [14]. Primary infection of swine usually occurs via the oral or nasal route. The virus replicates in the oropharynx and spreads through the cranical nerves within the central nervous system. Clinical symptoms in adult pigs include respiratory and gastrointestinal affections, fever, growth arrest, and CNS symptoms. While the mortality rate of mature swine is low, young pigs exhibit a high mortality rate. Clinical symptoms include fever, vomiting, diarrhea and involvement of the CNS. Death occurs about 6–8 days after infection. Infections of pregnant sows result in death and resorption of the embryo or the delivery of stillbirths and mummified fetuses.

In surviving pigs, PrV can establish latent infections in sensory ganglia, tonsils and lymphoid tissue. Latently infected animal sporadically shed small amounts of virus even when they show no signs of disease.

Secondary hosts may be infected by ingestion of dead piglets or infected placentas of infected pigs. The virus may also be transmitted to cattle through secretions of infected pigs. Most secondary hosts excrete virus for only 1 to 2 days before death occurrs.

PrV in men

Reports of Aujesky's disease in humans have been rare. Mravak and co-authors described three patients with a variety of neurological symptoms [13]. All three patients had close contact with cats or other domestic animals. Sera from these patients were tested in our institute for the presence of neutralizing antibodies to PrV and exhibited titers ranging from 1:8 to 1:16. Two sera from one patient were further characterized in immunoprecipitation assay with radioactively labelled PrV envelope proteins. Both sera reacted with the major capsid protein and the gB-protein of PrV.

All three patients recovered completely and clinical and neurological tests were normal.

Relationship between PrV and herpes simplex virus

These data indicate that PrV is able to infect of humans, although infection seems to be very mild with minor clinical symptoms.

The relationship between the human pathogenic herpes simplex virus (HSV) and PrV was initially recognized by Sabin [17]. Both viruses belong to the subfamily of alphaherpesviruses and share several structural and biological properties. Like HSV, PrV is a neurotropic virus, capable of infecting both the peripherial and the central nervous system (for review see [3]).

Structure

Examined by electron microscopy, PrV is indistinguishable from other members of the herpesvirus family. The virion consists of a core structure, surrounded by the icosadeltahedral capsid. The capsid is surrounded by an envelope containing at least 10 virus-encoded glycoproteins. The amorphous space between the capsid and the envelope is called „tegument" The envelope proteins of herpesviruses possess important functions in the reproduction cycle. They are involved in adsorption to the cell surface, entry of the virus into the cytoplasm, egress of progeny virus from infected cells and spread of the virus through the nervous system [3, 21, 24, 26]. To date, 11 glycoproteins are identified in PrV, all of which share amino acid similarity to glycoproteins of HSV, although homology may be low. Robbins et al. [16] calculated only 22.3% homology between PrV- and HSV-gC.

Reproduction

The wide range of host specificity is also reflected in tissue culture systems, since PrV infects a variety of cells including cells of human origin like WI-38

(ATCC CCL-75), Intestine 407 (ATCC CCL-6) and HEp-2 cells [19]. Binding to the cell surface and entry into the cytoplasm are the initial steps for a virus to establish a lytic or latent infection. Different viral and cellular structures are involved in these steps. There is evidence that PrV and HSV use similar or even identical cellular structures during adsorption and entry. Electron microscopic studies gave evidence that PrV reproduction shows no differences to replication of HSV [12, 15, 19].

Adsorption

Both PrV and HSV bind initially to heparan sulfate residues of cell surface proteoglycans [18]. The interaction with heparan sulfate is mediated by the gC-proteins of both viruses, although there is no conserved heparan sulfate-binding site. While three heparan sulfate-binding sites have been identified that mediate independently interaction with cellular heparan sulfate residues [4, 20, 22], amino acid residues critical for interaction between HSV-1 gC and cell surface heparan sulfates were localized to two separate regions of the protein [23]. Although PrV and HSV both interact with heparan sulfate, no data are available yet if both viruses bind to identical heparan sulfate structures. It was shown recently that the envelope proteins of herpes simplex virus type 1 and type 2 interact with different affinities for different structural features of the glycosaminoglycan heparin [5].

Entry accessary factor

Fusion of the viral envelope with the cell membrane and entry into the host cell requires at least one additional yet uncharacterized receptor. It was shown previously that chinese hamster ovary (CHO) cells are resistant to infection by PrV, because these cells lack a structure required for virus entry [19]. Since these cells are also resistant to infection by herpes simplex virus [25], it can be concluded that both viruses interact with the same entry-accessary factor.

Gene-regulation

The genes of pseudorabies virus during lytic infection are expressed as three groups, immediate-early (IE), early and late. Like gene expression of herpes simplex virus, expression of the PrV genes is highly regulated.

While the herpes simplex virus genome contains five IE genes, the PrV genome codes for only one IE protein of about 180 kDa. This protein is required to initiate transcription of other PrV genes [7]. The DNA sequence of the PrV IE gene is highly homologous with the IE proteins of other alphaherpesviruses, such as ICP4 of HSV and p140 of varicella-zoster virus [1].

Protection against the immune response of the host

Cellular and humural immune response are important factors in controlling infections by herpesviruses. This is demonstrated in particular by herpesvirus infections of newborns [8] and immunocompromised persons, e.g. AIDS patients [9]. Herpesviruses have evolved several mechanisms to protect infected cells from the immune response of the host organisms. Both PrV and HSV express glycoproteins that bind complement factors and antibodies of the IgG type.

Complement-binding activity of HSV-gC vs. PrV-gC

Cells infected by herpesviruses express viral glycoproteins on their surface. The gC-protein of HSV binds the third component of complement (C3b) and thereby protects the infected cell from lysis by the immune system [11].

PrV-gC binds C3b of porcine, but not human complement [6]. Therefore, expression of PrV-gC on the surface of a human cell would not protect the cell from complement-mediated lysis by the human immune system.

IgG-Fc-binding activity

The glycoproteins gE and gI of herpes simplex and pseudorabies virus form a non-covalently linked complex that is expressed on the surface of an infected cell [27]. The HSV-gE/gl complex binds the Fc-portion of human IgG and protects HSV-infected cells from the immune response [2]. In contrast, it was reported by Zuckermann and co-authors [27] that the PrV-gE/gl complex does not bind human IgG.

PrV as a potential risk in immunocompromised patients

Although only limited data are available, infection of humans with PrV seems possible. Infected persons may only exhibit minor or no clinical symptoms. Transmission of the virus may occur by contact with infected animals, transmission of the virus via food seems unlikely.

In contrast, use of porcine tissues or organs (e.g. livers) for human patients may pose a risk especially for immunocompromised patients. Animals used as a source for organs should therefore be examined very carefully for the presence of PrV and other pathogens that may pose a health risk for humans, like influenza A and B virus and rotavirus.

If these pigs are vaccinated against PrV, only inactivated vaccines should be used. Live attenuated vaccine strains of PrV are able to establish life long latent infections in vaccinated pigs, and may cause infections in immunocompromised patients. In our institute, animals that are used as a source for liver perfusion of patients in the Rudolf-Virchow-University Clinic, are routinely tested for antibodies directed against wild type and vaccine strain (gE-negative) PrV, as well as for the presence of antibodies directed against other human pathogenic viruses.

Note added in proof

The entry accessory factor that mediates entry of herpes simplex virus was recently identified by Montgomery and co-authors (Cell 87, 427–436, 1996) as an yet uncharacterized member of the TNF/NGF receptor family. The authors reported that this protein does not support the entry of PrV.

References

1. Cheung AK (1989) DNA nucleotide sequence analysis of the immediate-early gene of the Indiana-Funkhauser strain of pseudorabies virus. J Virol 62: 4 763–4 766
2. Dubin G, Socolof E, Frank I, Friedman HM (1991) Herpes simplex virus type 1 Fc receptor protects infected cells from antibody-dependent cellular cytotoxicity. J Virol 65: 7 046–7 050
3. Enquist LW (1994) Infection of the mammalian nervous system by pseudorabies virus (PRV). Semin Virol 5: 221–231
4. Flynn SJ, Ryan P (1996) The receptor-binding domain of pseudorabies virus glycoprotein gC is composed of multiple discrete units that are functionally redundant. J Virol 70: 1 355–1 364
5. Herold BC, Gerber SI, Belval BJ, Siston AM, Shulman N (1996) Differences in the susceptibility of herpes simplex virus types 1 and 2 to modified heparin compounds suggest serotype differences in viral entry. J Virol 70: 3 461–3 469
6. Huemer HP, Larcher C, Coe NE (1992) Pseudorabies virus glycoprotein III derived from virions and infected cells bind to the third component of complement. Virus Res 23: 271–280
7. Ihara S, Feldman L, Watanabe S, Ben-Porat T (1983) Characterization of the immediate-early functions of Pseudorabies virus. Virology 131: 437–454
8. Kohl S (1989) The neonatal human's immune response to herpes simplex virus infection: a critical review. Pediatr Infect Dis J 8: 67–74
9. Mann SL, Meyers JD, Holmes KL, Corey L (1984) Prevalence and incidence of herpesvirus infections among homosexuelly active men. J Infect Dis 149: 1 026
10. Matthews REF (1982) Classification and nomenclature of viruses. Intervirology 17: 1–200
11. McNearny TA, Odell C, Holers VM, Spear PG, Atkinson JP (1987) Herpes simplex virus glycoprotein gC-1 and gC-2 bind to the third component of complement and provide protection against complement-mediated neutralization of viral infection. J Exp Med 166: 1 525–1 535
12. Morgan C, Rose HM, Mednis B (1968) Electron microscopy of herpes simplex virus I. Entry. J Virol 2: 507–516
13. Mravak S, Bienzle U, Feldmeier H, Hampl H, Habermehl KO (1987) Lancet 1: 501–502
14. Nara PL (1982) Porcine herpesvirus 1. In: Olson JR, Krakown S, Blakesless Jr JR (eds) Comparative pathobiology of viral disease. CRC Press, Boca Raton, pp 90–113
15. Nii S, Morgan C, Rose HM (1968) Electron microscopy of herpes simplex virus II. Sequence of development. J Virol 2: 517–536
16. Robbins AK, Watson RJ, Whealy ME, Hays WW, Enquist LW (1986) Characterization of a pseudorabies virus glycoprotein gene with homology to herpes simplex virus type 1 and type 2 glycoprotein C. J Virol 58: 339–347
17. Sabin AB (1934) The immunological relationships of pseudorabies (infectious bulbar paralysis, mad itch). Br J Exp Pathol 15: 372–380
18. Sawitzky D, Hampl H, Habermehl KO (1990) Comparison of heparin-sensitive attachment of pseudorabies virus and herpes simplex virus type 1 and identification of heparin-binding PrV-glycoproteins. J Gen Virol 71: 1 221–1 225
19. Sawitzky D, Hampl H, Habermehl KO (1990) Entry of pseudorabies virus into CHO cells is blocked at the level of penetration. Arch Virol 115: 309–316

20. Sawitzky D, Voigt A, Habermehl KO (1993) A peptide-model for the heparin-binding property of pseudorabies virus glycoprotein III. Med Microbiol Immunol 182: 285–292
21. Schreurs C, Mettenleiter TC, Zuckermann F, Sugg N, Ben-Porat T (1988) Glycoprotein gIII of pseudorabies virus is multifunctional. J Virol 62: 2251–2257
22. Trybala E, Bergström T, Spillmann D, Svennerholm B, Olofsson S, Flynn SJ, Ryan P (1996) Mode of interaction between pseudorabies virus and heparan sulfate heparin. Virology 218: 35–42
23. Trybala E, Bergström T, Svennerholm B, Jeansson S, Glorioso JC, Olofsson S (1994) Localization of a functional site on herpes simplex virus type 1 glycoprotein C involved in binding to cell surface heparan sulfate. J Gen Virol 75: 743–752
24. Whealy ME, Robbins AK, Enquist LW (1988) Pseudorabies virus glycoprotein gIII Is required for efficient virus growth in tissue culture. J Virol 62: 2512–2515
25. WuDunn D, Spear PG (1989) Initial interaction of herpes simplex virus with cells is binding to heparan sulfate. J Virol 63: 52–58
26. Zsak L, Mettenleiter TC, Sugg N, Ben-Porat T (1989) Release of pseudorabies virus from infected cells is controlled by several viral functions and is modulated by cellular components. J Virol 63: 5475–5477
27. Zuckermann FA, Mettenleiter TC, Schreuers C, Sugg N, Ben-Porat T (1988) Complex between glycoprotein gl and gp63 of pseudorabies virus: its effect on virus replication. J Virol 62: 4622–4626

Authors' address: Dr. D. Sawitzky, Institut für Klinische und Experimentelle Virologie, Universitätsklinikum Benjamin Franklin, FU Berlin, Hindenburgdamm 27, D-12203 Berlin, Federal Republic of Germany.

The role of veterinary public health in the prevention of zoonoses

K. Stöhr and **F. X. Meslin**

Division of Emerging and other Communicable Diseases Surveillance and Control, World Health Organization, Geneva Switzerland

Summary. Veterinary public health is a component of public health activities devoted to the application of professional veterinary skills, knowledge, and resources for the protection and improvement of public health. VPH activities involve a very diverse range of functions within public health which reflect the broad community of interests between veterinary and human medicine. Zoonoses continue to represent an important health hazard in most parts of the world, where they cause considerable expenditure and losses for the health and agricultural sectors. Although the situation is improving in the industrialized world, zoonoses prevention and control will remain an area of major concern in most developing countries. Recent observations in these countries show that expenses related to the prevention of zoonotic diseases in humans are likely to increase dramatically in the near future. Programmes for their control and eventual elimination in animal reservoirs are urgently needed. The technical knowledge exists to bring diseases such as brucellosis, rabies, and bovine tuberculosis under control during the first decade of the next century. To achieve this goal, constant efforts will be needed for the next 15 to 20 years. In addition, as trade in animal products and the movement of human populations continues to increase, the risk that zoonotic diseases will be introduced or reintroduced into certain areas is likewise increasing. Over the past five years, a number of zoonotic diseases have emerged as either new pathological entities or known agents appearing in new areas or as new strains. Through its coordinating and information gathering functions, the WHO Emerging Disease Surveillance and Control Division provides a source of both practical and technical guidance that can help solve these and other threats to human health posed by animals.

Veterinary Public Health

The expression *Veterinary Public Health* (VPH) merges two terms: public health and veterinary medicine/profession and is thus defined as: "a component of public health activities devoted to the application of professional veterinary skills, knowledge and resources for the protection and improvement of human health".

The reason for the emergence and present existence of the expression VPH in everyday professional language usage should be observed as an indication that the knowledge and experiences of veterinarians and their resources do have some impact on the protection of human health. It also implies that there may be a demand on the professional "market" for specific skills and proficiencies which may be met best by veterinarians.

This does not necessarily mean that the position of veterinarians in public health is an easy one. Their knowledge and position in the field of disease control strategies, epidemiology, population medicine, medical science and particularly infectious diseases is often underestimated. Veterinary students' terms in microbiology for example are usually more than 250 h while the medical curricula in many universities and colleges cannot accommodate more than 30–40 h on microbiology, including parasitology.

The main scientific and applied disease control areas where the unique knowledge and position of veterinary medicine in natural science has proved useful for the protection of human health are:

- foodborne diseases control-food safety
- zoonoses control
- communicable disease epidemiology and population medicine
- comparative biomedicine
- animal welfare
- environmental protection

This paper focuses on zoonoses control as one of the most important components of VPH and will also touch upon some aspects of food hygiene for zoonoses control as VPH is ultimately linked with foodborne disease control.

Zoonoses

Zoonoses in general are defined as infectious diseases which can be transmitted from animals to humans under natural conditions. In general, zoonotic pathogens can survive and multiply without the presence of man. They circulate in wildlife and domestic animal populations and, in most cases, man is only a dead-end in the infectious chain and not even necessary for the perpetuation of the infectious or parasitic agent.

Zoonoses continue to represent an important health hazard in most parts of the world, where they cause considerable expenditure and losses for the health and agricultural sectors. Looking only at a short list of emerging and re-emerging human communicable diseases in 1995 and 1996, it appears that zoonotic diseases gain even more importance as in the epidemiology of many of these emerging and re-emerging diseases animals play a paramount role:

- Venezuelan Equine Encephalitis and Dengue Fever in South America
- Lyme Borreliosis, Hanta-virus-infection, Reston-virus infection and Cryptosporidiosis in North America
- STEC-infections in North America, Europe and Japan
- Leptospirosis in Central America and East Asia (Korea)
- non-typhi Salmonellosis in Europe and Asia
- Brucellosis in the Mediterranean Region
- Ebola in Africa
- Equine Morbillivirus-infection in Australia.

The reasons for this increasing trend are complex but some of them have been identified as follows:

- alteration of the environment affecting the size and distribution of certain animal species, vectors and transmitters of infectious agents affecting humans
- increasing human populations favouring an increased level of contact between humans and infected/affected animals
- industrialization of the animal production sector
- changes in food processing and consumer nutritional habits.

The group of zoonoses includes more than 120 communicable diseases which can be classified according to the causative biological agent as bacterial, viral, rickettsial and parasitic zoonoses.

To someone involved in the control of zoonotic diseases and in the light of the discussion on how the veterinary profession can contribute to the protection of human health, the subdivision of zoonotic diseases according to the control methods applied might be an alternative. In general there are five main intervention strategies for the prevention and control of zoonotic diseases: diagnosis and treatment of cases in human beings (usually the most expensive way), prevention of human-to-human transmission, interruption of the pathways of transmission from animals to humans, interruption of transmission from wild to domestic animals and vice-versa, and elimination of the disease in the animal reservoirs.

Which single or combination of five intervention strategies is chosen will mainly depend upon the epidemiology of the disease and its pathways of transmission, general availability of control and intervention methods, and the prevailing epidemiological, economical, and political conditions.

Generally, those important zoonotic diseases where there is no, or apparently no, feasible control possibility at the animal reservoir level are dealt with by public health authorities. For example, prevention and control of most of the arbovirus infections is built almost exclusively on diagnosis and treatment in humans and control of the non-vertebrate host animals. Exceptions are Venezuelan Equine Encephalitis, Japanese Encephalitis, and Rift Valley Fever for which vaccination of the domestic animal reservoirs plays a significant role in the interruption of the infectious chain.

Some other zoonotic diseases do not rank very high on the priority list in public health as:

- medical science and public health have no tool for the efficient control of the diseases within their area of responsibility
- the diseases have no or only insignificant economic impact (a knowledge very often greatly deficient, at least for many zoonotic diseases)
- funds are not available to deal with the disease.

Examples of zoonotic diseases

Let me give you a few examples of zoonotic diseases research and control in which the zoonoses group in WHO is either involved or which illustrate the current situation as to zoonotic diseases in the world.

- *Rabies* is certainly one of the most important zoonotic diseases. This does not become obvious when looking at the major causes of death worldwide (Table 1). Rabies is present in large areas of the northern hemisphere in many different wild animal species (e.g. fox, skunk, raccoon, raccoon dog, mongoose). Although only a small number of human deaths are reported there, rabies remains a permanent public health hazard, represents a major obstacle to the free movements of people and their pets and is at the origin of important spending for the public health sector and the general public (costs of post-exposure treatments, days of work lost). Dog rabies is still present in 87 countries and territories (with a total population at risk of about 2.4 billion persons) where between 35 000 and 50 000 human deaths due to rabies are estimated to occur each year. In these areas a total of 6.5 million persons receive rabies post-exposure treatment. Total annual expenses related to the prevention and control of the disease in humans and animals is estimated to be about US$ 300 million. Comprehensive programmes for the control of the disease in dogs have been shown to cost 25–52% less than a programme for improved post-exposure delivery in human beings.

- *Echinococcosis/Hydatidosis* is an important problem in many parts of the world especially in countries of the middle-east and north Africa where the reported incidence varies from 2 to 89.5 cases per 100 000 inhabitants (Table 2). In Maghreb countries alone, minimum spending and losses in the human and animal health sectors are estimated at US$ 60 million annually.

- *Leptospirosis* is found worldwide and most of the cases are diagnosed in professionally-exposed groups such as slaughterhouse workers, veterinarians, livestock breeders, sewer workers, and in some countries in groups representing a large part of the total population, i.e. paddy-field and sugar-cane workers. Cases are also associated with recreational activities such as swimming and fishing. In countries where rice cultivation is common, large numbers of farmers involved in water rice paddy field work throughout the year are affected with leptospirosis. Prevalence in endemic areas can range

Table 1. Main causes of mortality worldwide 1993

A. *Communicable diseases*	
	(× 1000)
Acute respiratory infections	6 900
Diarrhoeal diseases	4 200
Tuberculosis	3 300
Malaria	1 500
Hepatitis B	1 500
Rabies	35
Measles	220
Meningitis bacteria	200
Schistosomiasis	200
Pertussis	100
Amoebiasis	70
Hookworm	55
Yellow fever	30
African trypanosomiasis	20
Neonatal tetanus	560
Total	17 400
B. *Non communicable diseases*	
Diseases of the circulatory system	11 931
Malignant neoplasma	5 121
External causes	3 466
Perinatal causes	2 556
Chronic obstruction pulmonary disease	2 888
Maternal causes	504
Other causes	5 413
Total	31 879
Grand total	50 769

Table 2. Human Echinococcosis incidence in Middle East and North Africa (cases/100 000 inhabitants)

Country	Cases/100 000
Egypt	0–6.2
Iran	4–45
Iraq	3
Jordan	5
Libya	20
Morocco	13–19.8
Pakistan	89.5
Tunisia	16 (mortality 3.3–5.9%)
Turkey	2

Table 3. Human brucellosis in the middle and near east and the Arabian peninsula

Country	Year	Human cases	Rate (per 100 000)
Iran	1988	71 051	132.4
Jordan	1991	730	24.3
Kuwait	1989	2 164	12.9
Oman	1998	224	224.0
Saudi Arabia	1990	9 025	69.5
Syria	1990	988	8.2

from 4 to 100 per 100 000 population. China has reported that incidence during outbreaks was as high as 1300 cases per 100 000 inhabitants in Anlu district, Hubei province in 1982.

- Animal and human *brucellosis* is widespread. It has been reported in 86 out of 175 countries (49.1%), representing a population of 2.7 billion in all regions. Eastern Mediterranean countries experienced an increase in the number of brucellosis cases in 1982–1990. In seven countries of the region, about 82 000 cases were reported in 1988 as compared to 2 873 in 1985. This represented an average incidence of 18.9 per 100 000 in 1988. Incidence rates varied from 8.2 to 132.4 per 100 000 people in six countries of this region (Table 3).
- Zoonotic Salmonellosis

Increase in prevalence

During the last decade, reports of salmonella cases in humans have increased in many parts of the world. Figs. 1 and 2 show how *Salmonella* prevalence has evolved in some countries of the world since 1980 (first year of reporting = 100). This augmentation of reported salmonella cases in humans, particularly noticeable in Europe, was also reflected in the morbidity of human salmonellosis in the countries concerned. Reported morbidity reached levels of 150–350 bacteriologically-confirmed salmonellosis cases per 100 000 inhabitants. Particularly high rates were reported in the Czech Republic, Poland, Slovak Republic, Austria and Germany. The former USSR ranges in the middle of the listed countries with about 75 cases per 100 000 (Fig. 3).

One important reason for the rise in salmonellosis cases is certainly the dramatic increase in the number of cases of *Salmonella enteritidis*, a poultry-associated agent which is mainly transmitted through eggs and broilers to humans. *S. enteritidis* is only rarely transmitted to poultry through feed in the poultry operations. Transmission occurs mainly vertically through hatching eggs and infected animals and is maintained in the environment and through animal vectors (primarily mice). These epidemiological characteristics require a change in the general approach to salmonellosis control and emphasis should be placed in poultry operations on cleaning, disinfection and vector control,

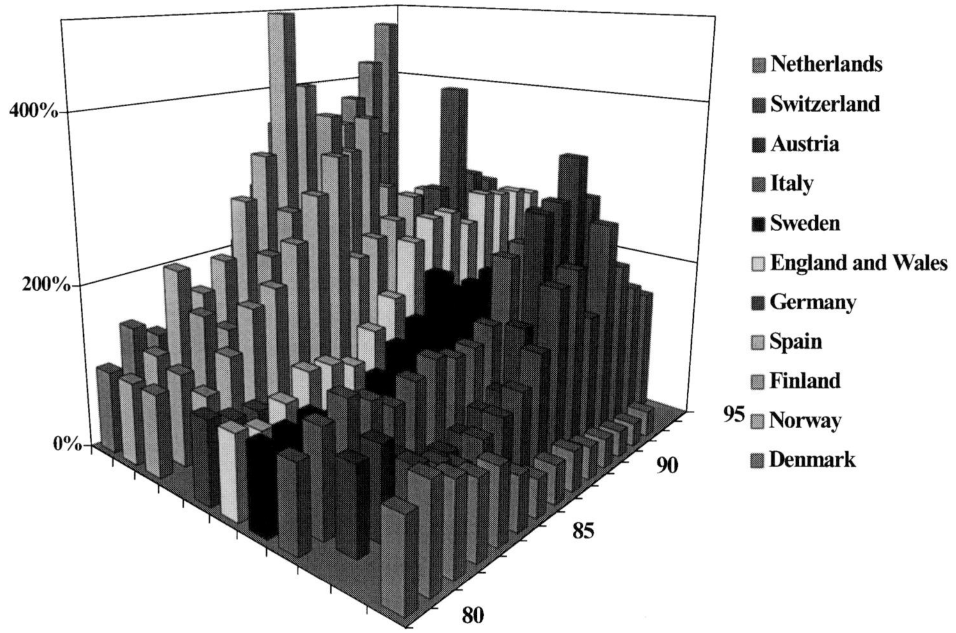

Fig. 1. Number of reported non-typhi *Salmonella* cases (1980=100%)

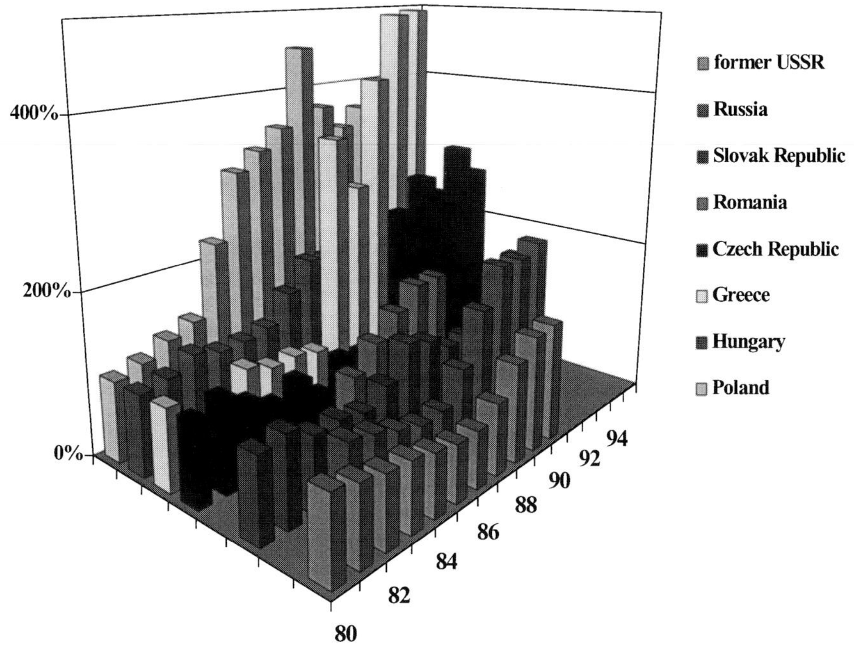

Fig. 2. Number of reported non-typhi *Salmonella* cases (1980=100%)

comprehensive monitoring systems throughout the poultry production chain, application of vaccines, competitive exclusion preparations and the tailored use of antimicrobials. WHO has developed guidelines covering each of the latter disease control methods.

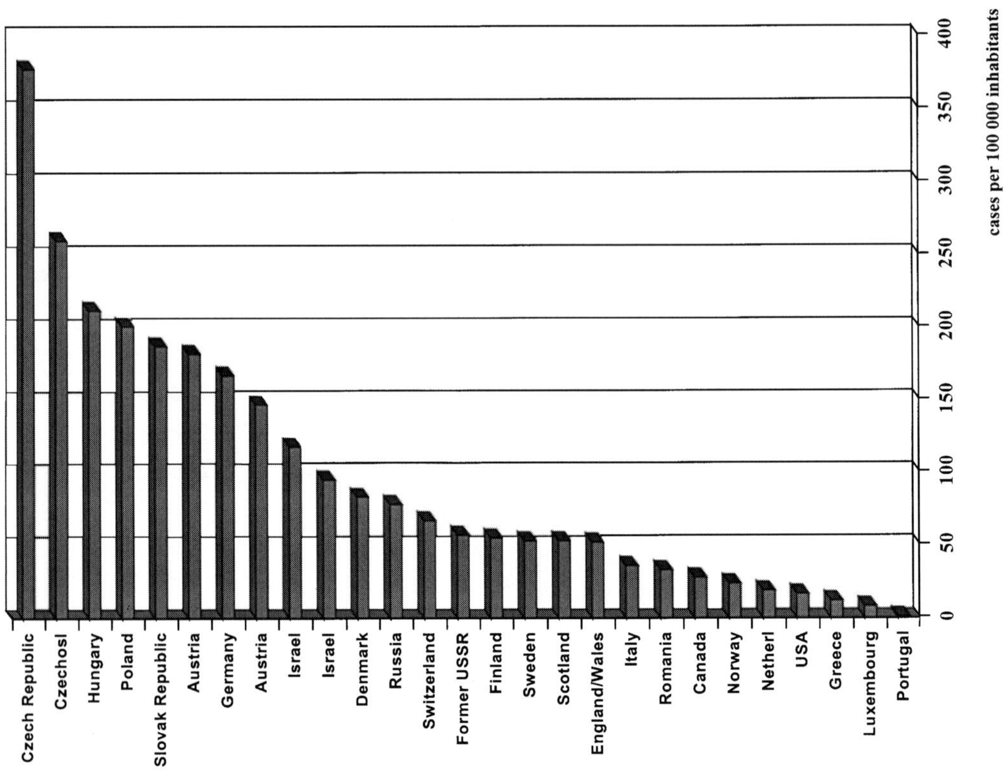

Fig. 3. *Salmonella* prevalence in various countries

Even though *Salmonella* strains (mainly *S. enteritidis* and *Salmonella typhimurium*) derived from poultry or eggs cause a significant proportion of human salmonellosis, other food commodities must not be neglected. Until the emergence of *S. enteritidis,* pork and beef were considered to be the most important sources for human salmonellosis cases in many developed countries. During recent investigations in Denmark, enteric salmonella were found in 6% of the pigs sampled in a national survey in 1363 pig herds in 1993/94. In 22% of the herds sampled, at least one *Salmonella* serotype was found. The predominant serotype was *S. typhimurium* (64% of all samples). Enteric *Salmonella* were also found in fresh pork (0.7%) and at the retail level (2.5% in fresh pork; 0.06% in heat-treated pork) during a continuous monitoring programme initiated in 1993. The distribution of serotypes in pork at slaughterhouse and in outlays of retailers largely *reflects the distribution of serotypes found in the pig herds.* In Denmark, pork was considered as the primary source of about 15% of all salmonellosis cases in humans in 1994 (20% poultry; 26% eggs, 14% travel, 26% unknown). The prevalence of salmonellosis in Denmark was 82.3 cases per 100 000 inhabitants in 1994.

Costs associated with salmonellosis

At least three factors have directed the attention of economists and others to the financial and social impact of salmonella infections:

– the dramatic increase in the recorded number of salmonella infections
– evidence which suggests that costs associated with foodborne diseases, particularly salmonella, are high
– attention placed on containing public health expenditure and pressure on resources in this sector.

The range of costs has been estimated in only a few individual *outbreaks* of salmonellosis from an economic point of view:

- 10 outbreaks in food service establishments in the USA – US$ 57 423 to US$ 699 400
- salmonellosis outbreak at a European Summit Conference – US$ 317 000
- two outbreaks due to manufactured food in the UK – £ 379 000 and £ 14.6 million

These figures are already indicative of the possible financial consequences of salmonellosis outbreaks involving many human beings. Interesting though the figures might appear, the number of salmonella cases occurring during outbreaks is comparatively low. Recent studies have shown that *sporadic, individual cases* account for the overwhelming majority of reported salmonellosis cases. These single, family cases, which are much less spectacular for the public and the media, often elude our attention though they generate most of the costs of salmonella illness in the public sector and individuals.

The economical losses due to salmonellosis in humans occur largely in the following sectors: public health, costs to individuals and society, and costs to industry. These costs do not completely reflect the impact of reported salmonella episodes on public health and economics. Intangible factors such as pain and suffering are very difficult to evaluate in financial terms.

Table 4 shows the costs of a single salmonellosis case in different countries. Figure 4 depicts the calculated annual costs for salmonellosis (in US$) based on the costs for a single reported salmonella case (Table 4), the reported prevalence, a fixed exchange rate (DM 1:52; £ 1:0.63; CA$: 1.37) and an annual inflation rate of 2%. For Germany, with a rather well established surveillance system, calculated costs would have amounted to approximately US$ 150

Table 4. Cost for reported human *Salmonella* cases

England/Wales	UK£ 788	Socket et al. (1991)
Canada	CA$ 1 350	Todd (1989)
Germany	DM 1 250	Krug et al. (1983)
USA	US$ 1 350	Todd (1989)

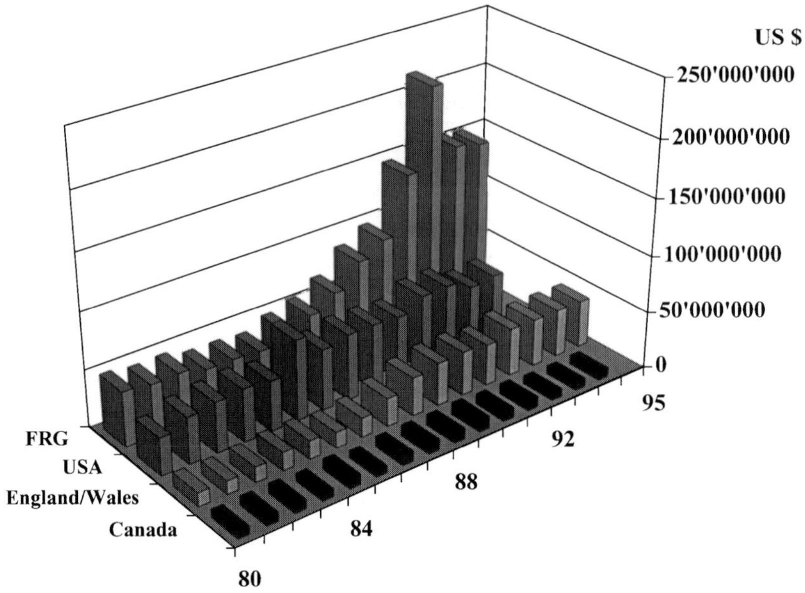

Fig. 4. Calculated annual costs for salmonellosis

million in 1994 to the public health sector, individuals and society. The data on which the analyses were made include *reported cases* only, but only a small proportion of foodborne disease episodes ever come to the attention of public health authorities. This also holds true for countries where surveillance and reporting systems function well. Probably only one out of ten episodes of salmonella diarrhoea is notified in industrialized countries. Under-reporting in developing countries is much higher and it is estimated that less than 1% of foodborne illness enters the public health statistics.

Although the *unreported salmonella episodes* usually do not need medical treatment, the costs they generate to society are substantial (about 1/5 of a reported case). An analysis in Germany revealed that costs for unreported salmonellosis cases account for about 2/3 of the total costs to the public sector.

Costs to the public and society are enormous. In the US it is estimated to amount to nearly 4 billion US$ which equates to an annual spending on salmonellosis of 160 US$ per inhabitant. This amount would represent 5.9% of the total annual health spending per capita in the USA.

- Emergence of multi-drug resistance Salmonella typhimurium DT104

Within five years the number of human cases due to this serotype have increased 10 fold in the UK. It is the second most important salmonella serotype in humans in England, Wales and Germany (where good surveillance systems exist). Almost 55% of all *S. typhimurium* cases are due to DT104 and almost 100% of DT104 is resistant to ampicillin, chloramphenicol, streptomycin, sulfonamides and tetracyclin. An increasing number of isolates are also resistant to trimethoprim and a few to ciprofloxacin. Infection with multi-drug resistant *S. typhimurium* DT104 is associated with hospitalization rates twice that of other zoonotic salmonella and with ten times higher case-fatality rates. Another worrying circumstance is that two, highly invasive salmonella serotypes might occur in parallel, something we have not yet experienced in epidemic proportions. *S. typhimurium* DT104 also causes considerable disease problems in sheep, pigs, horses and poultry. In Germany, 18% of all salmonellosis cases are due to *S. typhimurium* DT104. Pork and beef are found to be the most important sources of infection for humans.

- *Other significant zoonotic diseases* comprise Japanese encephalitis in many Asian countries where 30 000–40 000 cases are reported annually. Leishmaniasis are considered to be endemic in 82 countries (10 developed and 72 developing). A common estimate of the worldwide incidence per year is 600 000 newly reported clinical cases. Many other examples could be given, e.g. foodborne trematode infections, zoonotic tuberculosis, etc.

Other areas relevant to veterinary public health and zoonoses control are:

- Bovine Spongiforme Encephalopathie (BSE) and new variant Creutzfeld-Jacob-Disease (CJD)
- other zoonotic foodborne diseases (Campylobacteriosis, E. coli O157 and other enterotoxin-producing serotypes)
- monitoring of antimicrobial resistance in food animals and food of animal origin
- Lyme-Borreliosis
- Veterinary Urban Health
- Wildlife Reproduction Management

Recent observations show that expenses related to the prevention of zoonotic diseases in people are likely to increase dramatically in the near future if no programmes for their control in animal reservoirs are implemented. In the light of shrinking public health resources and the increase in the importance of communicable diseases it becomes more and more important to prioritize disease control, including zoonotic diseases. When dealing with the control and prevention of an individual (given) disease, the most significant questions are certainly:

- Are we tackling the right problem?
- In what terms can we identify the impact and consequences of the disease?
- What is the price society has to pay for the occurrence of this problem and its control?
- Do we have enough convincing arguments to compete successfully for public health spending?

Answers to these questions could lead to a consolidated appraisal of the economical impact of a disease. Cost-benefit analyses for its control and prevention of communicable diseases will become indispensable in the future as they will provide the rationale for the choice of preventive strategies at the most effective points of intervention. This applies even more for zoonotic diseases which often cause hidden, but frequently very costly problems in public health which do not enter the mortality statistics of public health.

The emergence of BSE and the new variant of CJD, drug-resistant S. typhimurium DT104, the general increase in Salmonellosis, outbreaks of STEC, Cryptosporidiosis and Campylobacteriosis and other recent developments in the field of zoonotic diseases make it even more obvious that the veterinary profession has to play a major role in the protection of human health from diseases which can be transmitted from animals to humans under natural conditions. As such disease problems are likely to appear again, animal disease and zoonoses surveillance will need to be reinforced and maintained at country level and internationally.

It is essential to recognize that health authorities alone cannot solve a number of zoonotic disease problems including foodborne diseases. It requires a national commitment and the collaboration of all parties concerned with health, agriculture, finance, planning, and commerce, as well as the agricultural and food industry, the biomedical and agricultural scientific community and the consumer. On the other hand veterinary medicine and science has its own animal production and animal health related problems and finance, manpower and other resources are lacking for veterinary public health activities including zoonoses control.

Salmonellosis is an example of how the implications of a zoonotic disease extend beyond the borders of a single sector of the society. Costs and benefits of foodborne diseases control are distributed among different sectors of the economy. The need for "responsibility sharing" is intrinsic to all programmes for the control of zoonotic diseases but this is obviously poorly understood in some countries. This appears to be one of the main reasons why zoonoses still have a significant impact on public health and economics in many parts of the world.

Control of diseases which are transmissible from animal to man require ability and willingness for intersectoral and interinstitutional cooperation. This is one of the decisive prerequisites not only for the control of Salmonellosis but also for all zoonoses control activities.

References

(References are available from the authors on request)

Authors' address: Dr. K. Stöhr, Division of Emerging and other Communicable Disease Surveillance and Control, WHO, CH-1211 Geneva 27, Switzerland.

Viral infections transmitted by food of animal origin: the present situation in the European Union

A. Stolle and **B. Sperner**

Institute for Hygiene and Technology of Food of Animal Origin, Veterinary Faculty, Ludwig-Maximilians-University, Munich, Federal Republic of Germany

Summary. The goal of this presentation was to clarify which foods are involved in viral diseases, which viruses are transmitted via food and how to evaluate the risk of a foodborne viral infection. Food items frequently identified as cause of viral disease outbreaks were shellfish harvested in sewage-contaminated water. Another common source of foodborne viral illness was cold food contaminated by infected food handlers. In the European Union the viruses most frequently associated with foodborne illness were hepatitis A virus and the SRSV's. A few isolated cases of foodborne hepatitis E were reported in Mediterranean countries. Compared to other foodborne diseases, those caused by viruses are less severe and seldom fatal. This might be a reason why the problem of viral contamination of food has been neglected. Yet, because many foodborne viral diseases are not recognized either as foodborne or as caused by viruses, the actual number of cases must be assumed to be significantly higher than the reported number. Consequently, food associated diseases of viral origin should be granted more attention.

Introduction

As concern about the possible health hazard of food increases among consumers, the question arises if and to what extent food is contaminated with viruses. In this regard the general public is insufficiently and sometimes even not correctly informed.

In Germany food is not routinely examined for the presence of viruses. Only if an outbreak of food poisoning is suspected to be of viral origin, virological examinations are performed in order to detect the source of infection. Therefore, no information about the actual extent of viral contamination of food is available. The goal of this presentation is to clarify which viruses are transmitted by food, which kinds of food are commonly associated with viral diseases and how to evaluate the risk of a foodborne viral infection. The information is based on the Medline Express data bases from 1966 to August 1996.

As many meetings deal exclusively with BSE, here we shall confine ourselves to diseases whose viral etiology is proven and which have been neglected on account of the paramount interest in BSE.

Contamination of food with viruses

First, the ways in which food can be contaminated with viruses shall be briefly explained. One has to discern *primary* and secondary contamination. The former takes place before harvest or slaughter either on a biological or a mechanical route. In the case of *biological* primary contamination viral replication takes place in the animal, i. e. the animal is infected. Contamination by this means happens especially in the case of asymptomatic infections or pre-symptomatic infections during the final stage of the incubation period. As the animals such affected do not show any clinical signs, the permission for slaughter is given. The primary biological contamination concerns only viruses infectious for animals, whereas the transmission of human viruses actually associated with foodborne infections is based on the following routes.

The first one is primary *mechanical* contamination, where no virus replication takes place. Viruses are rather just deposited in the animal or its product. Shellfish can be thus contaminated; this will be dealt with more thoroughly later. The transmission of viruses into egg or milk of infected animals is also considered as primary mechanical contamination.

Contrary to primary contamination, *secondary* contamination of food takes place during or after harvest or slaughter. It is an exclusively mechanical process. There is no virus replication, because viruses need living cells in order to replicate. Humans and animals represent the sources of contamination. Through virus-containing secretions or excretions they transmit viruses to food either directly by contact or indirectly via contamination of water, air, objects or surfaces.

Food associated with viral diseases

Cold food

Especially shellfish and various cold foods have been associated with viral foodborne infections. The latter include not only food eaten raw but also food heated at temperatures insufficient for the inactivation of viruses. Adequately heated food may also be contaminated if it is handled again after cooking. The crucial point is that no heating is carried out after contamination so that the viruses are not inactivated and remain infective.

Viruses reach food by way of secondary contamination through *food handlers* during processing, distribution or final preparation. As source of contamination overtly ill excretors can be identified who continue working despite their illness or who still shed viruses after cessation of clinical signs. Likewise, pre-symptomatic infected food handlers can already excrete viruses during incubation [27]. There might also be symptom-free carriers who shed viruses [31]. Furthermore mechanical transmission is suspected to be possible, which means that food handlers who are not infected themselves could for example transmit viruses from ill family members [22].

Fecal contamination as a result of poor personal hygiene is considered to be an important mode of contamination. For some viruses excretion in vomitus has been proven which implicates that viruses in aerosols produced by vomiting can directly reach food. On the other hand, viruses might be transmitted to food via surfaces or objects contaminated by one of the above-mentioned ways.

Aside from food handlers, *water* containing viruses is a possible source of contamination for cold food. The contact of such water with food is established either by directly adding it to the food in question or by using it for cleaning purposes.

The significance of viral contamination of cold food must not be underrated because disease outbreaks of this origin happen mostly in canteens and similar *institutions*. This implicates that large numbers of people are affected. Taking into consideration that in Germany about 3 million people are regularly supplied by such institutions, the potential health hazard associated with virus contaminated cold food can clearly be recognized.

Shellfish

Besides cold food, shellfish, especially bivalve molluscs, are commonly involved in outbreaks of foodborne viral diseases. They can acquire viruses by *secondary contamination* just as other cold food. But opposed to other food, molluscs are also affected by *primary mechanical contamination*. Shellfish are filter feeders; if shellfish-growing waters are polluted with human sewage, the shellfish also extract viruses infectious for humans. These viruses are concentrated in the gut, possibly even in the flesh of the molluscs. Depuration does not remove the viruses; they remain in the shellfish for several weeks. The difficulties in detecting viruses in shellfish pose further problems, as well as the fact that correlation between levels of bacteria indicator organisms and the extent of viral contamination is poor.

Oysters are most frequently associated with viral infections, because they are eaten raw. But other shellfish may also contain infective viruses if they are not sufficiently heated, as was the case in an outbreak of viral gastroenteritis caused by a seafood-cocktail consisting of pre-cooked mussels and cockles [15].

The mechanisms of foodborne infections are summarized in Fig. 1.

Viruses transmitted in food

All viruses which are of concern to human health and which were to this date detected in food emanate from the human digestive tract. Therefore, all viruses excreted with feces could theoretically contaminate food (Fig. 2). This comprises viruses of the genera *Enterovirus* and *Hepatovirus* (family *Picornaviridae*), of the genus *Rotavirus* (family *Reoviridae*) and members of

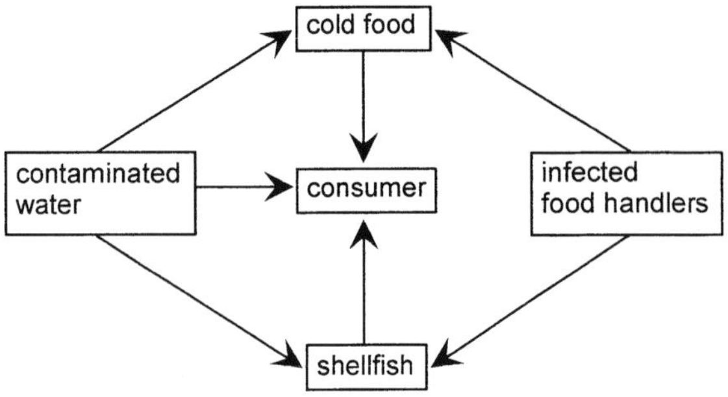

Fig. 1. Mechanisms of foodborne infections

the family *Astroviridae*. Furthermore, human feces may contain *Caliciviruses*, as which hepatitis E virus and small round structured viruses (SRSV's) were classified. Among the DNA-viruses members of the families *Adenoviridae*, *Herpesviridae*, *Parvoviridae* and *Papovaviridae* were detected in feces.

Only for some of these viruses transmission by food was actually proven, all of them uncoated, which can be explained by the greater stability of these viruses. Worldwide, several outbreaks of foodborne viral infections were reported, especially infections with SRSV's and hepatitis A virus. For this report however, foodborne viral diseases occurring in the EU were of interest (Fig. 3).

In several countries of the EU cases of foodborne hepatitis A and E were reported. Furthermore, reports about transmission via food exist concerning polioviruses, SRSV's and SRV's, which will probably be classified as *Parvoviridae*. It is not yet absolutely certain that *Rotaviruses* can be transmitted by contaminated food [16].

Picornaviridae
- Enterovirus
- Hepatovirus

Reoviridae
- Rotavirus

Caliciviridae
- Calicivirus

Astroviridae
- Astrovirus

Adenoviridae
- Mastadenovirus

Herpesviridae
- Betaherpesvirinae

Parvoviridae

Papovaviridae
- Polyomavirinae

Fig. 2. Viruses in human feces

Picornaviridae
  Enterovirus
    Poliovirus
  Hepatovirus
    Hepatitis A
Reoviridae
  Rotavirus
Caliciviridae
  Calicivirus
    SRSV
    Hepatitis E
Parvoviridae
  SRV (no final classification)

Fig. 3. Viruses transmitted in food (proven cases in the EU)

Poliovirus

The paralytic poliomyelitis caused by polioviruses develops after an incubation period of 2 to 35 days; the duration of the disease is variable. Raw milk is commonly implied as vehicle for the virus.

Nowadays poliomyelitis due to consumption of contaminated food only appears in developing countries. In developed countries like the ones of the EU no foodborne cases were reported since the 1950's. This is due to improvement of food hygiene, increased use of pasteurization of milk and also to the introduction of vaccination.

Hepatitis A

Like poliovirus hepatitis A virus has been classified as an *Enterovirus*. After an incubation period of 10 to 50 days hepatitis develops, accompanied by fever, nausea, vomiting and abdominal pain. Jaundice is observed in 70–80% of the infected adults, in children it is less frequent. There are no chronic cases, the disease lasts about 4 weeks, seldom up to 6 months. Only 1% of the patients die.

The vehicles of foodborne hepatitis A are shellfish and cold food. The contamination takes place by way of the already explained modes. It is important to bear in mind that the excretion of hepatitis A virus already starts 10 to 14 days before the onset of clinical signs. Consequently, infected persons can transmit viruses to food for quite some time before they are recognized as source of contamination on account of their illness. In addition, it is often difficult to link an infection to the consumption of contaminated food because of the long incubation period. It is even more demanding to identify the causative food item. As the virus is heat sensitive, infection can be avoided by adequate heat treatment. In 1988 the British Ministry for Agriculture, Fisheries and Food recommended the heating of shellfish for 1.5 min at 85–90 °C. From

Table 1. Reported outbreaks of foodborne Hepatitis A

Date of outbreak	Country	Vehicle	Reference
1963	UK	pork	[10]
1977	Netherlands	steak tatare	[10]
1984/1985	Italy	raw shellfish	[24]
1991	Germany	meat products	[5]

this time on no cases of hepatitis A occured after consumption of properly treated shellfish [4].

As far as the importance of hepatitis A is concerned, it is supposed that 30% of the cases of foodborne viral gastroenteritis are caused by hepatitis A [5]. According to investigations in the USA, 0.4% of the cases of hepatitis A are foodborne [6]. In England and Wales 23 outbreaks of food-associated hepatitis A were reported from 1978 to 1988. All except two of these outbreaks were caused by the consumption of shellfish.

Table 1 shows some reported outbreaks of foodborne hepatitis A.

Small round structured viruses

The SRSV's are also called "Norwalk-like viruses" and have to be distinguished from the small round viruses (SRV's). Yet, the latter cause diseases with similar clinical signs as those due to SRSV's. As there are furthermore only few reports about foodborne SRV-infections [1, 2, 26], these will not be discussed in further detail.

An infection with SRSV's is marked by a sudden onset of symptoms after an incubation of not more than 2 days. The most prominent sign of the disease is projectile vomiting, which appears in more than 50% of the cases. Besides, light diarrhoea and mild abdominal pain occur. In general, the disease is marked by mild symptoms and a duration of only 12 to 60 hours. Consequently, hospitalization or fluid therapy are seldom necessary.

SRSV's are transmitted either via shellfish or via cold food contaminated by food handlers. Shellfish-caused cases of SRSV-infections are sometimes mixed infections with hepatitis A in a way that gastroenteritis is followed by hepatitis after 3 or 4 weeks. Concerning the contamination via infected food handlers, it has to be pointed out that the SRSV's are excreted in vomitus as well as in feces. The excretion begins just a few hours before the onset of clinical signs and lasts only two to three days. Because of the short duration of virus excretion an etiological diagnosis via feces examination is not always possible. In this case infection with SRSV's can only be suspected based on epidemiological data. An important epidemiological feature is the high attack rate, which often exceeds 50%. Furthermore, many secondary cases occur because of person-to-person transmission characteristic for SRSV-infections. As the infective dose is very small, even light contamination of food is sufficient to cause SRSV-

Table 2. Reported outbreaks of foodborne SRSV-infections

Date of outbreak	Country	Vehicle	Author
1979	UK	chicken sandwiches	[27]
1983	UK	oysters	[14]
1983	UK	cold cooked ham	[29]
1985	UK	raw oysters	[18]
1987	UK	cold smoked trout fillets, mixed seafood	[28]
1991	UK	seafood cocktail	[15]
1991	UK	corned beef sandwich	[31]
1993	UK	raw oysters	[9]
1994	UK	turkey salad sandwiches, tuna salad	[22]

outbreaks. The SRSV's are heat sensitive and are inactivated at temperatures above 60 °C. Therefore, adequate heat treatment combined with good kitchen hygiene could prevent foodborne SRSV-infections.

The SRSV's are considered to be the main cause of foodborne illness of viral origin. SRSV's account for only 4% of the cases of sporadic gastroenteritis [34], but 55% of epidemic foodborne viral gastroenteritis are caused by SRSV's worldwide [5]. It is impossible to present accurate figures for the incidence of SRSV-infections, as many cases are not reported. This can be explained on the one hand by the mildness of the symptoms, on the other hand by diagnostic problems. It is suspected that many foodborne diseases with an unknown etiology are caused by SRSV's. For example, in Manchester eight foodborne SRSV-outbreaks were reported from 1982 to 1987. This represents 29% of all food-associated outbreaks of gastroenteritis during this time span. However, for the same number of outbreaks the causative agent could not be determined [34].

Because of the mild course and the short duration of the disease one may tend to neglect the importance of SRSV-infections. But outbreaks usually affect large numbers of people, taking into account that they occur mostly in institutions like schools, hospitals or hotels. Illness among staff may lead to organisational problems and to serious disruption of service in these institutions. Besides, the amount of working days lost may have considerable economic impact. Several outbreaks of foodborne SRSV-infections are listed in Table 2.

Hepatitis E

Like the SRSV's hepatitis E virus has been classified as a member of the *Caliciviridae*. After an incubation period of 15 to 40 days the disease presents itself with anorexia, nausea and vomiting. Hepatic coma develops in 5 to 10% of the cases.

As far as the transmission of hepatitis E via food is concerned, only a few cases have been reported. In these, shellfish was identified as the vehicle for the viruses. Most outbreaks, however, are caused by contaminated water.

Epidemics of hepatitis E occur only in tropical and subtropical developing countries. In the European Union foodborne hepatitis E is of little importance. Merely in some Mediterranean countries like Italy, Spain and Greece sporadic isolated cases were reported [7, 33].

Conclusions

Finally some particularities of foodborne viral diseases shall now be considered in respect to food hygiene.

Unlike bacteria contaminating food, viruses generally remain infective for a long time even in processed foods. Some methods of preservation like refrigeration, freezing and salting even prolong the persistence of viruses in food.

Foodborne viral diseases are generally less severe than other food-associated diseases. Moreover, they are seldom fatal. These are probably the reasons why viral contamination of food has been rather neglected. Viral foodborne diseases are marked by a small infective dose. Therefore, infection, especially of immunocompromised persons, is possible even after light contamination. SRSV's for example have an infective dose of $10–10^2$ particles/ml. This is significantly lower than the detectibility limit of $10^6–10^7$ particles/g for electron microscopy [19]. But not only the small number of viruses in a single food item and the low sensitivity of the common diagnostic methods, also the short duration of virus excretion often render an etiological diagnosis impossible. Furthermore, only some laboratories are properly equipped for virological examinations. Because of these diagnostic difficulties the viral etiology of many foodborne outbreaks of disease is not recognized. On the other hand, the link between illness and the consumption of contaminated food is often not established. All this leads to the conclusion that the true extent of foodborne viral infections appears to be considerably underestimated.

Therefore, definite assessment of the significance of foodborne viral diseases in the European Union is impossible. This implies the need for further research in this area and more thorough information of all parties involved.

References

1. Appleton H, Buckley M, Thom BT, Cotton JL, Henderson S (1977) Virus-like particles in winter vomiting disease. Lancet i: 409–411
2. Appleton H, Pereira MS (1977) A possible virus aetiology in outbreaks of food-poisoning from cockles. Lancet i: 780–781
3. Appleton H (1987) Small round viruses: classification and role in food-borne infections. Ciba Found Symp 128: 108–125
4. Appleton H (1990) Foodborne viruses. Lancet ii: 1 362–1 364

5. Becker B, Krämer J (1995) Übertragung humanpathogener Viren durch Lebensmittel. In: 36. Arbeitstagung des Arbeitsgebietes "Lebensmittelhygiene"/DVG, Garmisch-Partenkirchen 1995. Offset Köhler, Gießen, pp 203–209
6. Blackwell JH, Cliver DO, Callis JJ, Heidelbaugh ND, Larkin EP, McKercher PD, Thayer DW (1985) Foodborne viruses: their importance and need for research. J Food Prot 48: 717–723
7. Caredda F, Antinori S, Re T, Pastecchia C, Zavaglia C, Moroni M (1985) Clinical features of sporadic non-A, non-b hepatitis possibly associated with faecal-oral spread. Lancet ii: 48
8. Caul EO, Ashley C, Pether JVS (1979) "Norwalk"-like particles in epidemic gastroenteritis in the U.K. Lancet ii: 1 292
9. Chalmers JWT, McMillan JH (1995) An outbreak of viral gastroenteritis associated with adequately prepared oysters. Epidemiol Infect 115: 163–167
10. Cliver DO (1981) Gefahren einer Virusinfektion durch Fleisch. Fleischwirtsch 61: 432–434
11. Cliver DO (1990) Viruses. In: Cliver DO(ed) Foodborne diseases. Academic Press, San Diego, pp 275–292
12. Cooke EM (1990) Epidemiology of foodborne illness: UK. Lancet ii: 790–793
13. Fang G, Araujo V, Guerrant RL (1991) Enteric infections associated with exposure to animals or animal products. Infect Dis Clin North Am 5: 681–701
14. Gill ON, Cubitt WD, McSwiggan DA, Watney BM, Bartlett CLR (1983) Epidemic of gastroenteritis caused by oysters contaminated with small round structured viruses. Br Med J 287: 1 532–1 534
15. Gray SF, Evans MR (1993) Dose-response in an outbreak of non-bacterial food poisoning traced to a mixed seafood cocktail. Epidemiol Infect 110: 583–590
16. Greiser-Wilke I, Fries R (1994) Methoden zum Nachweis viraler Kontaminationen in Lebensmitteln tierischer Herkunft. Dtsch Tierarztl Wochenschr 101: 249–300
17. Hedberg CW, Osterholm MT (1993) Outbreaks of food-borne and waterborne viral gastroenteritis. Clin Microbiol Rev 6: 199–210
18. Heller D, Gill ON, Raynham E, Kirkland T, Zadick PM, Stanwell-Smith R (1985) An outbreak of gastrointestinal illness associated with consumption of raw depurated oysters. Br Med J 292: 1 726–1 727
19. Iversen AM, Gill M, Bartlett CLR, Cubitt WD, McSwiggan DA (1987) Two outbreaks of foodborne gastroenteritis caused by a small round structured virus: evidence of prolonged infectivity in a food handler. Lancet ii: 556–558
20. Kaplan JE, Gary GW, Baron RC, Singh N, Schonberger LB, Feldman R, Greenberg HB (1982) Epidemiology of Norwalk gastroenteritis and the role of Norwalk virus in outbreaks of acute nonbacterial gastroenteritis. Ann Int Med 96: 756–761
21. Larkin EP (1981) Food contaminants-viruses. J Food Prot 44: 320–325
22. Lo SV, Connolly AM, Palmer SR, Wright D, Thomas PD, Joyson D (1994) The role of the pre-symptomatic food handler in a common source of foodborne SRSV gastroenteritis in a group of hospitals. Epidemiol Infect 113: 513–521
23. Mayr A (1979) Tatsachen und Spekulationen über Viren in Lebensmitteln. Zentralbl Bakteriol B 168: 109–133
24. Mele A, Rastelli MG, Gill ON, Di Bisceglie D, Rosmini F, Pardelli G, Valtriani C, Patriarchi P (1989) Recurrent epidemic hepatitis A associated with consumption of raw shellfish, probably controlled through public health measures. Am J Epidemiol 130: 540–546
25. Papaevangelou G (1992) Epidemiology of hepatitis A in Mediterranean countries. Vaccine 10: S63–S66

26. Paver WK, Caul EO, Ashley CR, Clarke SKR (1973) A small virus in human faeces. Lancet i: 237–240
27. Pether JVS, Caul EO (1983) An outbreak of food-borne gastroenteritis in two hospitals associated with a Norwalk-like virus. J Hyg Camb 91: 343–350
28. Reid JA, Caul EO, White DG, Palmer SR (1988) Role of infected food handler in hotel outbreak of Norwalk-like viral gastroenteritis: implications for control. Lancet ii: 321–323
29. Riordan T, Craske J, Roberts JL, Curry A (1984) Food borne infection by a Norwalk like virus (small round structured virus). J Clin Pathol 37: 817–820
30. Schunck B, Rziha HJ (1994) Der Nachweis von Viren in Lebensmitteln tierischer Herkunft mit der Polymerase-Kettenreaktion: eine Übersicht. Tierärztl Prax 22: 502–505
31. Stevenson P, McCann R, Duthie R, Glew E, Ganguli L (1994) A hospital outbreak due to Norwalk virus. J Hosp Infect 26: 261–272
32. Tassopoulos NC, Alikiotis M, Limotirakis F, Nicolakakis P, Mela H, Paralogou-Ioannides M (1988) Acute sporadic non-A, non-B hepatitis in Greece. J Med Virol 26: 71–77
33. Torné J, Miralles R, Tomás S, Saballs P (1988) Typhoid fever and acute non-A non-B hepatitis after shellfish consumption. Eur J Clin Microbiol Infect Dis 7: 581–582
34. Viral Gastroenteritis Sub-Committe and the PHLS Virology Committee (1995) Outbreaks of gastroenteritis associated with SRSV's PHLS Microbiol Digest 10: 2–8
35. Wittman RJ, Flick GJ (1995) Microbial contamination of shellfish: prevalence, risk to human health, and control strategies. Annu Rev Public Health 16: 123–140

Authors' address: Prof. Dr. A. Stolle, Institute for Hygiene and Technology of Food of Animal Origin, Veterinary Faculty, Ludwig-Maximilians-University Munich, Veterinärstr. 13, D-80539 Munich, Federal Republic of Germany.

Viral zoonosis from the viewpoint of their epidemiological surveillance: tick-borne encephalitis as a model*

J. Süss, P. Béziat, and **Ch. Schrader**

Federal Institute for Health Protection of Consumers and Veterinary Medicine, Berlin, Federal Republic of Germany

Summary. Tick-borne encephalitis (TBE) is a vector borne and, more rarely, a food (milk, milk products) borne disease of humans. For further characterization of the virus activity in natural foci of TBE more than 32 000 unengorged wild ticks were caught in low and high virus active foci in Germany (Mecklenburg-Western Pomerania, Saxony, Brandenburg, Thuringia, Bavaria, Baden-Württemberg, Saarland). The ticks were examined by RT-PCR and Southern blot hybridization as well as by classical virological methods. The dynamics of such natural foci of TBE in the last 35 years were discussed. Also nucleotide sequence data of parts of the virus genome (5′- non coding region) of 16 European and some Far East subtype strains were compared.

Introduction

The *Flavivirus* genus (family *Flaviviridae*) comprises more than 70 classified members [10, 46], most of which are arthropod-borne [29]. Flavivirions are small, roughly spherical and enveloped, their genome RNA is single-stranded, of positive polarity and approximately 11 kb in length [33].

Tick-borne encephalitis virus (TBEV) is a member of the family *Flaviviridae* and pathogenic for humans. The tick-borne encephalitis virus complex consists of 14 antigenically closely related viruses, 8 of which cause human disease (for review see [26]).

In many countries in Europe and Asia TBE represents by far the most important viral arthropod-borne disease and it is of major public health interest.

Tick-borne encephalitis occurs in an endemic pattern over a wide area of Europe and the former Soviet Union. The annual incidence is several thousands of cases, with considerable variation from year to year. The highest morbidity in Europe has been recorded in Austria, Slovenia, Hungary, Latvia and the former

*In memory of Dr. Herbert Sinnecker (†1991), the founder of the Arbovirology in Germany.

Czechoslovakia. Several hundred cases were registered annually, in the former Soviet Union several thousands.

The disease occurs in foci characterized by ecological habitats [37] favorable for ticks, and very little variation in geographic distribution is observed [26]. The virus is maintained in nature in a cycle involving ticks and wild vertebrate hosts, especially small mammals.

Large mammals such as sheep, cattle and goats are important blood feeding hosts for adult ixodid ticks , but have low viremias and are not considered to be important sources of tick infection. However the virus is excreted in milk. Human infection may thus result from consumption of unpasteurized goat or sheep milk or fresh cheese. Small outbreaks, involving all age groups and often family groups, result from the consumption of raw sheep or goat milk or cheese [26].

Once the natural foci of TBE are recognized, humans can be protected through vaccination. We and others [31, 32, 41–43] have characterized such foci by registration of autochthonous clinical cases in geographic maps. In Germany, all of the 306 cases of TBE were registered in 1994 in the Federal Länder Bavaria and Baden-Württemberg [31, 32]. This was the highest morbidity since registration began. 296 of these cases were autochthonous. Of these, 242 cases were registered in Baden-Württemberg and 57 in Bavaria. In 1995, Kaiser et al. [20] collected data on 156 clinical cases of TBE in Baden-Württemberg.

In our opinion the identification and characterization of a TBE focus requires more complex studies. Sinnecker [37] demanded very early the following as criteria of a focus: prevalence of ticks in the area; presence of naturally infected ticks in the area; naturally infected local small mammals in the area; and (non vaccinated) antibody reactors or TBE in man.

There is a distinct difference between the epidemiological TBE-situations in the north-eastern and southern Länder of Germany. From 1960 to 1994 we carried out an intensive TBE surveillance programme in the northeastern Länder. In the sixties we registered hundreds of human cases annually in Mecklenburg-Western Pomerania, Thuringia, Saxony and Brandenburg [41]. From 1960 to 1990, isolations of TBE virus were made from ticks and small mammals and investigations were performed to detect TBE antibodies in small mammals, game and humans in these Länder of Germany. Confirmed TBE cases for which the respective site of exposure could be determined were also registered. As a result of these epidemiological and ecological investigations, a map was presented showing the natural foci of TBE [41]. TBE was clearly endemic in the sixties, with a morbidity of up to 0.7 per 100 000 inhabitants, which decreased in the seventies to 0.02 per 100 000 inhabitants.

A special surveillance programme in one natural focus was made from 1983 to 1989 to elucidate whether such foci had become extinct or whether they persisted below a detectable level. These investigations performed on the Island of Usedom using conventional virological and serological methods showed that this TBE focus had not disappeared but was in a state of endemic latency [41].

The aim of our present studies was to analyse by PCR and conventional virological techniques the actual epidemiological situation of TBE in the natural foci where the virus had formerly been highly active. Investigations were focused on the testing of unengorged ticks from well characterized old natural foci in the northeastern Länder of Germany.

In addition, ticks from present highly active natural foci in Bavaria and Baden-Württemberg as well as from a non-endemic area in the Saarland were examined. The TBE situation in the Saarland required an intensive study after the first autochthonous clinical case was described in 1994 [44].

We present here the nucleotide sequence of the 5′-terminal noncoding region of 16 different strains of TBE virus (European subtype), also of our present and some former isolates and compare them to other TBE virus strains (European and Far Eastern subtype) and to Powassan virus.

Materials and methods

Collection of ticks

Unengorged ticks (*Ixodes ricinus*) were collected from May to October (1992–94) by flagging the vegetation as described [1]. Pools of 10–200 individuals (nymphs and adults) were kept in glass test tubes at 4 °C until further examination.

Isolation of viruses from ticks

Prior to homogenization, the ticks were washed three times with cold phosphate buffered saline (PBS) (pH 7.2) supplemented with 0.02 mg/ml gentamycin. They were ground in a mortar with seasand and suspended in Earle's minimal essential medium (EMEM) (pH 7.2) with 0.1 mg/100 ml gentamycin and 10% foetal calf serum (FCS). Pools of about 200 nymphs were made up to a final volume of 2 ml; smaller pools were resuspended in appropriate volumes of diluent. Subsequently each of these suspensions were used for virus isolation attempts, the remaining fluids were stored at −80 °C.

Virus isolation in cell cultures

Chick embryo cells (CEC)

CEC were seeded with 1×10^6 cells in 2 ml EMEM (pH 7.2) supplemented with 10 % FCS and 0.1 mg/100ml gentamycin. After incubation for 24 h at 37 °C the medium was removed, 0.2 ml tick suspension was added to each of two tubes and absorption was carried out for 30–60 min. After that each tube was filled with 1.8 ml EMEM (pH 7.2) supplemented with 4% FCS and 0.1 mg/100 ml gentamycin and the cells were cultivated at 37 °C. Three days post infection (p. i.) the supernatant from 1 tube was repassaged in CEC cultures, the second tube was cultivated until 6 days p.i. Each tick suspension was grown on CEC cultures and passaged three times.

Indirect fluorescent antibody technique (IFAT)

CEC cultures infected with tick suspension for 18–24 h were washed and trypsinized, then washed with PBS twice and resuspended to 5×10^6 cells/ml in physiological saline supplemented with 5% FCS.

10 μl cell suspension was dropped on 15-well multitest slides, air-dried and fixed with acetone for 10 min. It was overlaid with rat serum for 30 min, washed for 20 min in 2 changes of PBS, rinsed with distilled water and finally treated with 10 μl anti-rat globulin marked with fluorescein isothiocyanate (FITC) for 30 min. The staining took place in a moist chamber at 37 °C. The following hyperimmune rat antisera were used:

Flaviviruses: TBEV strain Hypr, TBEV strain Sofjin, TBEV strains IZ 58/65 and IZ 77/66;

Alphaviruses: Semliki forest virus, Semliki like viruses IM 236/66, IM 299/68, IZ 352/70;

Uukuviruses: Uukuniemi virus, Uukuniemi virus IZ 18/87. The virus strains were obtained from the former Institute of Virology, Slovak Academy of Science, Bratislava, the Institute Ivanowski, Moscow, and the former Central Laboratory for Encephalitis Viruses, Berlin.

Neutralization test (NT)

The NT in CEC cultures was carried out as usual with constant serum dilution and decreasing virus dilution.

RNA extraction

Viral RNA was isolated by a modified form of the procedure described by Chomczynski and Sacchi [7]. Briefly, 100 μl tick homogenate was mixed with 300 μl of guanidine isothiocyanate lysis buffer (4 M guanidine isothiocyanate, 1 M tris-HCl, 1% 2-mercaptoethanol and 0.5% sarcosyl). The solution was sequentially mixed with 300 μl water-saturated phenol and 150 μl chloroform and was then mixed vigorously for at least 30 sec. After centrifugation at 13 000 rpm for 5 min the aqueous phase was removed and combined with 25 μl 3 M sodium acetate and an equal volume of isopropanol to precipitate the RNA. Precipitation took place overnight at −20 °C, then the RNA was collected by centrifugation at 14 000 rpm for 30 min at 4 °C. The resulting RNA pellet was washed with 70% ethanol, dried in an exsiccator and dissolved in 20 μl of sterile distilled water.

Primers

Seven primers were selected with high sequence homology to regions in the 5′-noncoding region of two TBE viruses, Central European encephalitis subtype, Neudoerfl strain (CEE Neudoerfl) [22] and Russian Spring and Summer encephalitis subtype, Sofjin strain (RSSE Sofjin) [28]. The sequences of the primers 1–6 have been published previously [30]. Primers 2, 4, and 5 were degenerate in one position each, thus increasing the sequence homology to both subtypes of the TBE virus. Primers were synthesized by TIB Molbiol, Berlin, Germany.

Reverse transcription

Target viral RNA was converted to a DNA copy (cDNA) prior to enzymatic DNA amplification by use of reverse transcriptase and the TBEV specific reverse primer 1. 2 μl dissolved RNA sample was mixed with 3 μl sterile distilled water. The sample was heated at 74 °C for 5 min, cooled on ice and added to a reaction mixture containing 4 μl reverse transcriptase buffer (supplied with the enzyme), 5 mM dithiothreitol, 20 units RNAsin, 0.5 mM of each deoxynucleotide triphosphate, 25 pmol of primer 1 and 100 units of

M-MLV reverse transcriptase (Gibco BRL) in a final volume of 20 µl. Primer annealing took place at 42 °C for 5 min, cDNA synthesis was carried out at 37 °C for 55 min. Finally the mixture was heated to 94 °C for 5 min and cooled down on ice.

Polymerase chain reaction

2 µl (= 10%) of the RT reaction mixture was used as template in the subsequent amplification which was performed in a final volume of 25 µl (2.5 µl Taq DNA polymerase buffer, 1.5 mM magnesium chloride, 0.1 mM each of the four deoxynucleotide triphosphates, 12.5 pmol of both sense and antisense primers used and 1 unit Taq DNA polymerase (Promega). An overlay of two drops of mineral oil was added. The amplification was performed in an automated thermal cycler (Biometra) with two cycles at 96 °C for 1 min, 37 °C for 1 min, 72 °C for 2 min, followed by 38 cycles of 92 °C for 1 min, 37 °C for 1 min and 72 °C for 2 min. Final extension was 72 °C for 5 min.

A second amplification (nested PCR) was initiated with 1 µl of the first amplification product. The reaction mixture contained all the components described for the initial amplification; primers 3 and 4 were used. Cycling was performed as usual. Amplified products were visualized by staining with ethidium bromide following electrophoresis in a 2% agarose gel.

Southern blot hybridization

Amplification products from agarose gel were transferred to a nylon membrane by capillary transfer using 0.4 M NaOH [38].

DNA was fixed to the membrane by UV irradiation for 5 min. An internal digoxigenin-labelled probe was synthesized by standard PCR using digoxigenin (DIG)-UTP (Boehringer) and primers 5 and 6.

TBEV strain IZ 58/65 [1] served as template for reverse transcription and PCR. The membrane was prehybridized with hybridization buffer at 68 °C for 1 h. 4 µl DIG-labelled probe was added to the hybridization buffer, and the membrane was incubated overnight at 68 °C. After the hybridization, the membrane was washed twice with 2 × sodium-sodiumcitrate (SSC)/0.1 % SDS at 68 °C for 15 min. The bound probes were detected by incubation with alkaline phosphatase-labelled antibody to DIG and Lumigen PPD (Boehringer) in accordance with the manufacturer's protocol. Visualization of bound probes was accomplished by exposing Agfa Curix XP film to the membrane for 2–5 min.

Nucleotide sequence determination

The second amplification (nested PCR) was initiated using a biotinylated sense-primer. DNA-strands were separated with streptavidin-coated magnetic beads (Dynabeads) according to the manufacturer's protocol.

The single-stranded amplification products were mixed with 25 pM of biotinylated primer and 2 µl annealing buffer, incubated at 65 °C for 2 min and cooled down to 30 °C over a period of 30 min. This solution was mixed with 1 µl dithiothreitol (DDT) (0.1 M), 2 µl labeling mix (containing dATP (0.4 nM), dCTP (0.4 nM), dGTP (0.4 nM), dTTP (0.26 nM), DIG-dUTP (0.14 nM)), 1 µl ATP (10 µM), 1 µl Mn buffer (0.1 M $MnCl_2$), and 2 µl Sequenase (26 units). After incubation at room temperature for 5 min, the material was divided into 4 identical aliquots and 2.5 µl of either ddATP, ddGTP, ddCTP, or ddTTP, respectively, were added to one of the tubes.

After incubation at 50 °C for 10 min the reaction was terminated by addition of 4 μl of stop solution. The labeled extended fragments were subjected to electrophoresis using standard sequencing polyacrylamide (6%) thin gels, then the DNA fragments were transferred to a nylon membrane with a Two step Direct Blotter (Hoefer Scientific Instruments, USA), and were fixed to the membrane by UV irradiation (3x 1200 μ Joule × 100). SEQ-LIGHT Detection Kit (SERVA) was used for visualization of the biotin-labeled probes.

Results

In order to study the virus activity in the former natural foci of TBE in north-east Germany we collected 18 760 ticks (*Ixodes ricinus*), subdivided them in 260 tick pools according to the collecting areas and screened them by conventional and molecular biological methods (Table 1). Using RT-PCR it was possible to detect TBEV RNA in 6 pools. The tick pools were collected in the regions Darß-Zingst (strains IZ 100/92, IZ 102/92) and Usedom (strains IZ 11/92, IZ 72/92) in Mecklenburg-Western Pomerania, in the Märkische Schweiz (strain IZ 44/92) in Brandenburg and in the Elster Valley near Gera (strain IZ 91/93) in Thuringia (Table 2). Another TBE virus strain (IZ 68/92) from ticks from the Island of Usedom could be detected by IFAT.

Comparative investigation of 3 553 ticks divided in 34 pools from well known and high virus active natural foci in Bavaria and Baden-Württemberg resulted in the isolation of 2 TBE virus strains (IZ 34/92, IZ 61/93, Bavarian Forest) in CEC cultures as well as their proof by PCR. The strain IZ 34/92 could also be isolated in suckling mice. Another virus strain (IZ 62/92, Black Forest) could be detected by RT-PCR (Table 2).

In Saarland 8.780 ticks were collected in 70 habitats from all the districts and analysed using the PCR in 21 pools. It was possible to detect TBEV RNA in two of the 21 tick pools; these ticks were collected in Perl/Mettlach/Schwemlingen and Saarlouis/Überherrn.

In 1994 in special foci in Thuringia 1 171 ticks (740 nymphs; 431 adults, 206 ♀♀, 225 ♂♂) were collected and divided in 31 pools. 7 of these pools were positive in the RT-PCR (Bad Berka 2×; Erlau; Dienstedt 2×; Luisenthal; Georgenthal).

In addition to these TBE virus strains it was also possible to isolate some other tick-borne virus strains in CEC cultures. In the tick pools from Mecklenburg-Western Pomerania we found 6 Uukuniemi and 2 Tribeč virus strains; besides there were 2 Uukuniemi virus strains in pools from Saxony and 3 Tribeč virus strains in pools from Thuringia.

The nucleotide sequences of the 5′-termini of 16 different strains of TBE virus, European subtype, were determined by non-radioactive dideoxy-chain terminating sequencing. The results can be seen in Fig. 1. Five of the virus strains were isolated from ticks collected in natural foci all over Germany (IZ 11/92, IZ 34/92, IZ 62/92, IZ 61/93, IZ 91/93). Other TBE virus strains were obtained from the former Central Laboratory for Encephalitis Viruses, Berlin, Germany.

Table 1. Origin of the ticks examined by cell culture (CEC) and RT-PCR 1992–1993

Federal Länder	Collecting area	Larva	Nymph	♀♀	♂♂	Total
Mecklenburg-Western Pomerania	Rügen		563	12	11	586
	Darß-Zingst	54	288	16	11	369
	Rostock		129	8	7	144
	Greifswald-Lubmin-Wolgast	6	327	7	7	347
	Island of Usedom	102	3650	113	93	4258
	Schwaan-Bützow-Neukloster		274	28	16	318
	Mecklenburgische Schweiz		268	38	35	341
	Ückermünde		35		2	37
	Schwerin-Gadebusch		140	7	3	150
	Lake area of Mecklenburg	124	808	71	53	1056
	Neubrandenburg-Burg Stargard		100	3	2	105
	Hagenow-Ludwigslust		72		2	74
Brandenburg	Märkische Schweiz		2235	21	30	2286
	Schorfheide		920	52	42	1014
	Gransee/Uckermark		269	16	12	297
Saxony	Sächsische Schweiz		2381	40	38	2459
	Chemnitz		98	4	2	104
Saxony-Anhalt	Delitzsch			20[a]	20[a]	40[a]
Thuringia	Kyffhäuser-Hainleite/Finne	2	235	95	95	427
	Eichsfeld		28	8	8	44
	Thuringian Basin		1065	554	445	2064
	Saale Barrage		240	191	180	611
	Neustadt		65	27	45	137
	Elster Valley-Gera-Altenburg		616	511	365	1492
Rhineland-Palatinate	Siebengebirge		700	22	26	748
Baden-Württemberg	Black Forest		340	29	27	396
Bavaria	Bavarian Forest		2253	64	65	2382
Hesse	Frankfurt/Main		17	7	3	27
Total		288	18416	1964	1645	22313

[a]*Dermacentor marginatus*

Sequence comparison indicated that the 5′-terminal non coding regions of the different TBEV isolates are almost identical. Alignment of the nucleotide sequences of different TBE virus 5′-non-coding regions with those of the TBE virus Neudoerfl strain showed that 13 TBEV 5′-non-coding region sequence differed by 3 and that 1 TBEV 5′-non coding region sequence differed by 4 nucleotides over the remainder of the 5′-non-coding region. The sequences of two strains (IZ 34/92, IZ 61/63) were identical to the 5′-terminus of TBEV European subtype, Neudoerfl strain.

Table 2. Origin of PCR/Southern blot positive tick homogenates (Table 1)

Virus isolate	Collecting area	Number of ticks (nymphs, females, males)
IZ 11/92	Island of Usedom	0 n/9 f/ 2 m
IZ 44/92	Märkische Schweiz	200 n/ 0 f/ 0 m
IZ 72/92	Island of Usedom	180 n/ 0 f/ 0 m
IZ 100/92	Darß-Zingst	10n/ 0 f/ 0 m
IZ 102/92	Darß-Zingst	24 n/ 1 f/ 0 m
IZ 91/93	Elster Valley	0 n/75 f/ 0 m
IZ 34/92	Bavarian Forest	200 n/ 0 f/ 0 m
IZ 62/92	Black Forest	0 n/ 15 f/ 5m
IZ 61/93	Bavarian Forest	145 n/ 10 f/ 1 m

The sequence homology of the 5′-non-coding region between the virus strains examined and European prototype TBEV Neudoerfl strain ranged between 95.2 % and 100 %. This high conservation was observed even for TBEV isolates collected over a period of 29 years (1965–1993) from distant geographic locations (former Soviet Union, Poland, Czechoslovakia, Eastern and Western Germany) and different sources (human blood; ticks, *I. ricinus*; mite; mice, *Clethrionomys glareolus*; sentinel rabbits; birds, sea swallow *Riparia riparia*) [43].

Sequence homology of the 5′-non-coding region of TBEV Neudoerfl strain and Far Eastern prototype Sofjin strain was 83.3%, but homology of tick-borne flaviviruses TBE (European subtype)and Powassan virus was found to be only 68.7% (Table 3).

Discussion

It is well known that the natural foci of TBE in north-east Germany were very active in the early sixties [41]. Later, the virus activity decreased distinctly [43]. In 1992 we examined sera of small mammals and game in order to detect TBEV antibodies [41]. Attempts were also made to culture virus from ticks and small mammals. Using the Island of Usedom as an example, we showed that this natural focus had not disappeared, but still showed endemic latency. In our present study from 1992 to 1994 by RT-PCR it was possible to detect TBEV RNA in 18 out of 346 tick pools; 13 out of 291 tick pools from the north-east Länder, 3 out of 34 pools from Bavaria and Baden-Württemberg and 2 out of 21 pools from the Saarland. These results are in good agreement with the actual morbidity rates in Germany, which are very low in eastern Germany with 1–3 autochthonous cases per year and rather high in southern Germany with hundreds of cases annually.

We were able to show that the natural foci of TBE in the so-called new Federal Länder of Germany are not extinguished. These results also indicate

```
N          CGGACAGCAT TAGCAGCGGT TGGTTTGAAA GAGATATTCT TTTGTTTCTA

IM 268 G         ---- ---------- ---------- --A------- -------T--
IKa18III      ------- ---------- ---------- --?------- -------T--
IV 122/L            - ---------- ---------- --A------- -------T--
IZ 131         ------ ---------- ---------- --A------- -------T--
IM 271 G               --------- ---------- --A-????-- -------T--
IZ 77a        ------- ---------- ---------- --A------- -------T--
IP 60              -- ---------- ---------- --A------- -------T--
Milben 1       ------ ---------- ---------- --A------- -------T--
IZ 77          ------ ---------- ---------- --A------- -------T--
IZ 58/65          --- ---------- ---------- --A------- -------T--
IZ 11/92       ------ ---------- ---------- -GA------- -------T--
IZ 34/92      ------- ---------- ---------- ---------- ----------
IZ 62/92       ------ ---------- ---------- --A------- -------T--
IZ 61/93      ------- ---------- ---------- ---------- ----------
IZ 91/93       ------ ---------- ---------- --A------- -------T--

N          CCAGTCGTGA ACGTGTTGAG AAAAAGACAG CTTAGGAGAA CAAGAGCTGG

IM 268     ---------- ---------- ---------- ---------- ----------
IM 268 G   ---------- ---------- ---------- ---------- ----------
IKa18III   ---------- ---------- ---------- ---------- ---------?
IV 122/L   ---------- ---------- ---------- ---------- ----------
IZ 131     ---------- ---------- ---------- ---------- ----------
IM 271 G   ---------- ---------- ---------- ---------- ----------
IZ 77a     ---------- ---------- ---------- ---------- ----------
IP 60      ---------- ---------- ---------- ---------- ----------
Milben 1   ---------- ---------- ---------- ---------- ----------
IZ 77      ---------- ---------- ---------- ---------- ----------
IZ 58/65   ---------- ---------- ---------- ---------- ----------
IZ 11/92   ---------- ---------- ---------- ---------- ----------
IZ 34/92   ---------- ---------- ---------- ---------- ----------
IZ 62/92   ---------- ---------- ---------- ---------- ----------
IZ 61/93   ---------- ---------- ---------- ---------- ----------
IZ 91/93   ---------- ---------- ---------- ---------- ----------

N          GGATGGTCAA GAAGGCCATC CTGAAAGGTA

IM 268     -------C-- ---------- ---
IM 268 G   -------C-- -------
IKa18III   -------C-- ---------- ---
IV 122/L   -------C-- ---------- ---
IZ 131     -------C-- ---------- -
IM 271 G   -------C-- ---------
IZ 77a     -------C-- ---------- -
IP 60      -------C-- ---------- ---
Milben 1   -------C-- ---------- ---
IZ 77      -------C-- ---------- ---
IZ 58/65   -------C-- ---------- -
IZ 11/92   -------C-- ---------- -----
IZ 34/92   ---------- ---------- ---
IZ 62/92   -------C-- ---------- ---
IZ 61/93   ---------- ---------- -----
IZ 91/93   -------C-- ---------- ----
```

Fig. 1. Nucleotide sequence determination of the 5′-terminals of the nested PCR products of the TBEV strains (European subtype) IM 286, IM 268 G, IKa 18 III, IV 122/L, IZ 131, IM 271 G, IZ 77a, IP 60, Milben 1, IZ 77, IZ 58/65, IZ 11/92, IZ 34/92, IZ 62/92, IZ 61/93 and IZ 91/93. For further characteristics of these strains see [43]. The TBEV sequences were compared to a TBEV sequence, subtype Neudoerfl [22]. The AUG codon initiating the C-protein is underlined. *N* Neudoerfl; – identity with the TBEV Neudoerfl strain; ? unknown strain nucleotides, underlined are the primers for nested PCR: sense primer 4 pos. 31–48, reverse primer pos. 142–158

Table 3. 5′-noncoding region sequence homology between different tick-borne flaviviruses and TBEV strain Neudoerfl

Virus strain	Number of nucleotides	Sequence homology (%)
IZ 34/92	99[a]	100,0
Hypr	84[a]	95,2
IM 268	98[a]	97,9
IZ 77a	98[a]	97,9
Sofjin	126[b]	83,3
205	131[b]	83,3
POW	111[b]	68,7

[a]Part of the 5′-noncoding region
[b]Entire 5′-noncoding region

that areas which have been characterized as natural foci of TBE for the past 30 years require further surveillance. The results in ticks of Thuringia from 1994 show the further existence of active microfoci of TBEV.

Until recently, the Saarland was considered to be free of TBE, because no case of TBE in the Saarland had ever been documented. In 1994, the first TBE case aquired in the Saarland (near Saarlouis) was documented [44]. In our Saarland study, we systematically collected ticks in all districts. Two tick pools were positive in RT-PCR, once of which came from the Saarlouis area.

Furthermore we did nucleotide sequence determinations and molecular epidemiological studies using sequence data of our virus isolates and those of published.

During the past few years the genomic sequences of a growing number of flaviviruses have been elucidated (for review see [6]). Most of the sequence data, however, are from mosquito-borne flaviviruses, whereas for tick-borne flaviviruses, complete genome sequences are available only for the following: TBE virus European subtype strain Neudoerfl [21, 22], Far Eastern subtype strain Sofjin [27, 28, 50], strain 205 [34] and Powassan virus [25].

Furthermore, partial sequence data have been reported for TBE virus strain Hypr [23], Kumlinge strain [48], Vasilchenko strain [13], strain K23 [19], turkish sheep encephalitis virus [49], Louping ill virus strain 369/T2 [36], strains 31, K, and NOR [12], strain SB 526 [46], Negishi virus [46], Langat virus strains TP 21, Yelantsev [24], and Omsk hemorrhagic fever virus [14].

Since most of these data give information on flaviviral protein coding regions, as yet little is known about the terminal non-coding regions of the flaviviral genome. These regions are thought to carry signal sequences and conserved secondary structures, which might be involved in the initiation and regulation of viral RNA synthesis [40]. Among related viruses, such signals are generally highly conserved.

In the flaviviral genome signal sequences are thought to be located in the short non coding regions flanking the long open reading frame [6, 25]. The 3′-terminal non-coding region of the genome RNAs of mosquito-borne flaviviruses contains conserved primary sequence motifs and stable secondary structures which seem to be essential for virus replication [3, 6]. For the 5′-terminal non-coding region, only short regions of conserved sequences were observed between flaviviruses of different subgroups, but a high degree of 5′-non coding region sequence conservation was observed between different members of the same subgroup. Such is known between Dengue Fever Virus-2 [9] and Dengue Fever Virus-4 [51] and between St. Louis Encephalitis Virus (SLEV), West Nile Fever Virus, and Murray Valley Encephalits Virus [4].

The sequence analysis of the 5′-terminal non-coding region of 16 TBEV strains (European subtype) indicated an almost complete conservation among the RNA in this part of the genome. The comparison with other tick-borne flaviviruses showed a significant gradual decrease of sequence homology with decreasing degree of relationship. It was hardly possible to find any similarities to the 5′-terminal non-coding region of mosquito-borne flaviviruses.

Our sequence data and published sequences of different tick-borne flaviviruses were used to assess the relationship between different TBEV strains and between TBEV and Powassan virus. By means of serological parameters, Powassan virus has previously been shown to be the most divergent member of the tick-borne serocomplex [5].

A very similar result for sequence homology of the 5′-non-coding region between virus strains and a high conservation was reported by Brinton and Dispoto [4] who determined the 5′-terminal RNA sequence for 7 strains of SLEV. Although the SLEV strains differed from each other in their virulence phenotype, the sequence of the entire 5′-non-coding region was highly conserved among them.

Flaviviruses were subdivided into eight serological complexes based on cross-neutralization [5, 8]. This classification was confirmed by other assays, such as complement-fixation and plaque reduction neutralization tests [45], peptide mapping, use of monoclonal antibodies [11, 16, 17, 39], RNA-DNA hybridization [35, 52], and amino acid sequence homology of the E protein [18]. Our results show that there is also a strong connection between the degree of non-coding region sequence conservation and the degree of relationship of different flaviviruses.

The results shown in Table 3 are in good agreement with Mandl et al. [25], who showed that Powassan virus is more distantly related to the other members of the tick-borne serocomplex than these are to each other. However, sequence identities within the tick-borne flavivirus group are at least 30 % higher than between any tick-borne and any mosquito-borne flaviviruses investigated [25]. Two short conserved RNA sequences, located 5′ to a putative 3′-terminal secondary structure, are shared by all mosquito-borne flaviviruses. One of these is complementary to a conserved sequence near the 5′ end of the viral RNAs, suggesting that mosquito-borne flavivirus RNAs can form cycles [15]. Comparison of the 5′ and 3′-non-coding region of the tick-borne flaviviruses to mosquito-borne flaviviruses revealed very little primary sequence conservation [25]. It thus appears that while primary nucleotide sequence conservation occurs among mosquito-borne flaviviruses and among tick-borne flaviviruses, it is very limited between the two vector groups. One possible explanation for this strict division was given by Blok et al. [2], who compared the genome sequences of different flaviviruses and produced dendrograms which supported the traditional groupings based on serological evidence. They suggested that the flaviviruses have evolved by divergent mutational change, and that the primary division of the flaviviruses was between those transmitted by mosquitos and those transmitted by ticks.

Acknowledgements

Parts of this work were supported by grants from the Grimminger-Stiftung für Zoonosenforschung, Stuttgart. We thanks Drs. W. Dorn, Jena, A. Haas, Homburg, H. P. Rohr, Saarbrücken, E. Schreier, Berlin, J. Treib, Homburg and B. Kriebel, Berlin for many helpful discussions and relevant informations and for collecting ticks. We wish to thank Dr. W.-P. vogt, Berlin, for critically reading the manuscript.

References

1. Apitzsch L, Sinnecker H, Wigand R, Berndt D (1968) Zeckencephalitis-Virusisolierungen in der DDR 1965/66 und einige Stammdifferenzierungen. Zbl Bakt I Orig 207: 429–434
2. Blok J, McWilliam SM, Butler HC, Gibbs AJ, Weiller G, Herring BL, Hemsley AC, Aaskov Jg, Yoksan S, Bhamarapravati N (1992) Comparison of a Dengue-2 virus and its candidate vaccine derivative: sequence relationships with the flaviviruses and other viruses. Virology 187: 573–590
3. Brinton MA, Fernandez AV, Dispoto JH (1986) The 3′-nucleotides of flavivirus genomic RNA form a conserved secondary structure. Virology 153: 113–121
4. Brinton MA, Dispoto JH (1988) Sequence and secondary structure analysis of the 5′-terminal region of flavivirus genome RNA. Virology 162: 290–299
5. Calisher CH, Karabatsos N, Dalrymple JM, Shope RE, Porterfield JS, Westaway EG, Brandt WE (1989) Antigenic relationships between flaviviruses as determined by cross-neutralization tests with polyclonal antisera. J Gen Virol 70: 37–43
6. Chambers TJ, Hahn CS, Galler R, Rice CM (1990) Flavivirus genome, organization, expression and replication. Ann Rev Microbiol 44: 649–688
7. Chomczynski P, Sacchi N (1987) Single-step method of RNA isolation by acid guanidium thiocyanate-phenol-chloroform extraction. Anal Biochem 162: 156–159
8. DeMadrid AT, Porterfield JS (1974) The flaviviruses (Group B Arboviruses): a cross-neutralization study. J Gen Virol 23: 91–96
9. Deubel V, Kinney RM, Trent DW (1986) Nucleotide sequence and deduced amino acid sequence of the structural proteins of Dengue type 2 virus, Jamaica genotype. Virology 155: 365–377
10. Francki RIB, Fauquet CM, Knudson DL, Brown F (eds) (1991) Classification and Nomenclature of Viruses. Fifth Report of the International Committee on Taxonomy of Viruses. Springer, Wien New York, pp 223–233 (Arch Virol [Suppl] 2)
11. Gao GF, Hussain MH, Reid HW, Gould EA (1993a) Classification of a new member of the TBE flavivirus subgroup by its immunological, pathogenic and molecular characteristics: Identification of subgroup-specific pentapeptides. Virus Res 30: 129–144
12. Gao GF, Jiang WR, Hussain MH, Venugopal K, Gritsun TS, Reid HW, Gould EA (1993b) Sequencing and antigenic studies of a Norwegian virus isolated from

encephalomyelitic sheep confirm the existence of Louping ill virus outside Great Britain and Ireland. J Gen Virol: 109–114
13. Gritsun TS, Frolova TV, Pogodina VV, Lashkevich VA, Venogopal K, Gould EA (1993a) Nucleotide and deduced amino acid sequence of the envelope gene of the Vasilchenko strain of TBE virus, comparison with other flavivirus. Virus Res 27: 201–209
14. Gritsun TS, Lashkevich VA, Gould EA (1993b) Nucleotide and deduced amino acid sequence of the envelope glycoprotein of Omsk haemorrhagic fever, comparison with other flaviviruses. J Gen Virol 74: 287–291
15. Hahn CS, Hahn YS, Rice CM, Lee E, Dalgarno L, Strauss EG, Strauss JH (1987) Conserved elements in the 3′ untranslated region of flavivirus RNAs and potential cyclization sequences. J Mol Biol 198; 33–41
16. Heinz FX, Kunz C (1981) Homogeneity of the structural glycoproteins from European isolates of tick-borne encephalitis virus: comparison with other flaviviruses. J Gen Virol 57: 263–274
17. Heinz FX, Berger R, Tuma W, Kunz C (1983) A topological and functional model of epitopes on the structural glycoprotein of Tick-borne encephalitis virus defined by monoclonal antibodies. Virology 126: 525–537
18. Heinz FX, Mandl CW, Guirakhoo F, Holzmann H, Tuma W, Kunz C (1990) The envelope protein E of tick-borne encephalitis virus and other flaviviruses: structure, functions and evolutionary relationships. Arch Virol [Suppl] 1: 125–135
19. Jacobs SC, Stephenson JR, Wilkinson GWG (1993) Sequence of the NS 1 gene of the K23 isolate of Tick-borne encephalitis virus and identification of conserved motifs. Arch Virol 129: 279–285
20. Kaiser R, Kern A, Fressle R, Steinbrecher A, Omram H, Malzacher V, Kügler D, Kampa D, Batsford S (1996) Zeckenvermittelte Erkrankungen in Baden-Württemberg. Münch Med Wochenschr 138: 647–652
21. Mandl CW, Heinz, FX, Kunz C (1988) Sequence of the structural proteins of tick-borne encephalitis virus (western subtype) and comparative analysis with other flaviviruses. Virology 166: 197–205
22. Mandl CW, Heinz FX, Stöckl E, Kunz C (1989) Genome sequence of tick-borne encephalitis virus (western subtype) and comparative analysis of non-structural proteins with other flaviviruses. Virology 173; 291–301
23. Mandl CW, Kunz, C, Heinz, FX (1991a) Presence of poly(A) in a flavivirus: significant differences between the 3′ non coding regions of the genomic RNAs of tick-borne encephalitis virus strains. J Virol 65: 4 070– 4 077
24. Mandl CW, Iacono-Connors L, Wallner G, Holzmann H, Kunz C, Heinz FX (1991b) Sequence of the genes encoding the structural proteins of the low-virulence tick-borne flaviviruses Langat TP 21 and Yelantsev. Virology 185: 891–895
25. Mandl CW, Holzmann H, Kunz C, Heinz FX (1993) Complete genomic sequence of Powassan virus: evaluation of genetic elements in tick-borne versus mosquito-borne flaviviruses. Virology 194: 173–184
26. Monath TP, Heinz FX (1996) Flaviviruses. In: Fields BN, Knipe DM, Howley PM (eds) Virology. Lippincott Raven, Philadelphia New York, pp 961–1 034
27. Pletnev AG, Yamshchikov VF, Blinov VM (1986) Tick-borne encephalitis virus genome. The nucleotide sequence coding for virion structural proteins. FEBS Lett: 317–321
28. Pletnev AG, Yamshchikov VF, Blinov VM (1990) Nucleotide sequence of the genome and complete amino acid sequence of the polyprotein of Tick-borne encephalitis virus. Virology 174: 250–263

29. Porterfield JS (1994) Encephalitis viruses. In: Webster RG, Granoff A (eds) Encyclopedia of virology, vol 1. Academic Press, London, pp 361–372
30. Ramelow C, Süss J, Berndt D, Roggendorf M, Schreier E (1993) Detection of tick-borne encephalitis virus RNA in ticks (Ixodes ricinus) by the polymerase chain reaction. J Virol Methods 45: 115–119
31. Roggendorf M, Lenz P (1994) Änderungen der Epidemiologie der FSME in Süddeutschland. Ellipse 11: 33–40
32. Roggendorf M, Lenz P, Kaiser R, Jäger G (1995) Epidemiologische Situation der FSME in Bayern und Baden-Württemberg. In: Süss J (ed) Durch Zecken übertragbare Erkrankungen FSME und Lyme-Borreliose. Weller, Schriesheim, pp 7–19
33. Rice CM, Strauss EG, Strauss JH (1986) Structure of the flavivirus genome. In: Schlesinger S, Schlesinger MJ (eds) The Togaviridae and Flaviviridae. Plenum Press, New York London, pp 279–326
34. Safronov PF, Netesov SV, Mikryukova TP, Blinov VM, Osipova EG, Kiseleva NN, Sandakhchiev LS (1991) Nucleotide sequence of genes and complete amino acid sequence of proteins of tick-borne encephalitis virus strain 205. Mol Gen Microbiol Virusol 4: 23–29
35. Shamanin VA, Pletnev AG, Rubin SG, Zlobin VI (1990) Differentiation of strains of tick-borne encephalitis virus by means of RNA-DNA hybridization. J Gen Virol 71: 1 505–1 515
36. Shiu SYW, Ayres MD, Gould EA (1991) Genomic sequence of the structural proteins of Louping ill virus: comparative analysis with tick-borne encephalitis virus. Virology 180: 411–415
37. Sinnecker H (1986) Tick-borne encephalitis – German Democratic Republic. In: Tick-borne encephalitis and haemorrhagic fever with renal syndrome in Europe – Report on a WHO meeting 1983. EURO Rep Studies 104: 19–23
38. Southern EM (1975) Detection of specific sequences among DNA fragments separated by gel electrophoresis. J Mol Biol 98: 503–517
39. Stephenson JR, Lee JM, Wilton-Smith PD (1984) Antigenic variation among members of the tick-borne encephalitis complex. J Gen Virol 65: 81–89
40. Strauss EG, Strauss JH (1983) Replication strategies of the single-stranded RNA viruses of eukaryotes. Curr Top Microbiol Immunol 105: 1–98
41. Süss J, Sinnecker H, Sinnecker R, Berndt D, Zilske E, Dedek G, Apitzsch L (1992) Epidemiology and ecology of tick-borne encephalitis in the eastern part of Germany between 1960 and 1990 and studies on the dynamics of a natural focus of tick-borne encephalitis. Zbl Bakt 277: 224–235
42. Süss J, Béziat P, Ramelow C, Berndt D (1995) Untersuchungen zur epidemiologischen Situation der FSME in den neuen Bundesländern von 1992–1994. In: Süss J (ed) Durch Zecken übertragbare Erkrankungen FSME und Lyme-Borreliose. Weller, Schriesheim, pp 25–32
43. Süss J, Béziat P, Ramelow C, Kahl O (1997) Tick-borne encephalitis virus (TBEV)-specific RT-PCR for characterization of natural foci of TBE and some other applications. Zbl Bakt 286: 125–138
44. Treib J (1994) First case of tick-borne encephalitis (TBE) in the Saarland. Infection 22: 368–369
45. Varelas-Wesley I, Calisher CH (1982) Antigenic relationships of flaviviruses with undetermined arthropod-borne status. Am J Trop Med Hyg 31: 1 273–1 284
46. Venugopal K, Buckley A, Reid HW, Gould EA (1992) Nucleotide sequence of the envelope glycoprotein of Negishi virus shows very close homology to Louping ill virus. Virology 190: 515–521

47. Westaway EG, Brinton MA, Gaidamovich SY, Horzinek MC, Igarashi A, Kääriäinen L, Lvov DK, Porterfield JS, Russell PK, Trent DW (1985) Flaviviridae. Intervirology 24: 183–192
48. Whitby JE, Jennings AD, Barrett ADT (1993a) Nucleotide sequence of the envelope protein gene of the tick-borne flavivirus, Kumlinge A52. Virus Genes 7: 145–149
49. Whitby JE, Whitby SN, Jennings AD, Stephenson JR, Barrett ADT (1993b) Nucleotide sequence of the envelope protein of a Turkish isolate of tick-borne encephalitis (TBE) virus is distinct from other viruses of the TBE virus complex. J Gen Virol 74: 921–924
50. Yamshchikov VF, Pletnev AG (1988) Nucleotide sequence of the genome region encoding the structural proteins and the NS1 protein of the tick-borne encephalitis virus. Nucleic Acids Res 16: 7750
51. Zhao B, Mackow E, Buckler-White A, Markoff L, Chanock RM, Lai CJ, Makino Y (1986) Cloning full-length Dengue type 4 viral DNA sequences: analysis of genes coding for structural proteins. Virology 155: 77–88
52. Zlobin VI, Shamanin VA, Pletnev AG, Drokin DA, Dzhioev YP, Voronko IV, Melnikova OV, Gusarova NA (1992) Specific reactivity of kDNA- and desoxyoligonucleotide probes complementary to the genome of tick-borne encephalitis virus with RNA of the strains of different geographical origin. Vopr Virusol 37: 248–252

Authors' address: Dr. J. Süss, Department for Viral Zoonosis, Federal Institute for Health Protection of Consumers and Veterinary Medicine, Diedersdorfer Weg 1, D-12277 Berlin, Federal Republic of Germany.

Strategies to avoid virus transmissions by biopharmaceutic products

W. Werz[1], **H. Hoffmann**[1], **K. Haberer**[2] and **J. K. Walter**[1]

[1] Dr. Karl Thomae GmbH, Boehringer Ingelheim, Department of Biotech Production, Biberach/Riss, Federal Republic of Germany
[2] Hoechst A.G., Department of Pharma Quality Control, Microbiology, Frankfurt/Main, Federal Republic of Germany

Summary. The use of biopharmaceutical products offers an opportunity for the treatment of many diseases. Biotechnical manufacturing using recombinant mammalian cell lines is the most appropriate method today for the production of biopharmaceutical protein drugs for the treatment of human and animal diseases. However, mammalian cell line derived products have a potential risk for virus transmission to patients who are treated with these biopharmaceutical products. The reliability that biological products are free of any viruses requires a combination of several strategies: The use of well-characterized cell bank systems and, if feasible, the avoidance of biological raw materials for the cultivation of these mammalian cell lines and the production of biopharmaceuticals. Further on, the purification process for biopharmaceuticals has to be validated for its ability to efficiently remove and inactivate any potential virus contamination and, where applicable, also unconventional transmissible agents, such as BSE. In addition, the biopharmaceutical product itself can be tested for the presence of viruses. Like other manufacturing processes, biotechnical production processes have to be performed in compliance with current Good Manufacturing Practices (cGMP).

Introduction

The use of novel recombinant technologies offers a new opportunity for the treatment of many diseases with biopharmaceutical products. Even large and complex structures can be produced in sufficient amounts by this technology. However, virus infection and replication is an inherent risk during cultivation of mammalian cells. This is desirable and can be applied for the production of virus vaccines but must be prevented for the manufacturing of biopharmaceuticals. Therefore, virus transmission is a potential risk for patients who are treated with pharmaceuticals that consist of, or are of human or animal origin (biological products). There are several categories of biological products which are of different risk for virus transmission. Products derived from human blood or other tissues or organs bear a high risk, since this material may be contaminated with highly infectious human viruses. Products derived from animal organs and other materials may contain viruses which can cause zoonoses. The associated risk may vary from case to case. Products derived from continuous cell lines may become contaminated from latent virus infection of the producer cells. This is independent from the question whether the cells are genetically modified or not. Animal derived products (raw materials) which are used during the production of biologicals can introduce viruses into the production process. Even though the virological risk is probably smaller in these cases, it cannot simply be neglected. The treatment with biopharmaceuticals should not be of any hazard for the patients. The confidence that biological products are free of any viruses needs a combination of strategies for manufacturing and safety testing of biopharmaceutical products because of the large diversity of viruses and the lack of a test system or marker to detect all viruses. The following strategies are generally employed:

1. The producer cells for active ingredients in biopharmaceuticals should be derived from a non human continuous cell line that exists as a Master Cell Bank (MCB) and/or Working Cell Bank (WCB). Cells out of these characterized MCB or WCB stocks should than be the origin for each biotechnical production process. These Cell Banks have to be extensively characterized for the presence of viruses.
2. The use of biological raw materials (i.e. serum, serum albumin, insulin, transferrin, trypsin) for the cultivation of the MCB or WCB cells should be avoided, if feasible. In cases where this is not possible, all biological additives should be adequately tested to avoid potential virus contamination from these sources.
3. The purification process for biopharmaceuticals has to be validated to remove and/or inactivate any potential virus contamination. During such a validation study, model viruses with defined properties are added at various stages of the purification process (spike experiments). By each spike experiment the capacity of the purification step to remove and/or inactivate the model virus is determined, considering statistical methods for evaluation of the data. The overall clearance factor has to be determined and the safety margin has to be calculated. The safety margin for virus clearance should be in the range of 10^3 to 10^5, considering detection limits, ratio of testing volume to total production volume and dose of the drug per patient.
4. Like other manufacturing processes, biotechnical production processes have to be performed according to current Good Manufacturing Practices (cGMP). The general concept of GMP is to guarantee the uniformity, safety, efficacy and high quality of the product. Due to the complexity of biopharmaceutical products the whole manufacturing process, including environment, buildings, equipment, personal documentation, etc., are considered in the principles of these GMP-regulations.

5. Although the Cell Banks have been characterized and the manufacturing process has been validated for virus removal/inactivation, the cell culture fluids of production batches need to be tested in addition for freedom of adventitious viruses. The necessary safety requirement and risk assessment should determine the nature of virus test(s) applied.

Transmission of bovine spongiform encephalopathy (BSE) has recently become of special concern. This is due to the grave consequences of an infection with spongiform encephalopathy, the lack of diagnostic tests for this agent, and the uncertainty whether the bovine disease is transmissible to human beings. A European guideline for minimizing the risk of transmission of spongiform encephalopathies has been issued by CPMP in 1992 [3]. This was extended in Germany by several detailed guidelines giving exact strategies how to estimate and minimize the risk [1]. Factors considered are source of animals, origin of materials, methods used to inactivate or remove BSE-like agents, quantity of animal source material required, number of daily doses, and route of administration.

The aim of this paper is to give an overview on the complex topic of virus safety of biopharmaceutical products, as well as examples of methods used and results in order to demonstrate safety. Each of the different areas addressed will be shortly introduced in each section.

Materials and methods

Virus validation of the downstream process

Virus removal or inactivation studies were performed for each process step in question by spike experiments. The validation of the purification process has been first performed in a down-scaled model according to the large technical scale using the same conditions as in the manufacturing process (i.e. pH, conductivity, osmolarity). To the product solution of each distinctive purification step a defined amount of virus was added and the remaining amount of virus was determined in the product containing fraction after this step (balance). Viruses required for the spike experiments were propagated and titrated in different specific virus susceptible cell lines (i.e. Vero, MDBK, 3T3). The virus suspension was added to the process fluid from selected purification steps in a dilution of 1:10 or 1:20 (vol./vol.). The cumulative reduction factor is the product of individual reduction factors for each step or the sum of logarithmic factors, respectively, of all purification steps tested. Only log clearances with values > 1.0 are considered to be significant. It is, however, important to only consider the cumulative factor of different steps, since the same purification principle may not be as effective when it will be used in repeated steps.

In downstream processing it has to be differentiated between virus removal and virus inactivation.

With respect to virus removal there are two principal methodologies – filtration and chromatography. Filtration using membranes of a distinctive pore size, e.g. nanofilters (30–50 nm) or ultrafilters (70 kD–300 kD) are selective to mechanically separate product from viral particles.

Chromatographic methods use different physico-chemical properties of the product and virus in order to obtain a significant separation. The design and optimization of such a chromatographic separation is dedicated and unique to the physical conditions of the fluid to be processed. Hence this way of virus removal is highly sensitive to any process deviation.

Virus inactivation using harsh physical conditions such as high temperatures, extreme pH-values or chemical treatment by organic solvents, detergents or chaotropic salts can be highly effective and allow for a precise determination of the inactivation efficacy by performing kinetic studies of the virus inactivation [8, 9]. The most advanced physical treatment in order to inactivate virus is ultrashort, ultrahigh heat treatment using microwave technology. The product solution is processed at a constant flowrate through a microwave chamber. At a residence time in the range of 2–5 ms temperatures up to 165 °C can be applied at flowrates up to 80 lh^{-1} [2, 4].

For validation purposes with unconventional transmissible agents (TSE) the hamsteradapted 263k strain of scrapie, isolated by Kimberlin and Walker was used as model for BSE [5]. This scrapie strain was passaged twice in Golden Syrian Hamster at the NIH laboratory and once in Prof. Pocchiari's laboratory by intracerebral inoculation as 10% scrapie-infected brain homogenate [6]. Scrapie-infected brain homogenates were frozen for the subsequent experiments and directly used as a spike in experiments, performed similar to the process validation using different model viruses.

Individual samples from the spike experiments were titrated in weaned female Syrian Hamsters. The hamsters were scored for clinical sign 6 days per week and observed for up to 18 months. Dead or sacrificed animals were examined microscopically for brain lesions and in some cases for the presence of PrP 27–30 prion protein by immunoblotting.

Measurement of oxygen uptake rate (OUR)

The investigation of the oxygen uptake rate (OUR) in virus-cell-systems was carried out in computer-controlled 3 l-airlift-fermenters containing wire springs as carriers for the adherent cell lines. The determination was performed according to the dynamic method (7). Referring to this method the OUR was calculated from the on-line signal of a pO_2-electrode by determining the decrease of dissolved oxygen in defined time intervals while the aeration was switched off [7].

Results

Cell bank system

Prevention of virus contamination is the most important strategy for drug safety. The most dominant source of virus is the origin of the biopharmaceutical product. Human blood or blood and tissue derived products carry the highest risk for transmission of human pathogens. This is the most important reason to use non human cell lines as origin for biopharmaceutical production. The concept of cell banking is to have homogeneous pool of characterized cells from which each production run will start. So far we have generated nearly hundred Cell Banks (MCB, WCB) and extensively characterized for the presence of viruses. The majority of these cell banks are murine hybridomas and recombinant CHO-cell lines. Nearly all the murine hybridomas were shown to carry retrovirus or retrovirus like particles, whereas only a small portion of the recombinant CHO-cell lines contain non-infectious retrovirus like particles.

Table 1. Tests suitable to detect infectious and non-infectious virus contaminants for the characterization of cell banks, biological raw materials, fermentation samples and biopharmaceutical products

Strategies to avoid virus transmission by biopharmaceutical products
Cell bank characterization
Test for virus contamination
• adventitious virus test
–human diploid cells
–murine embryo cells
–human cell line
–bovine cell line
–production cells
• in vivo test (different inoculation routes)
–suckling mice
–adult mice
–guinea pigs
–fertilized eggs
• specific virus tests
–MAP (HAP; RAP)
–different PCR
• retrovirus test
–co-cultivation
–R.T.-ase assay
–specific PCR
• electron microscopic examination

In all of the cell banks tested so far no other virus contamination could be detected by the test systems summarized in Table 1. Hence these cell banks are considered to be an acceptable source for biotechnical production.

Biological raw material

It is state-of-the-art to use serum free media for the manufaction of biopharmaceuticals. In many processes developed before the emergence of the presently available techniques, serum (i.e. FCS) has to be used. It is not easily feasible to reformulate processes on the basis of new technologies, since this can imply new registration of the product. However, the supplies of these serum batches underlay strong selection criteria and each batch of serum has been tested for the presence of bovine viruses (i.e. BVD, Pl-3, Parvo, Leukemina). This is also true if other biological raw materials have to be used (i.e. insulin, transferin, serum albumin). In addition, the manufacturing process for such biological raw materials should be validated for its ability to efficiently remove and inactivate potential virus contaminations. In case of biological raw material from bovine, sheep or goat origin, where no test for the presence of TSE-agent is available, sourcing of the material is of prime importance. No such material from countries with a high prevalence of TSE may be used for the production of pharmaceuticals. Also, the tissues of infected animals are not equally infectious. The potential infectiosity of the tissue used is, therefore, another important factor in risk estimation.

Validation of removal and inactivation of virus and transmissible agent in the course of the purification process

It is one major effort to determine the capability of the downstream processing to remove and inactivate any potential virus contamination. The purification process has to be designed in such a way that minimum numbers of process steps lead to a high purity of the pharmaceutical product, without loss of its biological activity but a robust and efficient removal and inactivation of impurities including viruses. Since pharmaceutical protein products exhibit similar physico-chemical properties as viruses, only a limited number of purification methods are applicable to remove and/or inactivate viruses, particularly highly resistant viruses (i.e. Parvo, SV-40). It is important for the use of murine hybridoma cells that retroviruses (i.e. MuLV) can easily be removed by chromatographic and filtration methods and inactivated effectively by harsh physical treatment. Small non-enveloped viruses (i.e. SV-40) are typically more resistant to such treatments (Table 2). The flow chart of a downstream process for the manufacturing of Fab-fragments of a monoclonal antibody and the results of the corresponding validation figures showed that for each model virus different clearance factors can be achieved for each individual purification step and consequently for the overall clearance factors (Table 2). This is even much more obvious by comparing the results of 15 different validation studies for 15 different products (Table 3).

In some cases where bovine tissue has to be used as origin for a bioproduct, and where sufficient security cannot be gained from other considerations the manufacturer has to take into consideration to perform a validation study to demonstrate the robustness of the purification process regarding removal and inactivation of unconventional agents (i.e. scrapie as a model for BSE). In the case described in this paper, bovine lung lavage was used for the isolation of natural surfactant for the treatment of acute respiratory distress syndrome. From spike experiments similar to the validation studies described in the previous chapter it could be demonstrated that not only viruses but also the scrapie agent could be removed and inactivated by this purification process (Table 4). The reduction of infectivity following three independent spikes accounts for a clearance factor of more than 18 logs for SV 40, 22 logs for VSV and 21 logs for scrapie (Table 4).

Ultrashort-ultrahigh temperature virus inactivation

The uptake of the microwave energy correlates with the amount of water molecules associated with the target molecule or particle, i.e. the denaturating

Table 2. Flow chart of a generic downstream process for the manufacturing of Fab-fragments of a monoclonal antibody and corresponding results of the virus validation study

Hold step	Process step	PI->3	SV-40	REO-3	eMuLV	x MuLV
	30 kD					
	↓ Protein G Sepharose FF	7,0	3,3	2,0	6,5	>5,65
	↓ Virus Inactivation					
	↓ 30 kD					
hold step +4°C	↓ 40 nm Filtration	3,75	0,55	0,2	>4,9	>5,2
	↓ Fragmentation					
	↓ Q-Sepharose FF	>5,75	>6,9	>6,75	>5,7	4,1
	↓ 5 kD					
hold step Bulk freeze −20°C	↓ 100 kD	3,15	3,35	4,05	3,9	2,9
	↓ 5 kD					
	↓ Superdex 75 pg	2,9	2,65	2,65	3,33	n.d.
	↓ Formulated Bulk					
	Clearance Factor	>22,55	>16,2	>14,45	>24,33	>17,85

All values are given in log 10!

effect is dependent on the size of the respective target: the larger a virus particle the more effectively it can be destructed, but also proteins can be denaturated respective to their molecule size (Fig. 1).

Risk assessment

For each validation study a risk assessment has to be calculated under worst case assumptions. In case of fermentation dependent bioproducts, a realistic potential virus titer estimation of the cell culture fluid should be done. This theoretical titer should be in relation to the reduction factor, the dose needed for the patient and the relation of amount of starting material per dose (i.e. [titer starting material × amount of starting material × dose per starting material]/ in relation to the clearance factor). The figure obtained by this calculation should be smaller 5 log 10 units to be on the safe side (i.e. to be in "compliance with the statistics for sterility testing").

For blood and tissue derived products in addition to this calculation, the incidence of a specific virus burden and efficacy of the application route should be accounted for.

Table 3. Summary of 15 individual validation studies from 15 independent purification processes for 15 different biopharmaceuticals

Chromatography	SV-40	Reo3	Pl-3	MuLV
Affinity				
Protein G	2,0->6,0	2,0->6,0	3,0->6,0	3,0->6,0
Ion exchange				
Cation ($-SO_3$)	<1,0-2,0	<1,0->6,6	<1,0-3,2	<1,0->3,0
Anion (−DEAE)	<1,0-4,8	3,0->6,0	<1,0->6,0	3,0->6,0
Anion ($-N(CH_3)_3$)	<1,0-2,7	3,0->6,0	1,5->6,8	3,0->6,0
HIC				
Phenyl	<1,–1,8	<1,0	<1,0-2,3	<1,0-2,3
Filtration				
Ultrafiltration				
Omega 200kD	>5,6	>6,8	>7,0	>6,8
Omega 300kD	<1,0	2,8	>5,2	4,0
Ultipor 40	<1,0	2,9	>3,7	>5,2
Nano filtration DV 50	>5,8	>7,0	>6,8	n.d
Acid treatment				
pH-value				
pH 4.0	<1,0	<1,0	1,2	5,0
pH 3.5	<1,0	n.d.	n.d.	n.d.

Figures for reduction factors in log 10 units
n.d. Not done

Table 4. Virus- and scrapie (as model for BSE) validation study of a bovine lung derived biopharmaceutical product

Starting material: bovine lung lavage
spike:

VSV:	10,5	(log10 $TCID_{50}$/ml)
SV-40	8,3	(log 10 $TCID_{50}$/ml)
Scrapie:	8,5	(log 10 LD_{50})

Model agent / Purification step	VSV	SV-40	scrapie
Balance 1: Density gradient	1,3	>4,0	>7,7
Balance 2: Bligh & Dyer extraction (Chloroform/Methanol)	>7,1	>4,8	7,1
Balance 3: Homogenous phase B&D extraction filtraion	>7,4	>5,2	7,0
Balance 4: Methylenchloride precipitation filtration	>7,0	>4,8	n.d.
Overall reduction titer	>22,8	>18,8	>21,8

The main inactivation and removal steps of the purification process of bovine lung derived surfactant have been validated using VSV, SV-40 and scrapie as model for BSE. The individual reduction factors are given in log 10 units. The overall reduction factor is the sum of the individual reduction factors
n.d. Not done

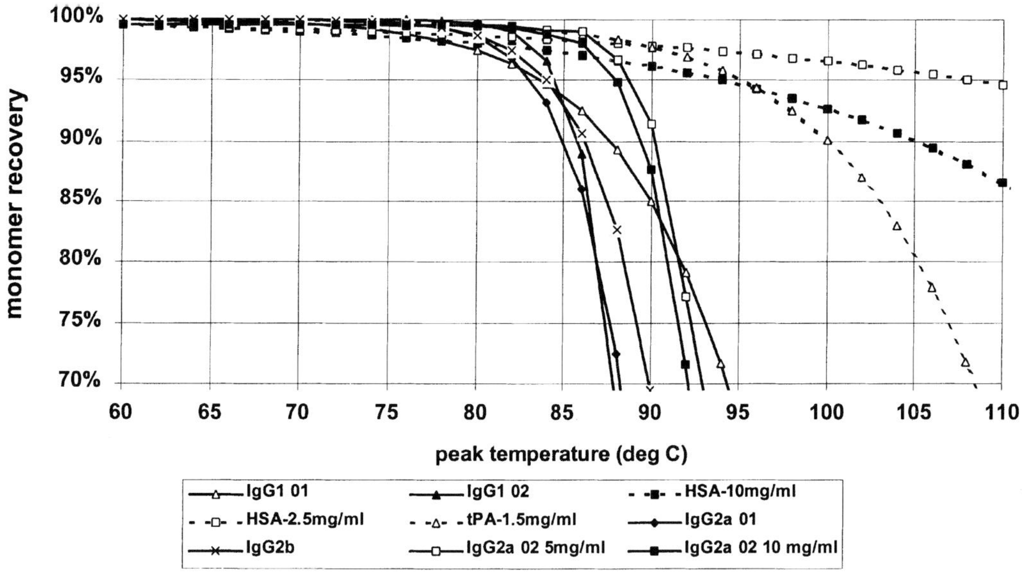

Fig. 1. Heat resistance of pharmaceutical proteins during ultrashort, ultrahigh microwave heating. Different proteins of pharmaceutical grade have been processed at a constant flowrate of 60 lh^{-1} through a 50″ coil, the residence time was about 4 ms. Heat resistance is strongly related to the size and physico-chemical properties of the protein molecules (UltraTherm, UT5AB1, Charm Bioengineering, Malden, MA, USA)

Current good manufacturing practice (cGMP)

The biotechnical production processes and operations have to be performed in compliance with cGMP. One of the scopes of cGMP is the prevention of adventitious virus contaminations. The employment of trained and skilled personnel, the use of appropriate equipment, as well as defined buildings and production areas are important pre-requisites to prevent virus contaminations. In all production areas a clean environment and a high level of hygiene by the operators is requested. The personnel and material flows should be restricted and regulated, and the process should be performed in closed systems. Only ultraclean water systems should be used, and where applicable the airflow should be filtered. The environment of the manufacturing areas should be classified and monitored accordingly. All operations including testing should be performed according to documented Standard Operation Procedures (SOP's) in order to guarantee the quality, safety and efficacy of each individual biopharmaceutical product.

The manufacturing process itself should be controlled and regulated. All relevant process parameters should be recorded and regulated by a process control system. In experimentally infected fermentation runs we searched for on-line recorded (e.g. metabolic) process parameters that may serve as indicators for a virus infection. The only parameter where a correlation to the

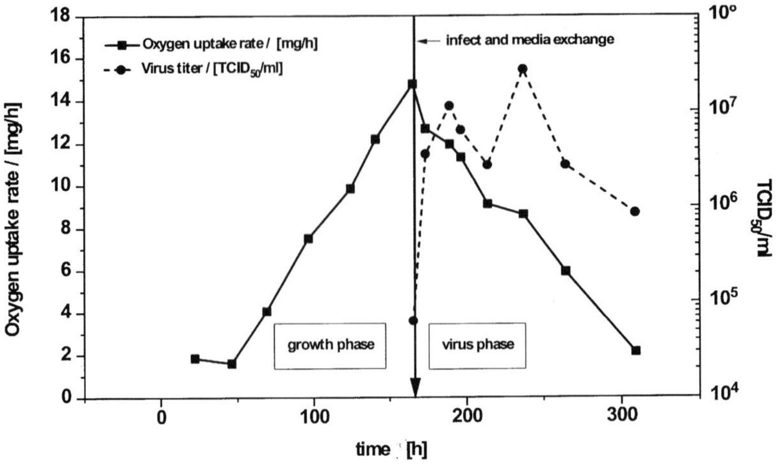

Fig. 2. Comparison of oxygen uptake rate (OUR) and titer of growing and virus producing cells. Experimental virus infection in small scale fermentation Vero/HSV-system. Vero cells have been cultivated adherent on wire springs in small scale fermenters [31] for 7 days and were then infected with HSV. The HSV propagation was recorded for further 7 days. The production for HSV by the Vero cells has been determined daily and was expressed at TCID 50/ml. During the cultivation and virus production phase the OUR was measured on-line

virus infection was seen was the oxygen uptake rate (OUR). During the growth and production phase, the cells have an increasing or constant OUR. After virus infection this OUR is changing (Fig. 2). Further experimental infection studies will show whether the change in OUR is really a general parameter in fermentation for virus contaminations (cytopathic vs. non-cytopathic viruses) or whether this is an observation relevant only for a few viruses-cell systems.

Testing of the purified drug product

Due to the detection limitations for residual virus in a highly purified biopharmaceutical product, it is not very meaningful to test the final drug substance after purification according to a validated purification process. Therefore, the most useful test point is at the end of the fermentation process, when the "product" is still in close contact with the producer cells. The assay employed to detect virus contaminations are identical to the test systems applied for the characterization of the cell bank system. The extention of testing should be defined for each product and should be in compliance with international guidelines, which will be harmonized by the International Conference of Harmonization (lCH).

Discussion

No single test or unique marker systems can be applied to detect all viruses present in nature. Therefore, it is theoretically and practically impossible to monitor the presence (or removal) of unknown biological transmissible agents and viruses. As a consequence, a spectrum of different strategies has to be applied to ensure safety of biopharmaceutical products. This includes the testing for viruses, in combination with control of the manufacturing process and with prevention methods. For testing, a broad spectrum of different test systems has to be used to look at the starting point (cell bank system, raw material) and end point (end of fermentation) for nonspecific and specific virus contaminations. The most important strategy for prevention is the validation of the purification process for the capacities to remove and inactivate known, e.g. in the case of murine hybridoma derived retroviruses, and unknown virus contaminations. For the difficult goal to demonstrate the clearance of unknown viruses, model viruses have to be used. These model viruses should represent the overall virus properties e.g. enveloped/non enveloped, small/large size, RNA/DNA genome, and resistance against different treatments to assess the worst case conditions. For each individual biopharmaceutical product an individual purification regime has to be developed and validated. For a defined purification step in distinct processes different clearance factors can be measured for the same model virus. The reasons for the variations in the reduction or clearance factors obtained at different purification steps for the same or for different viruses, is that individual purification steps need dedicated conditions (e.g. pH, osmolarity, conductivity) to purify and recover the bioactive biopharmaceutical product.

The GMP-regulations are also of relevance for prevention of virus contamination and for control of the manufacturing process. It would be of great advantage, if on-line parameters could be identified and applied as indicators for virus contamination during fermentation. We have analyzed different metabolic process parameters and could identify so far only the oxygen uptake rate (OUR) as an indicator for a virus contamination. In further experiments it has to be evaluated if the change in OUR after a virus infection is in general a marker or if this holds true only for few virus examples.

To guarantee virus freedom a virus barrier into the manufacturing process has to be introduced (e.g. heat, gamma-irradiation, chemical treatment). For the destruction of resistant non-enveloped viruses strong conditions are necessary under which the biopharmaceutical product could be denaturated. Ultrashort, ultrahigh microwave heat treatment is an innovative method that may be applied for many biopharmaceutical products or additives (e.g. HSA) to destroy viruses without inactivating the biopharmaceutical product. This method has to be evaluated for each individual bioproduct to find the optimum between minimal loss of biological activity of the product and maximum virus destruction. Another approach compromises the selection of raw materials of animal origin to minimize the risk of potential infections. This can be a geographic selection from countries where certain diseases are not endemic, selection of closed herds as well as selection of low risk animal materials where feasible.

In conclusion, safety of biotechnically manufactured products can be ensured by using different strategies for detection of virus contaminations and for protection and control of the manufacturing system. For modern biotechnical products no virus transmission has been reported until today. This demonstrates the success of these strategies. As a result biopharmaceuticals are of great benefit for many human diseases, can be applied to the patients without concerns on biological safety.

References

1. BGA (1994) Bekanntmachung der Sicherheitsanforderungen an Arzneimittel aus Körperbestandteilen von Rind, Schaf oder Ziege zur Übertragung von BSE bzw. Scrapie. Bundesanzeiger Nr. 40, 26. Februar 1994
2. Charm SE, Landan S, Williams B, Hovowitz B, Prince AM, Pascual D (1992) High-temperature short-time heat inactivation of HIV and other viruses in human blood pharma. Vox Sang 62: 12–20
3. CPMP (1992) Guidelines for minimizing the risk of transmitting agents causing spongiform encephalopathy via medicinal products. Biologicals 20: 155–158
4. Golden RF (1994) Microwave purification for viral inactivation and bioprocess validation. Gene Eng News, May 15,17
5. Kimberlin RH, Walker C (1977) Characteristics of a short incubation model of scrapie into the golden hamster. J Gen Virol 34: 295–304
6. Pocchiari M (1991) Methodological aspects of the validation of purification procedures of human/animal derived products to remove unconvential slow viruses. Dev Biol Standard 75: 87–95
7. Taguehi H, Humphrey A (1966) Dynamic measurement of the volumetric oxygen transfer coefficient in fermentation systems. J Ferment Technol 44: 881–889
8. Walter JK, Werz W, Berthold W (1992) Virus removal and inactivation. Concept and data for process validation of downstream processing. Biotech Forum Europe 9: 560–564
9. Walter JK, Werz W, Berthold W (1996) Process scale considerations in evaluation studies and scale-up. In: Brown F, Lubiniecki AS (eds) Viral safety and evaluation of viral clearance from biopharmaceutical products. S. Karger, Basel, pp 99–108

Authors' address: Dr. W. Werz, Dr. Karl Thomae GmbH, Boehringer Ingelheim, Department of Biotech Production, D-88397 Biberach/Riss, Federal Republic of Germany.